Signals and Systems

A PRIMER WITH MATLAB®

Signals and Systems

A PRIMER WITH MATLAB®

MATTHEW N. O. SADIKU
WARSAME H. ALI

CRC Press is an imprint of the
Taylor & Francis Group, an **informa** business

CRC Press
Taylor & Francis Group
6000 Broken Sound Parkway NW, Suite 300
Boca Raton, FL 33487-2742

First issued in paperback 2020

ISBN-13: 978-1-4822-6151-6 (hbk)
ISBN-13: 978-0-367-73777-1 (pbk)

Visit the Taylor & Francis Web site at
http://www.taylorandfrancis.com

and the CRC Press Web site at
http://www.crcpress.com

Dedicated to
My parents, Ayisat and Solomon Sadiku, and my wife, Kikelomo

Matthew N.O. Sadiku

My parents, Faduma Ibrahim Omer and Hassan Ali Hussein (Shigshigow), for their unconditional love and support, to my beloved wife, engineer Shukri Mahdi Ali, and my children, Mohamed, Faduma, Dahaba, Hassan, and Khalid, for their patience, and to all my siblings

Warsame H. Ali

Contents

Preface

The idea of signals and systems arises in different disciplines such as science, engineering, economics, politics, and medicine. Most students of electrical and computer engineering will deal with signals and systems in their professional life. This is why a course on signals and systems is an important part of most engineering curricula.

Some books on signals and systems are designed for a two-semester course sequence. Unfortunately, the field of electrical engineering has changed considerably and its curriculum is so crowded that there is no room for a two-semester course on signals and systems. This book is designed for a three-hour semester course on signals and systems. It is intended as a textbook for junior-level students in electrical and computer engineering curricula. The prerequisites for a course based on this book are knowledge of standard mathematics (including calculus and differential equations) and electric circuit analysis.

This book intends to present continuous-time and discrete-time signals and systems to electrical and computer engineering students in a manner that is clearer, more interesting, and easier to understand than other texts. This objective is achieved in the following ways:

- We have included several features to help students feel at home with the subject. Each chapter opens with a historical profile or career talk. This is followed by an introduction that links the chapter with the previous chapters and states the chapter objectives. The chapter ends with a summary of key points and formulas.
- All principles are presented in a lucid, logical, step-by-step manner. As much as possible, we have avoided wordiness and giving too much detail that could hide concepts and impede overall understanding of the material.
- Thoroughly worked examples are liberally given at the end of every section. The examples are regarded as part of the text and are clearly explained without asking the reader to fill in missing steps. They give students a good understanding of the solution and the confidence to solve problems themselves. Some of the problems are solved in two or three ways to facilitate an understanding and a comparison of different approaches.
- To give students practice opportunity, each illustrative example is immediately followed by a practice problem along with its answer. Students can follow the example step by step to solve the practice problem without flipping pages or looking at the end of the book for answers. The practice problems are also intended to test if students understand the preceding example. It will reinforce their grasp on the material before they move on to the next section.
- Toward the end of each chapter, we discuss some application aspects of the concepts covered in the chapter. The material covered in the chapter is applied to at least one or two practical problems or devices. This helps the students see how the concepts are applied to real-life situations.

- Ten review questions in the form of multiple-choice objective items are provided at the end of each chapter with answers. The review questions are intended to cover the little "tricks" that the examples and end-of-chapter problems may not cover. They serve as a self-test device and help students determine how well they have mastered the chapter.

In recognition of the requirements by ABET (Accreditation Board for Engineering and Technology) on integrating computer tools, the use of MATLAB® is encouraged in a student-friendly manner. We have introduced MATLAB in Appendix C and applied it gradually throughout this book. MATLAB has become a standard software package in electrical engineering curricula. To avoid confusing learning MATLAB with the main subject (signals and systems), we have delayed MATLAB till the end of each chapter.

Matthew N.O. Sadiku
Warsame H. Ali

Acknowledgments

We are indebted to Dr. Sudarshan Nelatury at the Penn State University at Erie, who provided us with a lot of end-of-chapter problems. We thank Dr. Saroj Biswas of Temple University, Philadelphia, for helping with MATLAB. Special thanks to our colleagues Drs. Siew Koay, Lijun Qian, and Annamalai Annamalai, who reviewed the manuscript, and to our graduate students, Mamatha Gowda, Savitha Gowda, Tekena Abibo, and Mahamadou Tembely, for going over the manuscript and pointing out some errors. We are also grateful to Dr. Pamela Obiomon, the interim head of the Department of Electrical and Computer Engineering, and Dr. Kendall Harris, the dean of the College of Engineering, for providing a sound academic environment at Prairie View A&M University, Prairie View, Texas.

Finally, we express our profound gratitude to our wives and children, without whose cooperation this project would have been difficult if not impossible. We appreciate feedback from students, professors, and other users of this book. We can be reached at sadiku@ieee.org and whali@pvamu.edu.

Authors

Dr. Matthew N.O. Sadiku received his BSc in 1978 from Ahmadu Bello University, Zaria, Nigeria, and his MSc and PhD from Tennessee Technological University, Cookeville, Tennessee, in 1982 and 1984, respectively. From 1984 to 1988, he was an assistant professor at Florida Atlantic University, where he did graduate work in computer science. From 1988 to 2000, he was at Temple University, Philadelphia, Pennsylvania, where he became a full professor. From 2000 to 2002, he worked with Lucent/Avaya, Holmdel, New Jersey, as a system engineer and with Boeing Satellite Systems as a senior scientist. At present, he is a professor at Prairie View A&M University.

He is the author of over 240 professional papers and over 50 books, including *Elements of Electromagnetics* (Oxford University Press, 6th ed., 2015), *Fundamentals of Electric Circuits* (McGraw-Hill, 5th ed., 2013, coauthored with C. Alexander), *Numerical Techniques in Electromagnetics with MATLAB®* (CRC Press, 3rd ed., 2009), and *Metropolitan Area Networks* (CRC Press, 1995). Some of his books have been translated into French, Korean, Chinese (and Chinese long form in Taiwan), Italian, Portuguese, and Spanish. He was the recipient of the McGraw-Hill/Jacob Millman Award in 2000 for outstanding contributions in the field of electrical engineering. He was also the recipient of the Regents Professor award for 2012–2013 given by the Texas A&M University System.

His current research interests are in the areas of numerical modeling of electromagnetic systems and computer communication networks. He is a registered professional engineer and a fellow of the Institute of Electrical and Electronics Engineers (IEEE) "for contributions to computational electromagnetics and engineering education." He was the IEEE Region 2 Student Activities Committee chairman. He was an associate editor for *IEEE Transactions on Education*. He is also a member of Association for Computing Machinery (ACM).

Dr. Warsame H. Ali received his BSc from King Saud University Electrical Engineering Department, Riyadh, Saudi Arabia, and his MS from Prairie View A&M University, Prairie View, Texas. He received his PhD in electrical engineering from the University of Houston, Houston, Texas. Dr. Ali was promoted to associate professor and tenured in 2010. Dr. Ali joined NASA, Glenn Research Center, in the summer of 2005, and Texas Instruments (TI) in 2006.

Dr. Ali has given several invited talks and is also the author of 80 research articles in major scientific journals and conferences. Dr. Ali has received several major NSF, NAVSEA, AFRL, and DOE awards. At present, he is teaching undergraduate and graduate courses in the Electrical and Computer Engineering Department at Prairie View A&M University. His main research interests are the application of digital PID controllers, digital methods to electrical measurements, and mixed signal testing techniques, power systems, HVDC power transmission, sustainable power and energy systems, power electronics and motor drives, electric and hybrid vehicles, and control systems.

NOTES TO STUDENTS

Although electrical engineering is an exciting and challenging discipline, a course on signals and systems may intimidate you. This book was written to prevent that. A good textbook and a good professor are an advantage—but you are the one who does the learning. If you keep the following ideas in mind, you will do very well in this course:

- This course provides a foundation for other courses in the electrical engineering curriculum. For this reason, put in as much effort as you can. Study the course regularly.
- Signals and systems is a problem-solving subject, learned through practice. Solve as many problems as you can. Begin by solving the practice problems following each example and then proceed to the end-of-chapter problems. The best way to learn is to solve a lot of problems. An asterisk in front of a problem indicates a challenging problem.
- *MATLAB* is a software that is very useful in signals and systems and other courses you will be taking. A brief tutorial on *MATLAB* is given in Appendix C to get you started. The best way to learn *MATLAB* is to start with it once you know a few commands.
- Each chapter ends with a section on how the material covered in the chapter can be applied to real-world situations. The concepts in this section may be new and advanced for you. No doubt, you will learn more of the details in other courses. We are mainly interested in making you gain a general familiarity with these ideas.
- Attempt the review questions at the end of each chapter. They will help you discover some "tricks" not revealed in the class or in the textbook.
- It is very tempting to sell your book after taking this course. However, our advice to you is, *Do not sell your engineering books!* When we were students, we did not sell any of our engineering books and were very glad we did not. We found that we needed most of them throughout our career.

A short review of the mathematical formulas you may need are covered in Appendix A while answers to odd-numbered problems are given in Appendix D.

Have fun!

1 Basic Concepts

Society is never prepared to receive any invention. Every new thing is received, and it takes years for the inventor to get people to listen to him and years more before it can be introduced.

—Thomas Alva Edison

GLOBAL POSITIONING SYSTEM

Artist interpretation of GPS satellite. (Image courtesy of NASA, Washington, DC; Global Positioning System. http://en.wikipedia.org/wiki/Global_Positioning_System.)

The global positioning system (GPS) is a typical illustration of what signals and systems are all about. GPS is a satellite-based navigation system made up of a network of 24 satellites placed into orbit by the U.S. Department of Defense. GPS was originally designed for military use, but in the 1980s, the government made the system available for civilian use.

The 24 satellites that make up the GPS space segment are orbiting the earth about 12,000 miles above us. These satellites travel at speeds of roughly 7,000 miles/h. GPS satellites transmit two low power radio signals. The signals travel by line of sight, meaning they will pass through clouds, glass, and plastic but will not go through most solid objects, such as buildings and mountains.

A GPS signal contains three different bits of information—a pseudorandom code, ephemeris data, and almanac data. The pseudorandom code identifies which satellite is transmitting information. You can view this number on your GPS unit's satellite page. Ephemeris data contains important information about the status of the satellite, current date, and time. This part of the signal is essential for determining a position. The almanac data tells the GPS receiver where each GPS satellite should be at any time throughout the day. Each satellite transmits almanac data showing the orbital information for that satellite and for every other satellite in the system.

A GPS receiver calculates its position by precisely timing the signals sent by GPS satellites. The receiver uses the messages it receives to determine the transit time of each message and computes the distance to each satellite. These distances along with the satellites' locations are used to compute the position of the receiver.

1.1 INTRODUCTION

The idea of signals and systems arises in different disciplines such as science, engineering, economics, politics, and medicine. Scientists, mathematicians, financial analysts, cardiologists, and engineers all use the concepts of systems and signals because they are the foundation on which we build many things in our daily lives. Typical examples of **systems** include radio and television, telephone networks, radar systems, computer networks, wireless communication, military surveillance systems, and satellite communication systems. The theory of signals and systems provides a solid foundation for control systems, communication systems, power systems, and networking, to name a few.

Our objective in this book is to present an introductory, yet comprehensive, treatment of signals and systems with an emphasis on computing using MATLAB®. The knowledge of a broad range of signals and systems is of practical value in describing human experience. It is also important because engineers must be familiar with signal and system concepts and apply the knowledge to analyze some specific signals and systems they will deal with in their professional life.

In this chapter, we begin by discussing some of the basic concepts in signals and systems. We introduce the mathematical representations of signals, their properties, and applications. We also discuss different systems and how the material covered in this chapter is used in some applications. We finally learn how to use MATLAB to process signals.

1.2 BASIC DEFINITIONS

To avoid any misconception, it is expedient that we define at the outset what we mean by signals and systems.

> A **signal** $x(t)$ is a set of data or function of time that represents a variable of interest.

A signal typically contains information about the nature of a phenomenon. Examples of signals include the atmospheric temperature, humidity, human voice, television images, a dog's bark, and birdsongs. More generally, a signal may be a function of more than one independent variable (time). For example, pictures are signals that depend on two independent variables (horizontal and vertical positions)

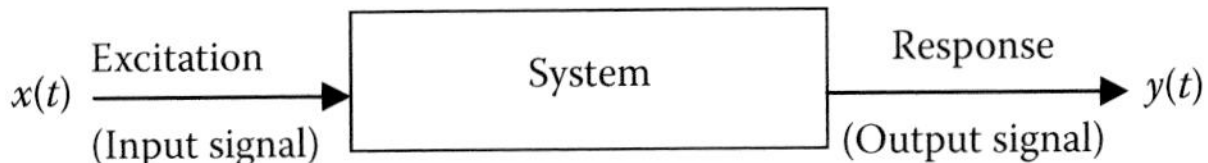

FIGURE 1.1 A simple system.

and may be regarded as two-dimensional signals. However, in this text, we will consider only one-dimensional signals with time as the independent variable.

A **system** is a collection of devices that operate on input signal $x(t)$ (or excitation) to produce an output signal $y(t)$ (or response).

A system may also be regarded as a mathematical model of a physical process that relates the input signal to the output signal. Examples of systems include electric circuits, computer programs, the stock market, weather, and the human body. A system may have several mathematical models or representations. The variables in the mathematical model are described as **signals**, which may be current, voltage, or displacement. In electrical systems, signals are often represented as currents and voltages. In mechanical systems, they are often represented as temperatures, forces, and velocities. In hydraulic systems, signals may be displacements and pressures.

Figure 1.1 illustrates the block diagram of a single-input single-output system. We classify the signals that enter the system as input signals, while the signals produced by the system as outputs. For example, we may regard voltages and currents as functions of time in an electric circuit as signals, while the circuit itself is regarded as a system. In engineering systems, signals may carry energy or information.

1.3 CLASSIFICATIONS OF SIGNALS

There are many ways of classifying signals: continuous-time or discrete-time, periodic or nonperiodic, energy or power, analog or digital, random or nonrandom, real or complex, etc.

1.3.1 Continuous-Time and Discrete-Time Signals

A signal $x(t)$ that is defined at all instants of time is known as a continuous-time signal.

A **continuous-time signal** takes a value at every instant of time t.

An example of a continuous-time signal $x(t)$ is shown in Figure 1.2a.

A **discrete-time signal** is defined only at particular instants of time.

A discrete-time signal is usually identified as a sequence of numbers, denoted by $x[n]$, where n is an integer. It may represent a phenomenon for which the

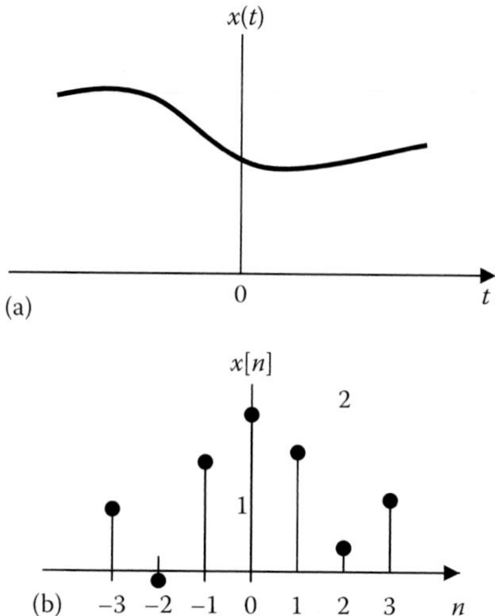

FIGURE 1.2 Typical examples of (a) continuous-time and (b) discrete-time signals.

independent variable n is inherently discrete. An example of discrete-time signal $x[n]$ is shown in Figure 1.2b.

Since time is naturally continuous, most physical systems are continuous-time systems. Discrete-time signals are often obtained from continuous-time signals through sampling.

As typically shown in Figure 1.3, the continuous-time signal $x(t)$ in Figure 1.3a is sampled uniformly with sampling period T to produce the discrete-time signal $x[n]$ in Figure 1.3b. We simplify notation by letting $x(kT) \triangleq x[k]$. A discrete-time signal

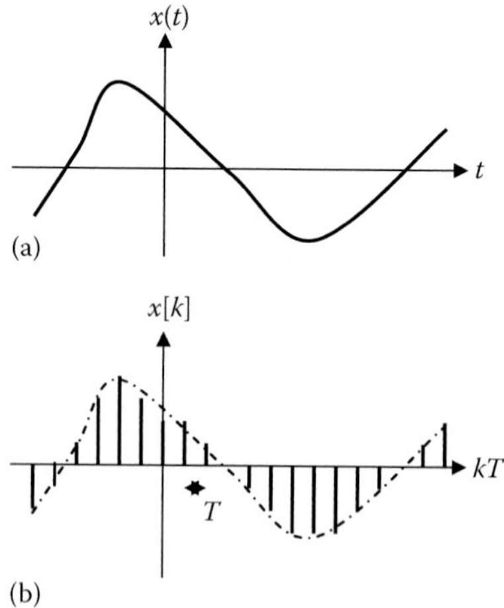

FIGURE 1.3 Obtaining x[k] from x(t) through sampling.

is equally spaced in time with sampling period *T*. Thus, discrete-time signals are samples of continuous-time signals, or they may exist naturally.

1.3.2 Periodic and Nonperiodic Signals

A periodic continuous-time signal satisfies

$$x(t)=x(t+nT) \tag{1.1}$$

where
  n is an integer
  T is the period of the signal

A *periodic signal* is one that repeats itself every T seconds.

A popular example of a periodic signal is the sinusoid

$$x(t)=A\sin(\omega t+\theta), \quad -\infty<t<\infty \tag{1.2}$$

where
  A is the amplitude of the signal
  $\omega(=2\pi f=2\pi/T)$ is the angular frequency in radians per second
  θ is the phase in radians

Another example of a periodic (nonsinusoidal) continuous-time signal is shown in Figure 1.4a. Any signal that does not satisfy the periodicity condition in Equation 1.1

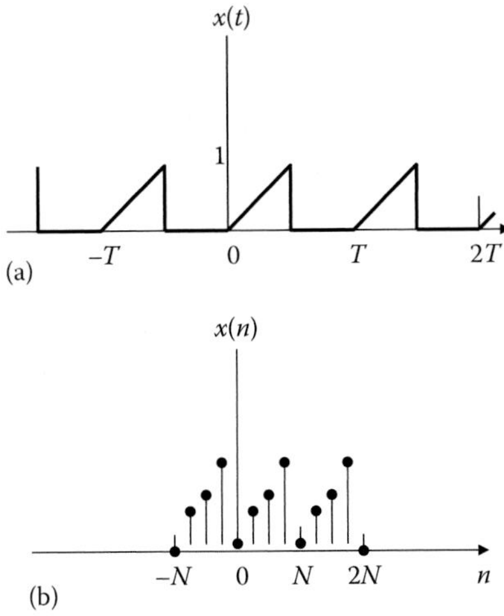

FIGURE 1.4 Examples of periodic (a) continuous-time and (b) discrete-time signals.

is called a *nonperiodic* signal. We will be dealing with nonperiodic signals (step functions, ramp functions, rectangular functions, etc.) later.

A discrete-time signal $x[n]$ is periodic with period N if it satisfies

$$x\left[n\right] = x\left[n+N\right] \tag{1.3}$$

The discrete-time sine and cosine signals are given in terms of complex exponential signals as

$$x[n] = \cos(\omega n) = \frac{1}{2}(e^{j\omega n} + e^{-j\omega n}) \tag{1.4}$$

$$y[n] = \sin(\omega n) = \frac{1}{2j}(e^{j\omega n} - e^{-j\omega n}) \tag{1.5}$$

Another example of a periodic discrete-time signal is shown in Figure 1.4b.

1.3.3 Analog and Digital Signals

If a continuous-time signal $x(t)$ can assume any value in the range $-\infty < t < \infty$, then it is called an analog signal. Although all analog signals are continuous-time signals, not all continuous-time signals are analog signals. If a discrete-time signal assumes only finite values, then it is called a digital signal.

An *analog signal* is a continuous-time signal in which the variation with time is analogous (or proportional) to some physical phenomena.

A *digital signal* is a discrete-time signal that can have a finite number of values (usually binary).

A digital signal can assume only a finite number of values. The difference between analog and digital signals is that analog is a continuous electrical signal, whereas digital is a discrete electrical signal. We live in an analog world and most signals are analog. Although some signals are inherently digital, most digital signals are obtained from analog signals by sampling or an analog-to-digital converter (ADC).

For example, an analog signal is taken straight from the microphone and recorded into a tape in its original form. The signal from the microphone is an analog signal, and therefore the signal on the tape is analog as well. Since data is sent using variable currents in an analog system, it is very difficult to remove noise and signal distortions during the transmission. For this reason, analog signals cannot perform high-quality data transmission. On the other hand, digital signals use binary data strings (0 and 1) to reproduce data being transmitted.

1.3.4 Energy and Power Signals

For continuous-time signal $x(t)$, the normalized energy E of $x(t)$ (assuming $x(t)$ is real) is

$$E = \int_{-\infty}^{\infty} x(t)^2 \, dt \tag{1.6}$$

If $x(t)$ is complex valued, Equation 1.6 can be generalized:

$$E = \int_{-\infty}^{\infty} |x(t)|^2 \, dt \tag{1.7}$$

where $|x(t)|$ is the magnitude of $x(t)$. The normalized power P for real $x(t)$ is

$$P = \lim_{T\to\infty} \frac{1}{T} \int_{-T/2}^{T/2} x(t)^2 \, dt \tag{1.8}$$

This can be generalized for complex value $x(t)$ as

$$P = \lim_{T\to\infty} \frac{1}{T} \int_{-T/2}^{T/2} |x(t)|^2 \, dt \tag{1.9}$$

Similarly, for a discrete-time signal $x[n]$, the normalized energy E of $x[n]$ is

$$E = \sum_{n=-\infty}^{\infty} |x[n]|^2 \tag{1.10}$$

while the normalized power P is

$$P = \lim_{N\to\infty} \frac{1}{2N+1} \sum_{n=-N}^{N} |x[n]|^2 \tag{1.11}$$

Based on the definitions of E and P in Equations 1.6 through 1.11, we define the following:

A signal $x(t)$ or $x[n]$ is an *energy signal* if and only if $0 < E < \infty$ and consequently $P = 0$.

A signal $x(t)$ or $x[n]$ is a *power signal* if and only if $0 < P < \infty$ and consequently $E = \infty$.

If a signal is a power signal, then it cannot be an energy signal or vice versa; power and energy signals are mutually exclusive. A signal may be neither a power nor an energy signal if the conditions in Equations 1.6 through 1.11 are not met. Almost all periodic functions of practical interest are power signals.

1.3.5 Even and Odd Symmetry

By definition, a signal is even if

$$x(t) = x(-t) \tag{1.12}$$

A function is even if its plot is symmetrical about the vertical axis; that is, the signal for $t < 0$ is the mirror image of the signal for $t > 0$. Examples of even signals are $\cos t$, t^2, and t^4.

By definition, a signal is odd if

$$x(t) = -x(-t) \tag{1.13}$$

The plot of an odd function is antisymmetrical about the vertical axis. Examples of odd functions are t, t^3, and $\sin t$.

An *even signal* $x(t)$ is one for which $x(t) = x(-t)$ and an *odd signal* $y(t)$ is one for which $y(t)= -y(-t)$.

Any signal $x(t)$ can be represented as the sum of even and odd signals as

$$x(t) = x_e(t) + x_o(t) \tag{1.14}$$

where
$x_e(t)$ is the even part
$x_o(t)$ is the odd part

Replacing t with $-t$ in Equation 1.14 and incorporating Equations 1.12 and 1.13, we get

$$x(-t) = x_e(-t) + x_o(-t) = x_e(t) - x_o(t) \tag{1.15}$$

Adding Equations 1.14 and 1.15 and dividing by 2,

$$\boxed{x_e(t) = \frac{1}{2}\left[x(t) + x(-t)\right]} \tag{1.16}$$

Subtracting Equation 1.15 from Equation 1.14 and dividing by 2,

$$\boxed{x_o(t) = \frac{1}{2}\left[x(t) - x(-t)\right]} \tag{1.17}$$

Thus,

$$x(t) = \underbrace{\frac{x(t) + x(-t)}{2}}_{even} + \underbrace{\frac{x(t) - x(-t)}{2}}_{odd} \tag{1.18}$$

Equation 1.16 shows that the two signals are added and scaled in magnitude to produce the even signal $x_e(t)$, while Equation 1.17 indicates that $x(-t)$ is subtracted from $x(t)$ and the result is amplitude-scaled by 0.5 to produce the odd signal $x_o(t)$. For a discrete-time signal $x[n]$, we can construct the even and odd parts using Equations 1.16 and 1.17.

Note the following properties of even and odd functions:

1. The product of two even functions is also an even function.
2. The product of two odd functions is an even function.
3. The product of an even function and an odd function is an odd function.
4. The sum (or difference) of two even functions is also an even function.
5. The sum (or difference) of two odd functions is an odd function.
6. The sum (or difference) of an even function and an odd function is neither even nor odd.

Each of these properties can be proved using Equations 1.12 and 1.13.

Example 1.1

Show that the signal in Equation 1.2 is periodic.

Solution

The period of the signal in Equation 1.2 is

$$T = \frac{1}{f} = \frac{2\pi}{\omega}$$

$$\begin{aligned} x(t+T) &= A\sin[\omega(t+T)+\theta] = A\sin\left[\omega\left(t+\frac{2\pi}{\omega}\right)+\theta\right] \\ &= A\sin(\omega t + 2\pi + \theta) = A\sin(\omega t + \theta) \\ &= x(t) \end{aligned}$$

showing that $x(t)$ satisfies Equation 1.1; hence it is periodic.

Practice Problem 1.1 Show that the signal $x(t) = A\cos(2\pi t + 0.1\pi)$ is periodic.

Answer: Proof.

Example 1.2

Determine whether the following signals are energy signals, power signals, or neither.

(a) $x(t) = \begin{cases} e^{-at}, & 0 < t < \infty, \quad a > 0 \\ 0, & \text{otherwise} \end{cases}$

(b) $x(t) = A\cos(\omega t + \theta)$

(c) $x[n] = 10e^{j2n}$

Solution

(a) The normalized energy of the signal is

$$E = \int_{-\infty}^{\infty} x(t)^2 dt = \int_0^{\infty} e^{-2at} dt = \left.\frac{e^{-2at}}{-2a}\right|_0^{\infty} = \frac{1}{2a} < \infty$$

confirming that $x(t)$ is an energy signal.

(b) The sinusoidal signal has period $T = 2\pi/\omega$. The normalized average power is

$$p = \frac{1}{T}\int_0^T x(t)^2 dt = \frac{\omega}{2\pi}\int_0^T A^2\cos^2(\omega t + \theta)dt$$

$$= \frac{A^2\omega}{2\pi}\int_0^{2\pi/\omega} \frac{1}{2}\left[1 + \cos(2\omega t + 2\theta)\right]dt$$

$$= \frac{A^2\omega}{2\pi}\frac{1}{2}\frac{2\pi}{\omega} = \frac{A^2}{2} < \infty$$

showing that $x(t)$ is a power signal. All periodic signals are generally power signals.

(c) $\left|x[n]\right| = \left|10e^{j2n}\right| = 10\left|e^{j2n}\right| = 10\,|\cos(2n) + j\sin(2n)|$

$= 10\sqrt{\left[\cos^2(2n) + \sin^2(2n)\right]} = 10$

The normalized average power is

$$P = \lim_{N\to\infty}\frac{1}{2N+1}\sum_{n=-N}^{N}|x[n]|^2 = \lim_{N\to\infty}\frac{1}{2N+1}\sum_{n=-N}^{N}10^2$$

$$= \lim_{N\to\infty}\frac{1}{2N+1}100(2N+1) = 100 < \infty$$

that is, $x[n]$ is a power signal.

Practice Problem 1.2 Determine whether the following signals are energy signals, power signals, or neither.

(a) $x(t) = \begin{cases} t, & 0 < t < \infty \\ 0, & \text{otherwise} \end{cases}$

(b) $x(t) = \begin{cases} A, & -a < t < a \\ 0, & \text{otherwise} \end{cases}$

(c) $x[n] = 5e^{-j4n}$

Answer: (a) Neither, (b) energy, (c) power.

Example 1.3

Determine the even and odd components of

(a) $x(t) = 10\sin t + 5\cos t - 2\cos t \sin t$

(b) $y(t)$ shown in Figure 1.5a

Solution

(a) We first find $x(-t)$ and then apply Equations 1.13 and 1.14. We should keep in mind that $\sin(-t) = -\sin t$ and $\cos(-t) = \cos t$.

$$\begin{aligned} x(-t) &= 10\sin(-t) + 5\cos(-t) - 2\cos(-t)\sin(-t) \\ &= -10\sin t + 5\cos t + 2\cos t \sin t \end{aligned}$$

Hence,

$$x_e(t) = \frac{x(t) + x(-t)}{2} = 5\cos t$$

$$x_o(t) = \frac{x(t) - x(-t)}{2} = 10\sin t - 2\cos t \sin t$$

(b) We first obtain $y(-t)$ as shown in Figure 1.5b. Note that

$$y(t) = \begin{cases} 0, & t < 0 \\ 1, & t > 0 \end{cases}$$

$$y(-t) = \begin{cases} 1, & t < 0 \\ 0, & t > 0 \end{cases}$$

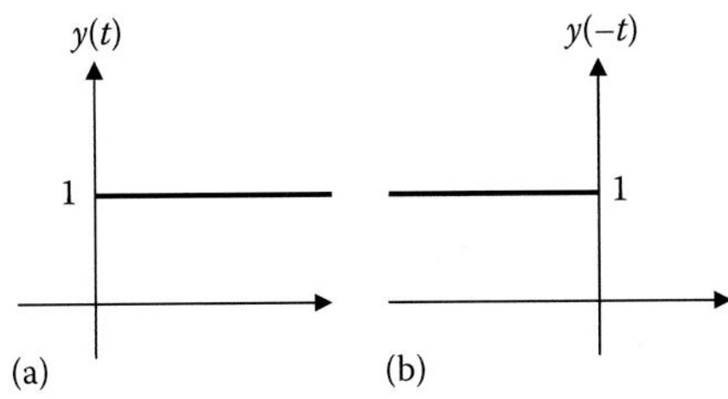

FIGURE 1.5 For Example 1.3(b).

Hence,

$$y_e(t) = \frac{y(t)+y(-t)}{2} = \begin{cases} \frac{1}{2}(0+1), & t<0 \\ \frac{1}{2}(1+0), & t>0 \end{cases} = \frac{1}{2}$$

$$y_e(t) = \frac{y(t)-y(-t)}{2} = \begin{cases} \frac{1}{2}(0-1), & t<0 \\ \frac{1}{2}(1-0), & t>0 \end{cases} = \begin{cases} -\frac{1}{2}, & t<0 \\ \frac{1}{2}, & t>0 \end{cases}$$

Both $y_e(t)$ and $y_o(t)$ are sketched in Figure 1.6.

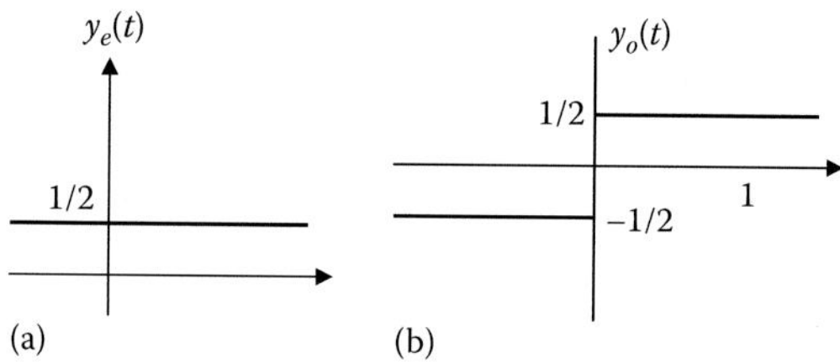

FIGURE 1.6 For Example 1.3(b).

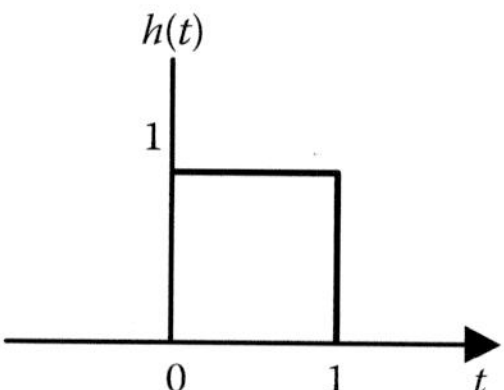

FIGURE 1.7 For Practice Problem 1.3b.

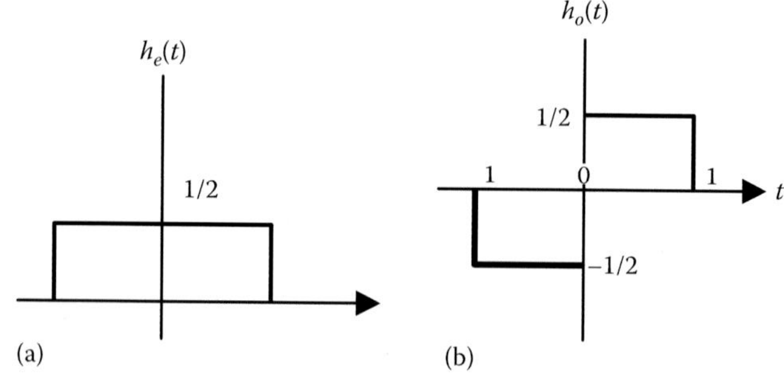

FIGURE 1.8 For Practice Problem 1.3b.

Practice Problem 1.3 Find the even and odd parts of the following signals:

(a) $z(t) = t^2 + 4t - 10$
(b) $h(t)$ shown in Figure 1.7

Answer: (a) $z_e(t) = t^2 - 10$, $z_o(t) = 4t$, (b) See Figure 1.8.

1.4 BASIC CONTINUOUS-TIME SIGNALS

We now consider some simple, standard continuous-time signals. These include the unit step function $u(t)$, the unit ramp function $r(t)$, and the impulse function $\delta(t)$. These three functions are called **singularity functions**. A singularity function is one that is discontinuous or has discontinuous derivatives. In addition to the singularity functions, we will also consider the unit rectangular function $\Pi(t)$, the unit triangular function $\Lambda(t)$, the sinusoidal function, and the exponential function.

1.4.1 Unit Step Function

The *unit step function* $u(t)$, also known as *Heaviside unit function*, is defined as

$$\boxed{u(t) = \begin{cases} 1, & t > 0 \\ 0, & t < 0 \end{cases}} \tag{1.19}$$

The **unit step function** $u(t)$ is 0 for negative values of t and 1 for positive values of t.

The unit step function is shown in Figure 1.9. We should note that $u(t)$ is discontinuous at $t = 0$; it is undefined at $t = 0$.

We use the step function to represent an abrupt change, like the changes that occur in the circuits of control systems and digital computers. It is used in representing a signal that is zero up to some time and finite thereafter. For example, the signal

$$x(t) = \begin{cases} \cos \omega t, & t \geq 0 \\ 0, & t < 0 \end{cases} \tag{1.20}$$

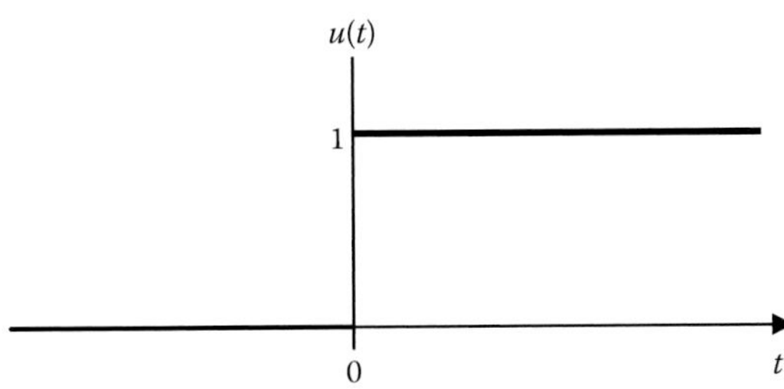

FIGURE 1.9 The unit step function.

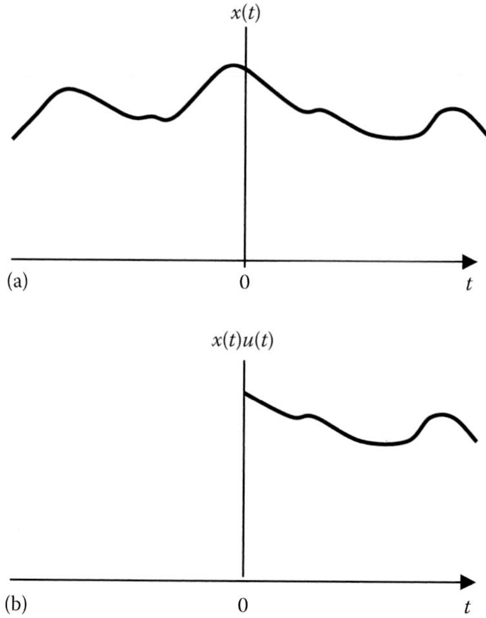

FIGURE 1.10 An arbitrary signal $x(t)$ in (a) is multiplied by $u(t)$ in (b).

can be written concisely as

$$x(t) = u(t)\cos\omega t \tag{1.21}$$

As another example, when we turn the key in the ignition system of a car, we are actually introducing a step voltage, the battery voltage. Given a continuous-time signal $x(t)$, the product $x(t)u(t)$ is given by

$$x(t)u(t) = \begin{cases} x(t), & t > 0 \\ 0, & t < 0 \end{cases} \tag{1.22}$$

This is illustrated in Figure 1.10. From this, we conclude that the multiplication of a signal $x(t)$ with the unit step function $u(t)$ eliminates the values of $x(t)$ for $t < 0$.

1.4.2 Unit Impulse Function

The derivative of the unit step function $u(t)$ is the **unit impulse function** $\delta(t)$, which we write as

$$\boxed{\delta(t) = \frac{d}{dt}u(t) = \begin{cases} 0, & t \neq 0 \\ \text{undefined}, & t = 0 \end{cases}} \tag{1.23}$$

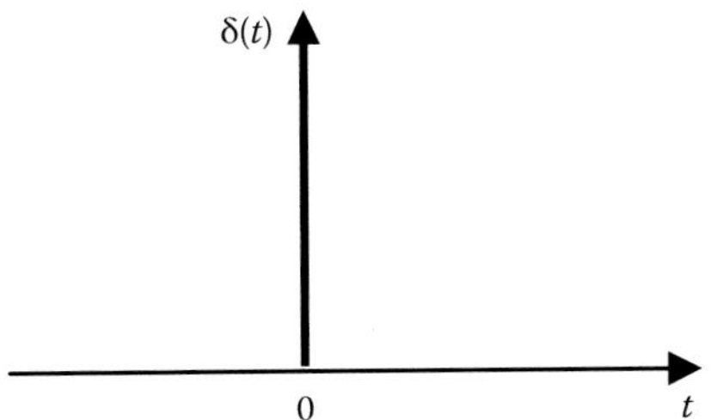

FIGURE 1.11 The unit impulse function.

The unit impulse function, also known as the *delta function*, is shown in Figure 1.11.

> The *unit impulse function* $\delta(t)$ is zero everywhere, except at $t = 0$, where it is undefined.

The unit impulse may be regarded as an applied or resulting shock. It may be visualized as a very short duration pulse of unit area. This may be expressed mathematically as

$$\int_{0^-}^{0^+} \delta(t)\,dt = 1 \tag{1.24}$$

where $t = 0^-$ denotes the time just before $t = 0$ and $t = 0^+$ is the time just after $t = 0$. For this reason, it is customary to write 1 (denoting unit area) beside the arrow that is used to symbolize the unit impulse function, as shown in Figure 1.12. For example, an impulse function $10\delta(t)$ has an area of 10. Figure 1.12 shows the impulse functions $5\delta(t + 2)$, $10\delta(t)$, and $-4\delta(t - 3)$.

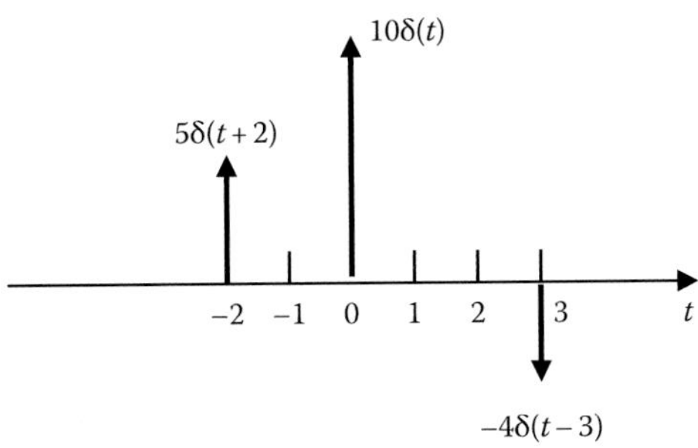

FIGURE 1.12 Three impulse functions.

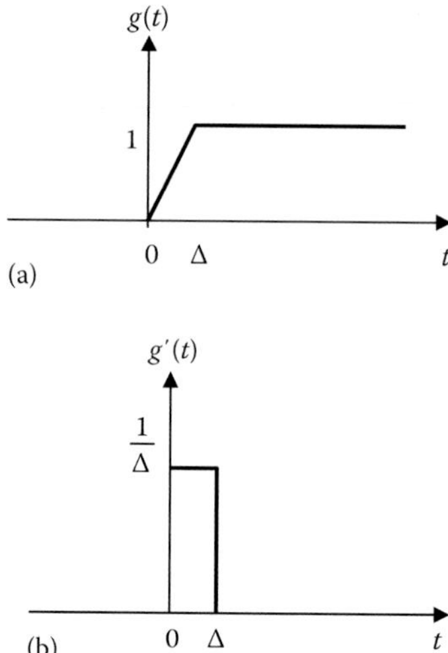

FIGURE 1.13 Functions that approach the unit step and unit impulse as $\Delta \to 0$.

To better understand the delta function, consider the function shown in Figure 1.13a. The function $g(t)$ becomes a unit step as $\Delta \to 0$.

$$u(t) = \lim_{\Delta \to 0} g(t) \tag{1.25}$$

The derivative $g'(t)$ of $g(t)$ is shown in Figure 1.13b, where we notice that as $\Delta \to 0$, $g'(t)$ becomes the unit impulse function.

$$\delta(t) = \lim_{\Delta \to 0} g'(t) \tag{1.26}$$

Thus, we can consider the delta function $\delta(t)$ as a large spike or impulse at the origin.

To illustrate how the impulse function affects other functions, let us evaluate the integral

$$\int_a^b f(t)\delta(t - t_o)\,dt \tag{1.27}$$

where $a < t_o < b$. Since $\delta(t - t_o) = 0$ except at $t = t_o$, the integrand is zero except at t_o.

Thus,

$$\int_a^b f(t)\delta(t - t_o)\,dt = \int_a^b f(t_o)\delta(t - t_o)\,dt = f(t_o)\int_a^b \delta(t - t_o)\,dt = f(t_o)$$

or

$$\boxed{\int_a^b f(t)\delta(t-t_o)dt = f(t_o)} \tag{1.28}$$

This shows that when a function is integrated with the impulse function, we obtain the value of the function at the point where the impulse occurs. This is a useful property of the impulse function known as the *sampling* or **sifting property**. The special case of Equation 1.28 is for $t_o = 0$. Then, Equation 1.28 becomes

$$\int_{0^-}^{0^+} f(t)\delta(t)\,dt = f(0) \tag{1.29}$$

If the integration interval does not include the impulse function, then the integration in Equation 1.23 is zero. For example,

$$\int_{-\infty}^{-2} \delta(t)dt = 0, \quad \int_{3}^{5} \delta(t)dt = 0 \tag{1.30}$$

This and other properties of the impulse functions are presented in Table 1.1.

1.4.3 Unit Ramp Function

Integrating the unit step function $u(t)$ results in the **unit ramp function** $r(t)$; we write

$$r(t) = \int_{-\infty}^{t} u(\lambda)\,d\lambda = tu(t) \tag{1.31}$$

TABLE 1.1
Properties of the Unit Impulse Function

1. $f(t)\delta(t-t_o) = f(t_o)\delta(t-t_o)$

2. $\int_{-\infty}^{\infty} f(t)\delta(t-t_o)dt = f(t_o)$

3. $\delta(t-t_o) = \frac{d}{dt}u(t-t_o)$

4. $u(t-t_o) = \int_{-\infty}^{t} \delta(\lambda - t_o)\,d\lambda = \begin{cases} 1, & t > t_o \\ 0, & t < t_o \end{cases}$

5. $\delta(-t) = \delta(t)$, that is, $\delta(t)$ is an even function

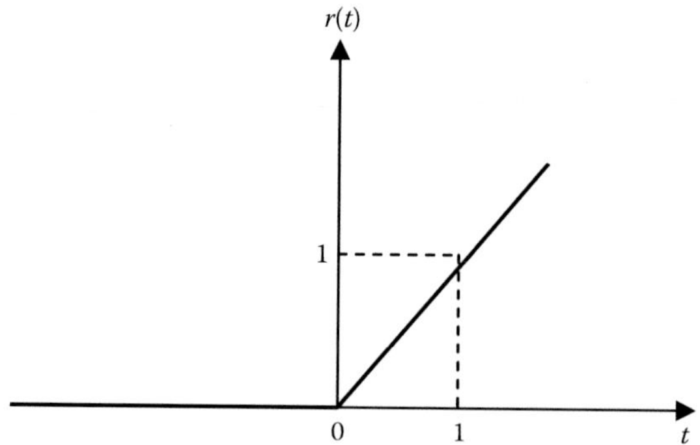

FIGURE 1.14 The unit ramp function.

or

$$\boxed{r(t) = \begin{cases} 0, & t < 0 \\ t, & t \geq 0 \end{cases}} \tag{1.32}$$

The **unit ramp function** is zero for negative values of t and has a unit slope for positive values of t.

Figure 1.14 shows the unit ramp function. In general, a ramp is a function that changes at a constant rate.

We should keep in mind that the three singularity functions (step, impulse, and ramp) are related by differentiation as

$$\delta(t) = \frac{du(t)}{dt}, \quad u(t) = \frac{dr(t)}{dt} \tag{1.33}$$

or by integration as

$$u(t) = \int_{-\infty}^{t} \delta(\lambda)\, d\lambda, \quad r(t) = \int_{-\infty}^{t} u(\lambda)\, d\lambda \tag{1.34}$$

1.4.4 Rectangular Pulse Function

The **unit rectangular pulse function** is defined as

$$\Pi\left(\frac{t}{\tau}\right) = \begin{cases} 1, & |t| < \tau/2 \\ 0, & \text{otherwise} \end{cases} = \begin{cases} 1, & -\tau/2 < t < \tau/2 \\ 0, & \text{otherwise} \end{cases} \tag{1.35}$$

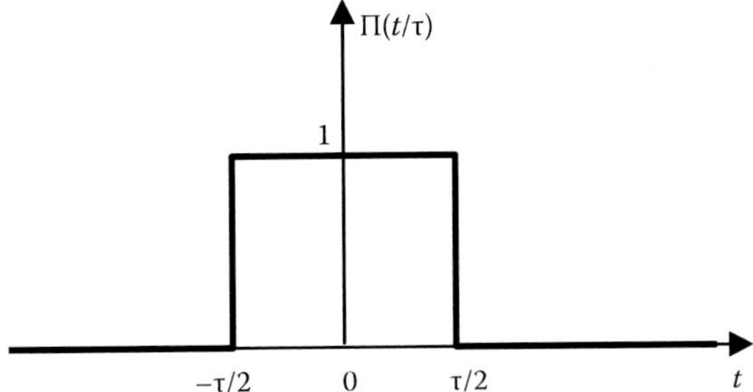

FIGURE 1.15 The unit rectangular pulse function.

It is centered at the origin and has unit height and width τ. Figure 1.15 shows the rectangular pulse function. It is evident from the figure that we can express the rectangular pulse function in terms of the unit step function as

$$\Pi\left(\frac{t}{\tau}\right) = u\left(t + \frac{\tau}{2}\right) - u\left(t - \frac{\tau}{2}\right) \tag{1.36}$$

The rectangular function results from an on–off switching operation of a source. It is used in extracting part of a signal.

1.4.5 Triangular Pulse Function

The unit triangular function is defined as

$$\Lambda\left(\frac{t}{\tau}\right) = \begin{cases} 1 - \frac{|t|}{\tau}, & -\tau < t < \tau \\ 0, & \text{otherwise} \end{cases} = \begin{cases} 1 + \frac{t}{\tau}, & -\tau < t < 0 \\ 1 - \frac{t}{\tau}, & 0 < t < \tau \\ 0, & \text{otherwise} \end{cases} \tag{1.37}$$

This means that the unit triangular pulse is centered at the origin, has unit height, and width 2τ. The triangular function is shown in Figure 1.16.

1.4.6 Sinusoidal Signal

Perhaps the most important and useful of all the standard signals is the sine wave or *sinusoidal signal*. It can take the form of

$$x(t) = \sin \omega t, \quad x(t) = \cos \omega t \tag{1.38}$$

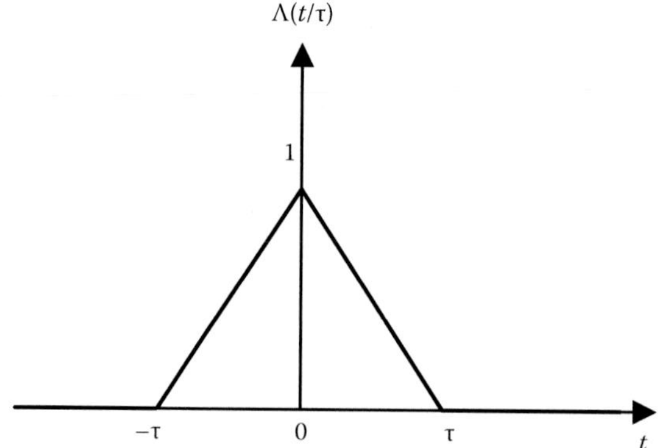

FIGURE 1.16 The unit triangular pulse function.

or by combining both to form a complex exponential signal

$$x(t) = e^{j\omega t} = \cos\omega t + j\sin\omega t \tag{1.39}$$

where ω is the frequency of the sinusoid. We already came across sinusoids in Equation 1.2.

1.4.7 Exponential Signal

The continuous-time exponential function is defined as

$$x(t) = Ae^{at} \tag{1.40}$$

where A and a are constants, which may be complex in general. If we assume that A and a are real, the exponential function is shown in Figure 1.17 for both positive and negative values of a. The signal $x(t)$ in Equation 1.39 is a complex exponential signal,

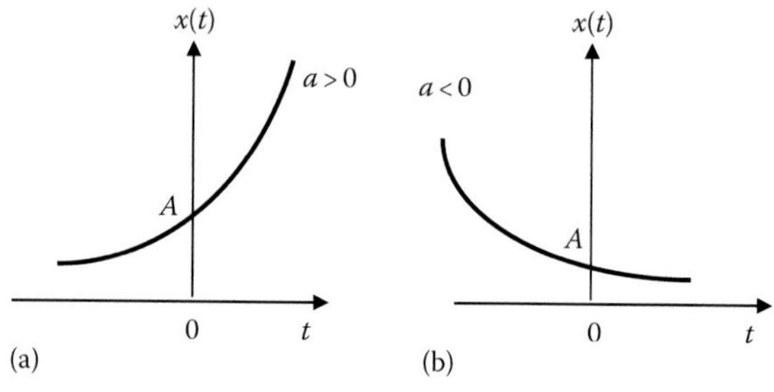

FIGURE 1.17 The exponential signal: (a) $a > 0$ and (b) $a < 0$.

which will play a major role in our treatment of signals and systems. One reason that causes the signal to arise in many applications is its rate of change.

$$\frac{d}{dt}x(t) = aAe^{at} = ax(t) \tag{1.41}$$

This means that the derivative of $x(t)$ at any time is proportional to its value at that time.

Example 1.4

Evaluate the following:

(a) $\int_{-\infty}^{\infty} \sin 3t\delta\left(t - \frac{\pi}{2}\right) dt$

(b) $\int_{-\infty}^{\infty} t^4\delta(t-2)\, dt$

Solution

(a) $\int_{-\infty}^{\infty} \sin 3t\delta\left(t - \frac{\pi}{2}\right) dt = \sin 3t\Big|_{t=\pi/2}$

$$= \sin\frac{3\pi}{2} = -1$$

(b) $\int_{-\infty}^{\infty} t^4\delta(t-2)\, dt = t^4\Big|_{t=2} = 2^4 = 16$

Practice Problem 1.4 Evaluate the following:

(a) $\int_{-\infty}^{\infty} \left[\sin\left(t^3 + \frac{\pi}{2}\right)\right]\delta(t)\, dt$

(b) $\int_{0}^{10} (t^2 + 4t - 2)\delta(t-1)\, dt$

Answer: (a) 1, (b) 3.

Example 1.5

Express the pulse (or rectangular) in Figure 1.18 in terms of the unit step. Calculate its derivative and sketch it.

Solution

The signal in Figure 1.18 is also called the **gate function**. It may be regarded as a step function that switches on at $t = 2$ s and switches off at $t = 5$ s. As shown in Figure 1.19a, it consists of two step functions. From the figure, we get

$$v(t) = 10u(t-2) - 10u(t-5) = 10[u(t-2) - u(t-5)]$$

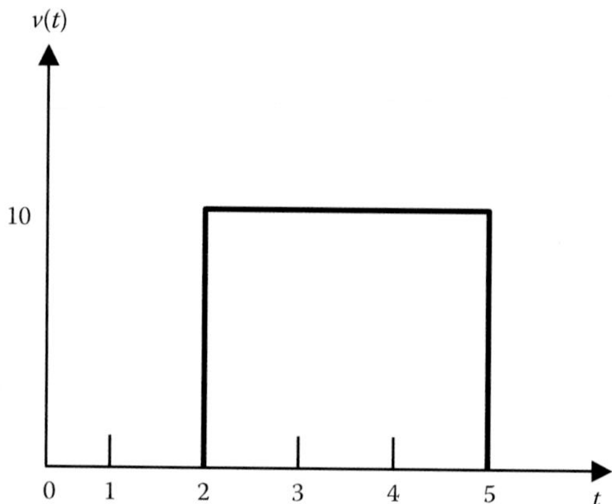

FIGURE 1.18 For Example 1.4.

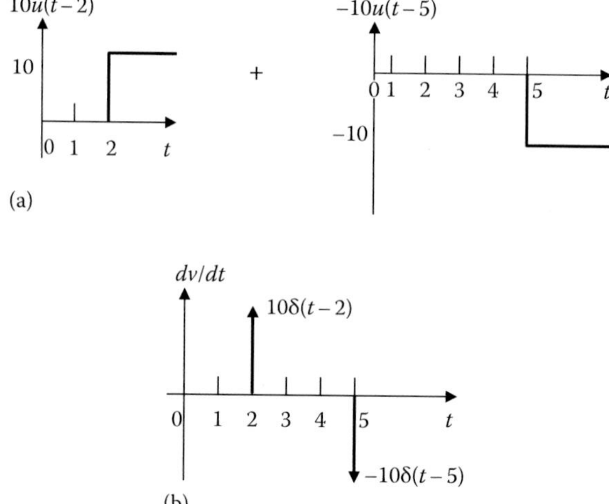

FIGURE 1.19 For Example 1.5: (a) decomposition of the pulse in Figure 1.18 and (b) derivative of the pulse in Figure 1.18.

Taking the derivative of this gives

$$\frac{dv}{dt} = 10[\delta(t-2) - \delta(t-5)]$$

which is shown in Figure 1.19b.

Practice Problem 1.5 Express the current pulse in Figure 1.20 in terms of the unit step function. Find its integral and sketch it.

Answer: $10\left[u(t)-2u(t-2)+u(t-4)\right]$, $10\left[r(t)-2r(t-2)+r(t-4)\right]$. See Figure 1.21.

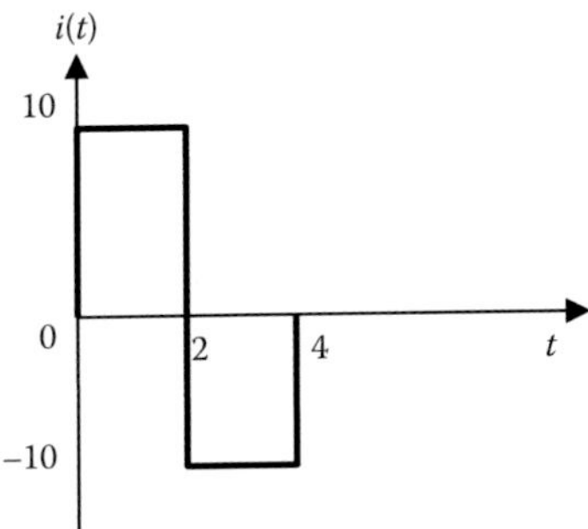

FIGURE 1.20 For Practice Problem 1.5.

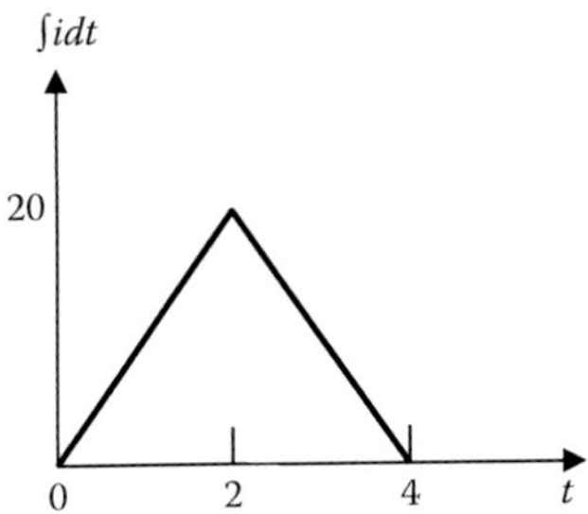

FIGURE 1.21 For Practice Problem 1.5.

Example 1.6

Sketch the following signals:

(a) $x(t) = 3\Pi\left(\dfrac{t-4}{2}\right)$

(b) $y(t) = 2\Lambda\left(\dfrac{t-3}{2}\right)$

Solution

(a) The signal $x(t)$ is a rectangular pulse centered at $t = 4$ with magnitude 3 and has width 2. It is sketched in Figure 1.22a. To be sure that the sketch is correct, we use Equation 1.35. Replace every t with $t - 4$ and set $\tau = 2$. We obtain

$$x(t) = 3\Pi\left(\frac{t-4}{2}\right) = 3\begin{cases} 1, & -2/2 < t-4 < 2/2 \\ 0, & \text{otherwise} \end{cases}$$

$$= \begin{cases} 3, & -1 < t-4 < 1 \\ 0, & \text{otherwise} \end{cases}$$

$$= \begin{cases} 3, & 3 < t < 5 \\ 0, & \text{otherwise} \end{cases}$$

which is exactly what we have in Figure 1.22a.

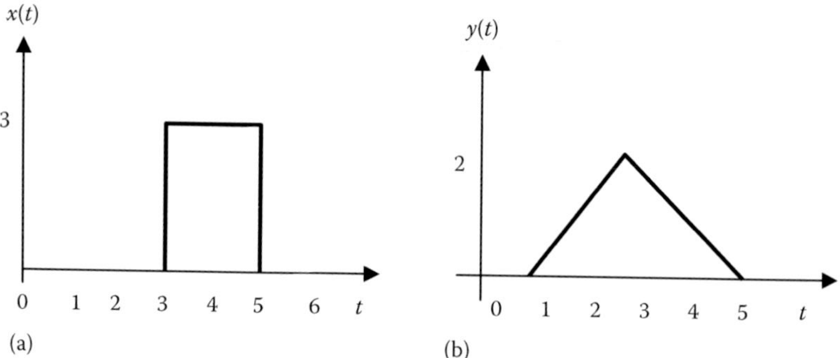

FIGURE 1.22 For Example 1.6.

(b) This is a triangular pulse centered at $t = 3$ with magnitude 2 and with $\tau = 2$ so that the width is $2\tau = 4$. Thus, $y(t)$ is sketched in Figure 1.22b. To confirm this, we use Equation 1.37. Replace every t with $t - 3$ and set $\tau = 2$. We obtain

$$y(t) = 2\Lambda\left(\frac{t-3}{2}\right) = 2\begin{cases} 1+\dfrac{t-3}{2}, & -2 < t-3 < 0 \\ 1-\dfrac{t-3}{2}, & 0 < t-3 < 2 \\ 0, & \text{otherwise} \end{cases}$$

$$= \begin{cases} 2+t-3, & 1 < t < 3 \\ 2-t+3, & 3 < t < 5 \\ 0, & \text{otherwise} \end{cases}$$

$$= \begin{cases} t-1, & 1 < t < 3 \\ 5-t, & 3 < t < 5 \\ 0, & \text{otherwise} \end{cases}$$

which confirms the sketch in Figure 1.22b.

Practice Problem 1.6 Sketch the following signals:

(a) $x(t) = 5\Pi\left(\dfrac{t-2}{4}\right)$

(b) $y(t) = 3\Lambda\left(\dfrac{t-2}{3}\right)$

Answer: See Figure 1.23.

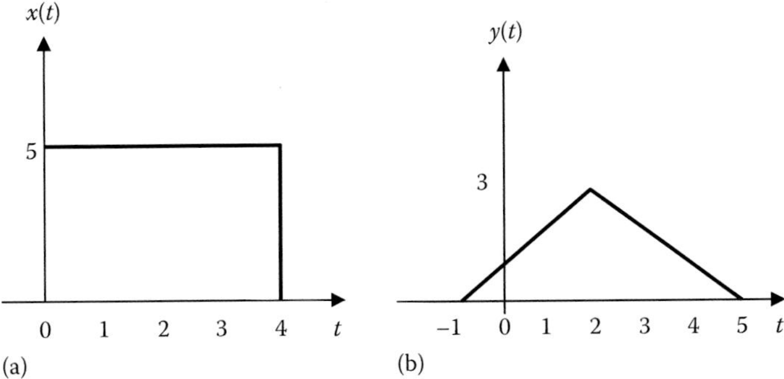

FIGURE 1.23 For Practice Problem 1.6.

1.5 BASIC DISCRETE-TIME SIGNALS

Here we consider some simple, standard discrete-time signals.

1.5.1 Unit Step Sequence

Like the continuous-time unit step, we define the unit step sequence $u[n]$ as

$$u[n] = \begin{cases} 0, & n < 0 \\ 1, & n \geq 0 \end{cases} \tag{1.42}$$

As shown in Figure 1.24, $u[n]$ is a sequence of 1s starting at the origin. Notice that $u[n]$ is defined at $n = 0$ unlike $u(t)$, which is not defined at $t = 0$.

1.5.2 Unit Impulse Sequence

In discrete time, we define unit impulse sequence as

$$\delta[n] = \begin{cases} 0, & n \neq 0 \\ 1, & n = 0 \end{cases} \tag{1.43}$$

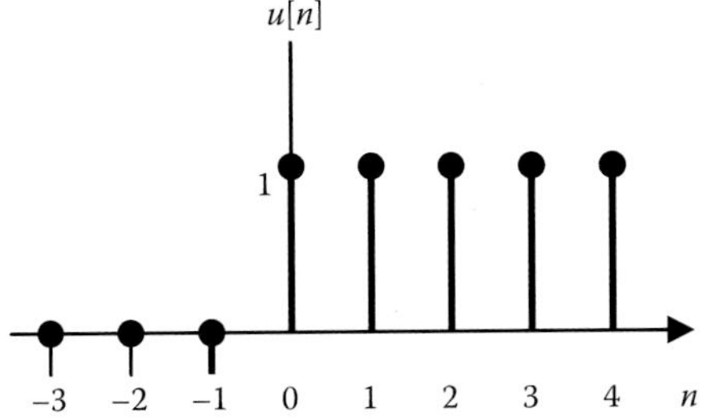

FIGURE 1.24 The unit step sequence.

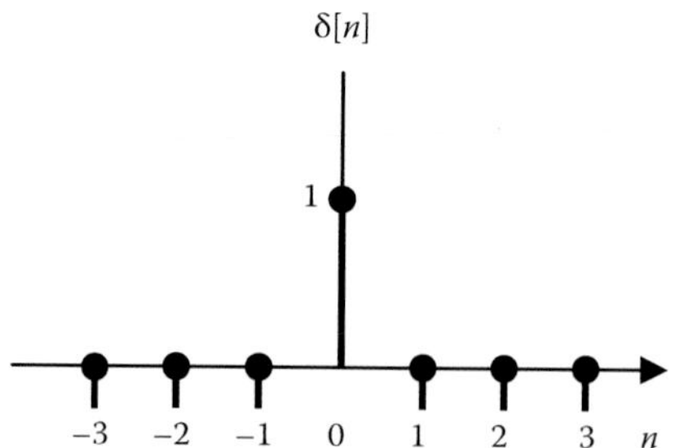

FIGURE 1.25 The unit impulse sequence.

TABLE 1.2
Properties of the Unit Impulse Sequence

1. $x[n]\delta[n] = x[0]\delta[n]$
2. $x[n]\delta[n-k] = x[k]\delta[n-k]$
3. $\delta[n] = u[n]-u[n-1]$
4. $u[n] = \sum_{k=-\infty}^{n} \delta[k]$
5. $x[n] = \sum_{k=-\infty}^{\infty} x[k]\delta[n-k]$

This is illustrated in Figure 1.25. Notice that we do not have difficulties in defining $\delta[n]$ unlike $\delta(t)$. Some properties of the unit impulse sequence are listed in Table 1.2.

1.5.3 Unit Ramp Sequence

The **unit ramp sequence** is defined as

$$r[n] = \begin{cases} n, & n \geq 0 \\ 0, & n < 0 \end{cases} \tag{1.44}$$

The sequence is shown in Figure 1.26. The relationships between unit impulse, unit step, and unit ramp sequences are

$$\delta[n] = u[n] - u[n-1] = \begin{cases} 1, & n = 0 \\ 0, & n \neq 0 \end{cases} \tag{1.45}$$

$$u[n] = \sum_{m=-\infty}^{n} \delta[m] \tag{1.46}$$

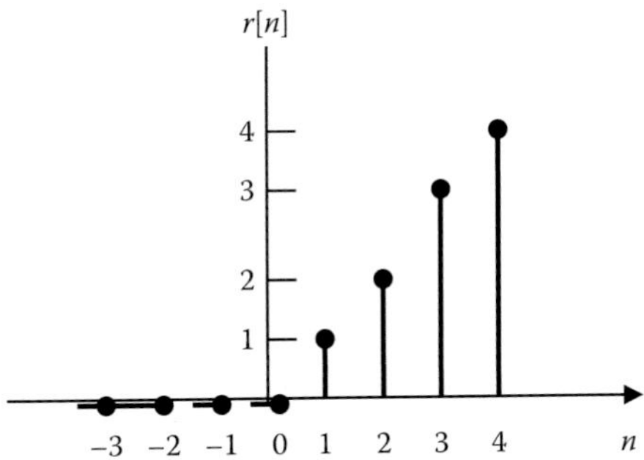

FIGURE 1.26 The unit ramp sequence.

$$u[n] = r[n+1] - r[n] \tag{1.47}$$

$$r[n] = \sum_{m=-\infty}^{n-1} u[m] \tag{1.48}$$

1.5.4 Sinusoidal Sequence

The sinusoidal sequence or a discrete-time sinusoid is given by

$$x[n] = A\cos\left(\frac{2\pi n}{N} + \theta\right) = \text{Re}\left[Ae^{j(2\pi n/N+\theta)}\right] \tag{1.49}$$

where

A is a positive real number and is the amplitude of the sequence
N is the period
θ is the phase
n is an integer

A typical sinusoidal sequence for $A = 1$, $N = 12$, and $\theta = 0$ is shown in Figure 1.27.

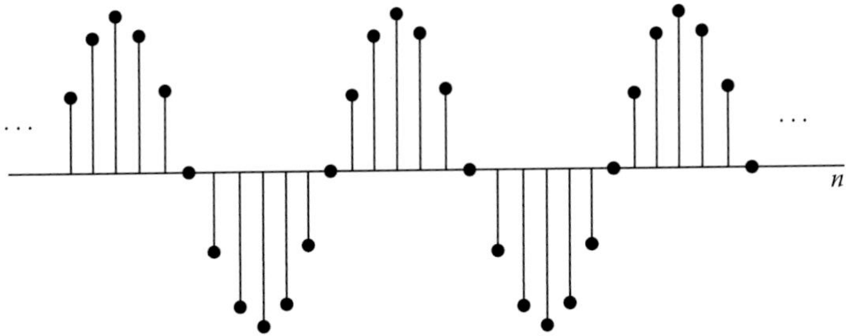

FIGURE 1.27 A discrete-time sinusoidal sequence, $x[n] = \cos(\pi n/6)$.

1.5.5 Exponential Sequence

If we sample a continuous-time exponential function $x(t) = Ae^{-at}$ with sampling period T, we obtain the sequence $x[n] = Ae^{-anT} = A\alpha^n$, with $\alpha = e^{-aT}$. Thus, the exponential sequence is given by

$$x[n] = A\alpha^n \tag{1.50}$$

where

A and α are generally complex numbers
n is an integer

A typical discrete-time exponential sequence is shown in Figure 1.28. For the signal shown in Figure 1.28, both A and α are real numbers.

Example 1.7

If $r[n] = nu[n] = \begin{cases} n, & n \geq 0 \\ 0, & n < 0 \end{cases}$, Find $y[n] = 2r[1-n]$

Solution

To obtain $y[n]$, we replace every n in $r[n]$ with $-n + 1$.

$$y[n] = 2r[-n+1] = 2(-n+1)u[-n+1] = \begin{cases} -2n+2, & -n+1 \geq 0 \\ 0, & -n+1 < 0 \end{cases}$$

$$= \begin{cases} -2n+2, & n \leq 1 \\ 0, & n > 1 \end{cases}$$

Practice Problem 1.7 Given $r[n]$ in Example 1.7, obtain $z[n] = r(n + 2)$.

Answer: $z[n] = \begin{cases} n+2, & n \geq -2 \\ 0, & n < -2 \end{cases}$

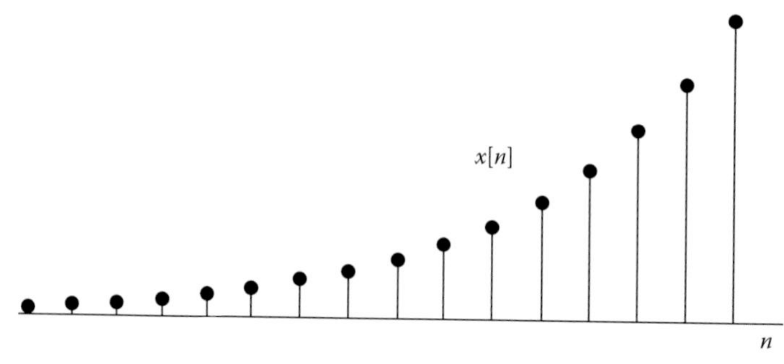

FIGURE 1.28 A discrete-time exponential sequence, $\alpha > 1$.

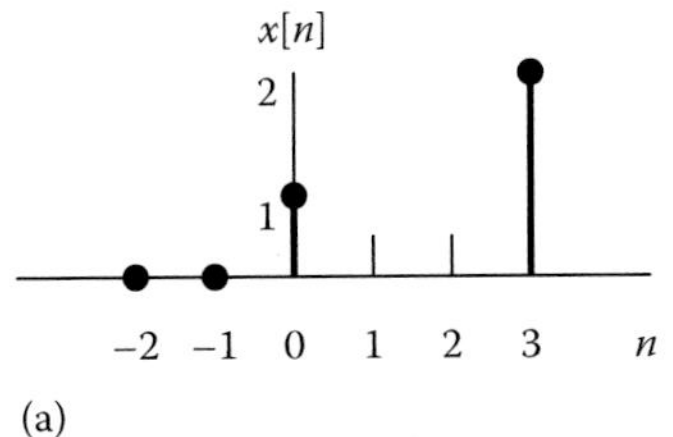

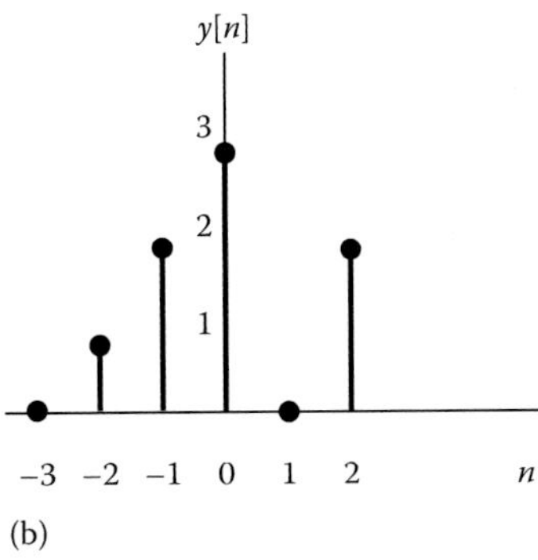

FIGURE 1.29 For Example 1.8.

Example 1.8

Write down expressions for the sequences shown in Figure 1.29.

Solution

We use item 5 in Table 1.2 as a general way of expressing any discrete signal.

$$x[n] = \sum_{k=-\infty}^{\infty} x[k]\delta[n-k]$$

(a) For the discrete-time signal in Figure 1.29a,

$$x[n] = \delta(n) + 2\delta(n-3)$$

(b) Similarly, for $y[n]$ in Figure 1.29b,

$$y[n] = \delta[n+2] + 2\delta[n+1] + 3\delta[n] + 2\delta[n-2]$$

Practice Problem 1.8 Write down expressions for the sequences shown in Figure 1.30.

Answer:

(a) $x[n] = \delta[n+1] + 2\delta[n-1]$
(b) $y[n] = 2\delta[n+2] + \delta[n+1] + \delta[n-2] + 2\delta[n-2]$

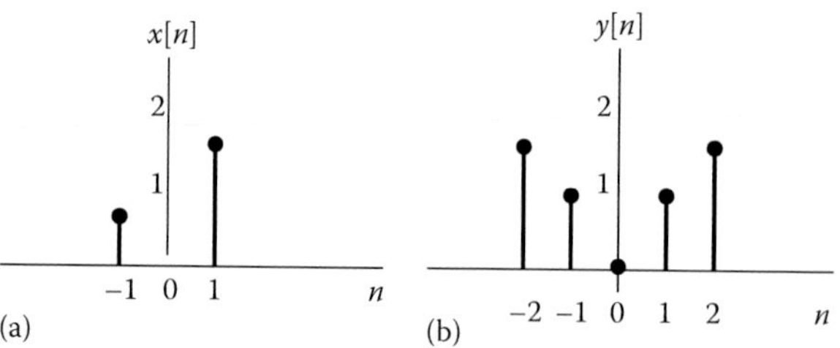

FIGURE 1.30 For Practice Problem 1.8.

1.6 BASIC OPERATIONS ON SIGNALS

We have limited ourselves to considering signals with one independent variable time (t) or integer [n]. We consider six basic operations or transformations on real function. The first three operations have to do with time, while the second three deal with transformations in amplitude. The combinations of these transformations make it possible to obtain complex signals from the basic, standard signals. Signals that are produced by these transformations are useful in sonar, radar, signal processing, and communication systems.

1.6.1 Time Reversal

Given a signal $x(t)$, its time reversal is $x(-t)$. The reversed signal $x(-t)$ is obtained as a reflection of $x(t)$ about the $t = 0$, that is, we perform a reflection about the vertical axis. (Reflection about the horizontal axis results in $-x(t)$.) An example of this is shown in Figure 1.31a. In the figure, we get

$$x(t) = \begin{cases} -1, & -1 < t < 0 \\ 1 - t, & 0 < t < 2 \\ 0, & \text{otherwise} \end{cases} \tag{1.51}$$

From this, we obtain $x(-t)$ by replacing every t with $-t$ in Equation 1.51. Hence,

$$x(-t) = \begin{cases} -1, & -1 < -t < 0 \\ 1 - (-t), & 0 < -t < 2 \\ 0, & \text{otherwise} \end{cases} = \begin{cases} -1, & 0 < t < 1 \\ 1 + t, & -2 < t < 0 \\ 0, & \text{otherwise} \end{cases} \tag{1.52}$$

as evident from Figure 1.31b.

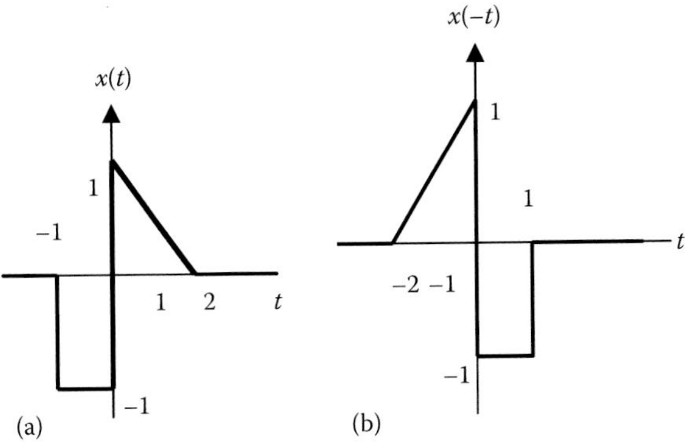

FIGURE 1.31 Time reversal of a signal.

1.6.2 Time Scaling

Time scaling involves the compression or expansion of a signal in time. Given a continuous-time signal $x(t)$, the time-scaled form of $x(t)$ is $x(at)$, where a is a constant. The scaled signal $x(at)$ will be compressed if $|a| > 1$ or expanded if $|a| < 1$; a negative value of a yields time reversal as well as compression or expansion. Again, we can obtain $x(at)$ from the signal in Equation 1.51 by replacing every t with at and simplifying. We show an example in Figure 1.32. Notice that the time reversal discussed in Section 1.6.1 can be considered a special case of time scaling with the scaling factor $a = -1$.

1.6.3 Time Shifting

Given a continuous-time signal $x(t)$, the time-shift form of $x(t)$ is $x(t - t_o)$, where t_o is a constant. If $t_o > 0$, then the signal $x(t - t_o)$ is *delayed* and the signal is shifted to the right relative to $t = 0$. When $t_o < 0$, the signal $x(t - t_o)$ is an *advanced* replica of $x(t)$, with $x(t)$ shifted to the left. Examples of $x(t - 2)$ and $x(t + 1)$ are shown in Figure 1.33.

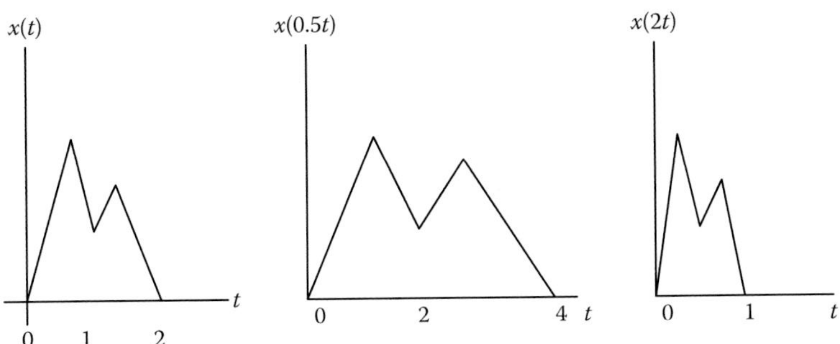

FIGURE 1.32 An example of time scaling.

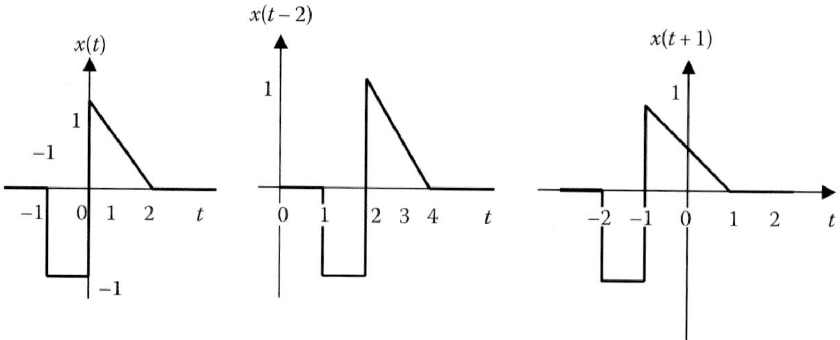

FIGURE 1.33 Time-shifted signals.

For any signal $x(t)$, the transformation $at + b$ on the independent variable can be performed as follows:

$$x(at+b) = x\left(a\left(t+\frac{b}{a}\right)\right) \tag{1.53}$$

where a and b are constants. This involves two steps:

1. Scale by factor a. If a is negative, reflect $x(t)$ about the vertical axis.
2. If b/a is negative, shift $x(t)$ to the right. If b/a is positive, shift $x(t)$ to the left.

1.6.4 Amplitude Transformations

The previous three transformations deal with the independent variable, t. Equivalent transformations of the amplitude of the signal will now be discussed. Given a signal $x(t)$, amplitude transformations take the general form

$$y(t) = Ax(t) + B \tag{1.54}$$

where A and B are constants. For example, let

$$y(t) = -2x(t) + 4 \tag{1.55}$$

We notice that $A = -2$ means amplitude reversal ($-x(t)$ implies reflection about the horizontal axis) and amplitude scaling ($|A| = 2$). Also, $B = 4$ shifts vertically the amplitude of the signal.

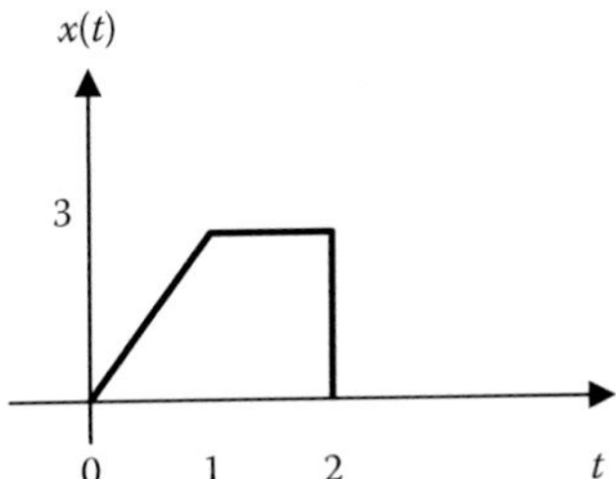

FIGURE 1.34 For Example 1.9.

Example 1.9

A continuous-time signal is shown in Figure 1.34. Sketch each of the following signals:

(a) $x(2t-6)$, (b) $x(t/2 + 1)$, (c) $y(t) = -1 + 2x(t)$

Solution

(a) We can do this in two ways:

Method 1: We can write this as $x(2(t-3))$, indicating that $x(t)$ is compressed by a factor of 2 and shifted to the right by three units. Hence, $x(2t-6)$ is as sketched in Figure 1.35a.

Method 2: From Figure 1.34,

$$x(t) = \begin{cases} 3t, & 0 < t < 1 \\ 3, & 1 < t < 2 \\ 0, & \text{otherwise} \end{cases} \tag{1.9.1}$$

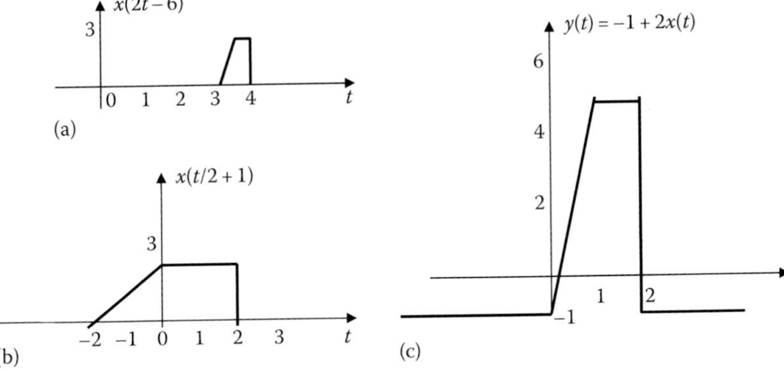

FIGURE 1.35 For Example 1.9.

Replacing every t with $2t - 6$ gives

$$x(2t-6) = \begin{cases} 3(2t-6), & 0 < 2t-6 < 1 \\ 3, & 1 < 2t-6 < 2 \\ 0, & \text{otherwise} \end{cases}$$

$$= \begin{cases} 6t-18, & 3 < t < 3.5 \\ 3, & 3.5 < t < 4 \\ 0, & \text{otherwise} \end{cases}$$

which is what we have in Figure 1.35a.

(b) *Method 1*: We write $x(t/2 + 1) = x(1/2(t + 2))$ indicating that $x(t)$ is expanded by a factor of 2 and advanced or shifted left by 2 units. The signal $x(t/2 + 1)$ is sketched in Figure 1.35b.

Method 2: Replace every t in Equation 1.9.1 by $t/2 + 1$.

$$x\left(\frac{t}{2}+1\right) = \begin{cases} 3(t/2+1), & 0 < t/2+1 < 1 \\ 3, & 1 < t/2+1 < 2 \\ 0, & \text{otherwise} \end{cases}$$

$$= \begin{cases} 1.5t+3, & -2 < t < 0 \\ 3, & 0 < t < 2 \\ 0, & \text{otherwise} \end{cases}$$

which agrees with the sketch in Figure 1.35b.

(c) This deals with amplitude amplification; there is no transformation of time t.

Method 1: The signal $x(t)$ is amplified by 2 and lowered by -1, as shown in Figure 1.35c, giving $-1 + 2x(t)$.

Method 2:

$$y(t) = -1 + 2x(t) = \begin{cases} -1+6t, & 0 < t < 1 \\ -1+6, & 1 < t < 2 \\ -1, & \text{otherwise} \end{cases}$$

$$= \begin{cases} -1+6t, & 0 < t < 1 \\ 5, & 1 < t < 2 \\ -1, & \text{otherwise} \end{cases}$$

which agrees with what we have in Figure 1.35c.

Practice Problem 1.9 With the continuous-time signal in Figure 1.34, sketch each of the following signals:

(a) $x(2t + 4)$, (b) $-2x(t-1)$, (c) $2 + x(-t)$.

Answer: See Figure 1.36.

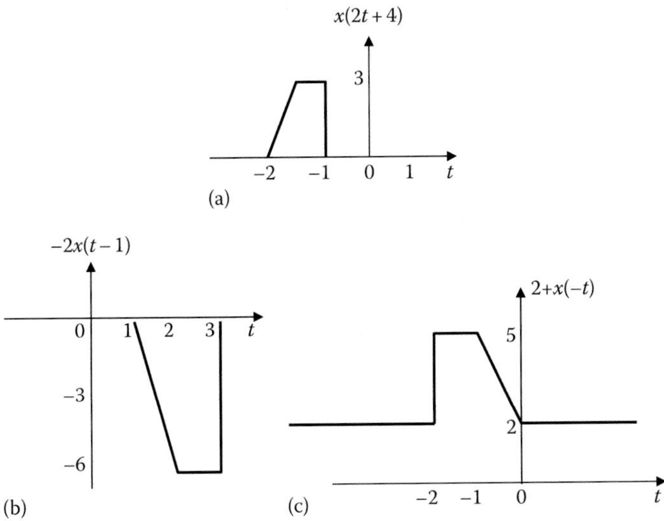

FIGURE 1.36 For Practice Problem 1.9.

Example 1.10

A discrete-time signal is shown in Figure 1.37. Sketch each of the following signals: (a) $x[n-3]$, (b) $x[-n + 3]$, (c) $x[2n]$.

Solution

(a) This involves shifting $x[n]$ three units to the right, as shown in Figure 1.38a.
(b) We can write $x[-n + 3] = x[-(n - 3)]$, meaning that we perform a reflection about the vertical axis to get $x[-n]$ and we then shift these three units to the right. This is shown in Figure 1.38b.
(c) This involves compressing $x[n]$ by a factor of two, as shown in Figure 1.38c.

Practice Problem 1.10 With the discrete-time signal shown in Figure 1.37, sketch each of the following signals: (a) $x[-n]$, (b) $x[n + 2]$, (c) $x[n/2]$

Answer: See Figure 1.39.

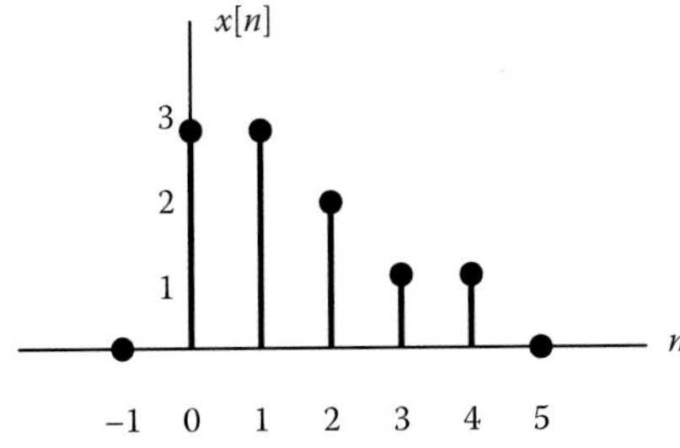

FIGURE 1.37 For Example 1.10.

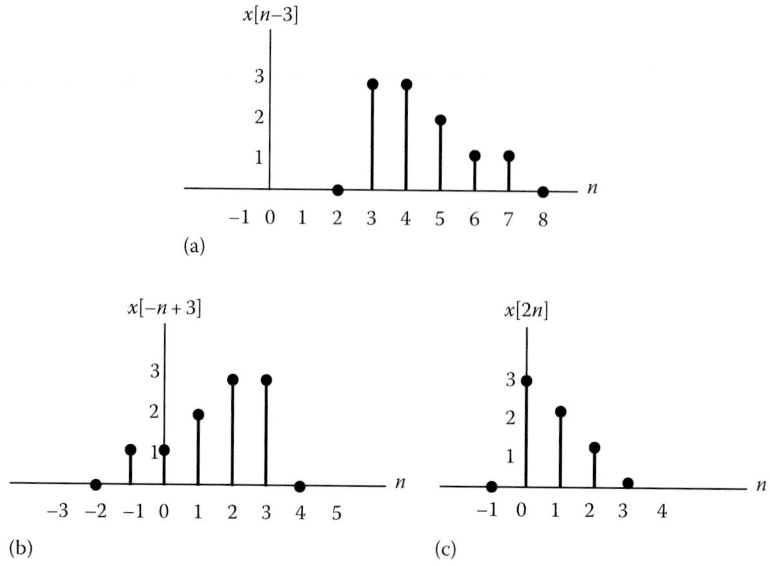

FIGURE 1.38 For Example 1.10.

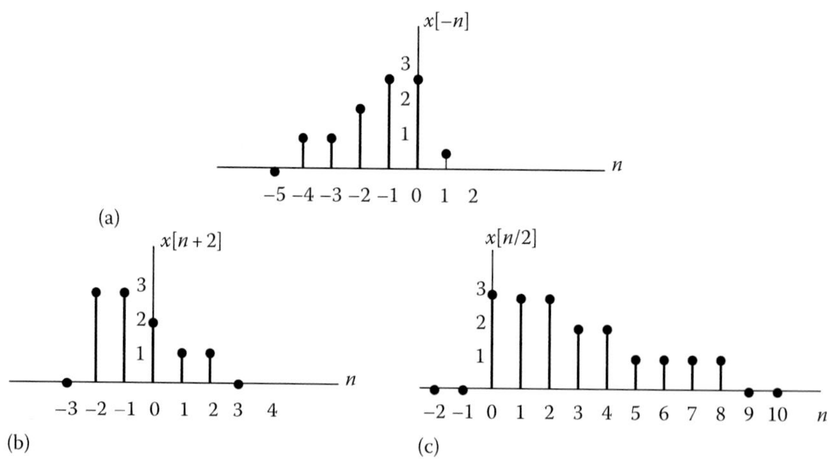

FIGURE 1.39 For Practice Problem 1.10.

1.7 CLASSIFICATIONS OF SYSTEMS

As mentioned in Section 1.2, a system may be regarded as a mathematical model of a physical process that relates the input signal $x(t)$ to the output signal $y(t)$. This relationship is illustrated in Figure 1.40a. Although a system may have many input and output signals as shown in Figure 1.40b, we will focus our attention on the single-input single-output case in this book.

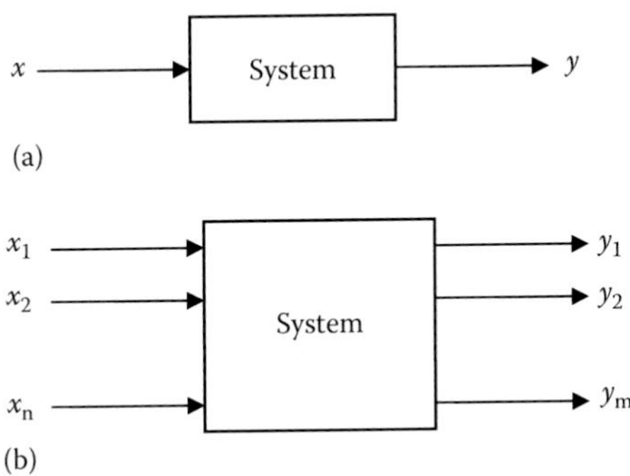

FIGURE 1.40 Block diagram representation: (a) system with single-input and single-output and (b) system with many inputs and outputs.

If x is the input signal and y is the output signal, they are related through a transformation

$$y = \mathbf{T}x \tag{1.56}$$

where **T** is an operator transforming x into y.

1.7.1 Continuous-Time and Discrete-Time Systems

A system is continuous-time if the input and output signals are continuous-time. It is discrete-time if the input and output signals are discrete-time. In continuous-time systems, time is measured continuously. Some continuous-time systems are described by ordinary differential equations, algebraic equations (resistive circuits), polynomial equations (diodes), integral equations (op amp integrator circuits), etc. For discrete-time systems, time is defined only at discrete instants and the systems are described by difference equation and any other way the input–output properties of the system may be specified. Continuous-time and discrete-time systems are illustrated in Figure 1.41.

1.7.2 Causal and Noncausal Systems

The output of a system may depend on the present and past inputs. A system that has this property is known as *causal system*. A causal (or nonanticipatory) system is one whose output $y(t)$ at the present time depends only on the present and past values of the input $x(t)$.

A **causal system** is one whose present response does not depend on the future values of the input.

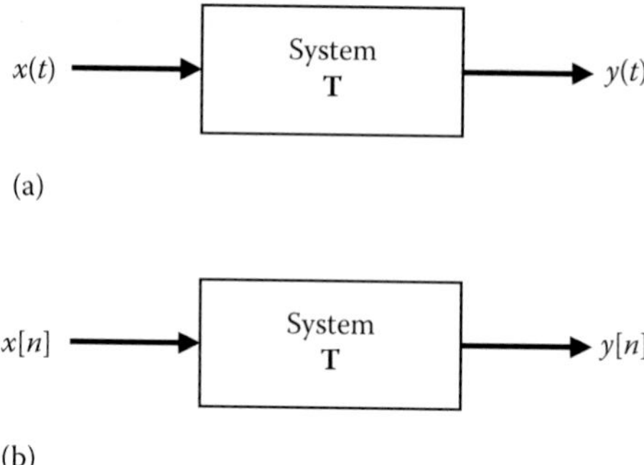

FIGURE 1.41 Block diagram representing (a) a continuous-time system and (b) a discrete-time system.

Examples of causal systems are

$$y(t) = x(t-1) \tag{1.57}$$

$$y[n] = x\left[\frac{n}{2}\right] \tag{1.58}$$

Every physical system is causal. The motion of a car is causal since it does not expect the future actions of the driver. Causality is a necessary condition for a system to be realized in the real world we live in. Causal systems are physically realizable.

If a system is not causal, it is said to be *noncausal* or anticipatory. Typical examples of **noncausal systems** are

$$y(t) = x(t+2) \tag{1.59}$$

$$y[n] = x[1-n] \tag{1.60}$$

The system in Equation 1.59 is not causal because the output at, say $t = 0$, is equal to input at $t = 2$. This is not physically realizable. The same logic applies to the system in Equation 1.60.

An ideal filter is noncausal and is not physically realizable; it cannot be built in practice. It should be noted that causality is not often an essential constraint in applications in which the independent variable is not time, such as in image processing.

1.7.3 Linear and Nonlinear Systems

Linearity is the property of the system describing a linear relationship between input (cause) and output (effect). The property is a combination of both homogeneity (scaling) property and the additivity property.

The homogeneity property requires that if the input is multiplied by any constant k, then the output is multiplied by the same constant, that is,

$$\mathbf{T}\{kx\} = ky \tag{1.61}$$

The additivity property requires that the response to a sum of inputs is the sum of the responses to each input applied separately. If $\mathbf{T}x_1 = y_1$ and $\mathbf{T}x_2 = y_2$, then

$$\mathbf{T}\{x_1 + x_2\} = y_1 + y_2 \tag{1.62}$$

We can combine Equations 1.61 and 1.62 as

$$\mathbf{T}\{k_1x_1 + k_2x_2\} = k_1y_1 + k_2y_2 \tag{1.63}$$

where k_1 and k_2 are constants.

A *linear system* meets the two conditions in Equations 1.61 and 1.63, that is, a system is linear if it satisfies homogeneity and additivity properties.

A linear system may be regarded as one in which all the interrelationships among the quantities involved are expressed by linear equations, which may be algebraic, differential, or integral. A very important consequence of linearity is that *superposition principle* applies. If an input consists of the weighted sum of several signals, then the output is the weighted sum of the responses of the system to each of those signals.

An example of a linear system is a circuit that contains resistors, capacitors, and inductors, since these are linear elements. Devices such as rectifiers, diodes, and saturating magnetic devices are nonlinear. A system having even one such device is treated as nonlinear. Examples of nonlinear systems are

$$y = \sin x \tag{1.64}$$

$$y = x^3 \tag{1.65}$$

1.7.4 Time-Varying and Time-Invariant Systems

When one or more parameters of the system vary with time, the system is said to be time-varying.

A *time-varying system* is one whose parameters vary with time.

For continuous-time, time-varying systems are described by time-varying differential equations. For discrete-time, time-varying systems are described by time-varying difference equations. If all system parameters are constant with time, the system is said to be time-invariant or fixed.

In a *time-invariant system*, a time shift (advance or delay) in the input signal leads to the time shift in the output signal.

For a continuous-time system, the system is time-invariant when

$$T\{x(t-\tau)\} = y(t-\tau) \tag{1.66}$$

where τ is a constant. Consider the input $x(t)$ of a linear system yielding an output $y(t)$. For the system to be time-invariant, a shifted input $x(t-\tau)$ should result in a shifted output $y(t-\tau)$. Figure 1.42 shows an example of time-invariant system, while Figure 1.43 gives an example of time-varying system.

For a discrete-time system, the system is time-invariant when

$$T\{x[n-m]\} = y[n-m] \tag{1.67}$$

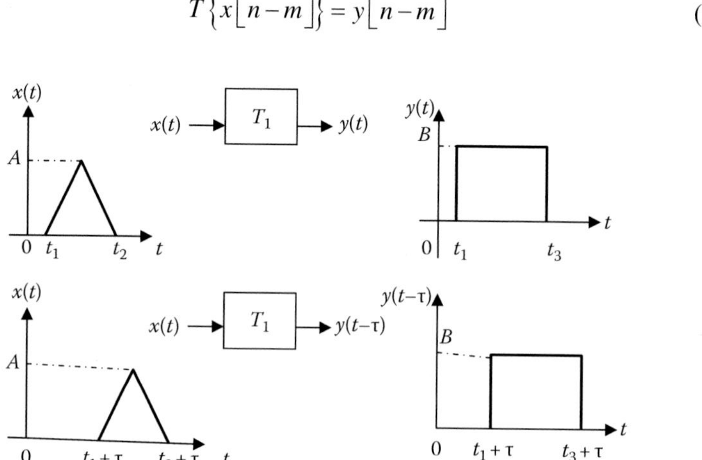

FIGURE 1.42 An example of a time-invariant system.

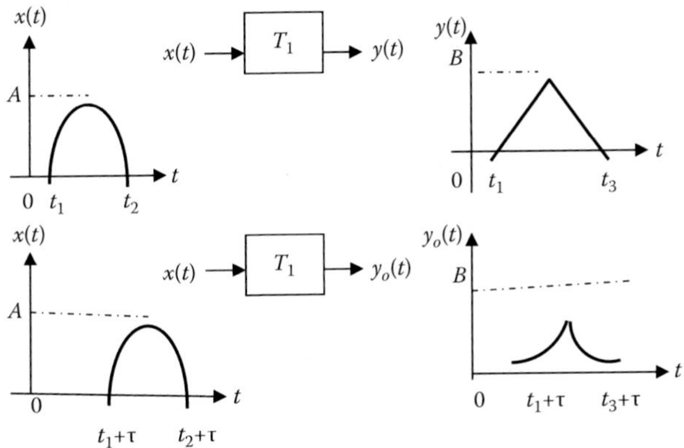

FIGURE 1.43 An example of a time-varying system.

for any integer *m*. A continuous-time/discrete-time system that does not satisfy Equation 1.63 or 1.64 is time-varying. Such a system has parameters that vary with time. We will confine our efforts to linear time-invariant (LTI) systems in this book.

1.7.5 Systems with and without Memory

When the output of a system depends on the past and/or future input, the system is said to have a memory. For example, a system described by

$$y(t) = x(t) - x(t-1) + x(t+2) \tag{1.68}$$

has a memory since the output $y(t)$ requires the past input $x(t-1)$, the current input $x(t)$, and the future input $x(t+2)$.

If the output of a system does not depend on the past and/or future input, the system is memoryless (or instantaneous).

A *memoryless system* is one in which the current output depends only on the current input; it does not depend on the past or future inputs.

A system with a memory is also called a *dynamic system.* A memoryless system is called a *static system.* Many practical systems such as resistive networks, amplifiers, operational amplifiers, and diodes are usually modeled as memoryless systems.

Example 1.11

Show that the system represented by

$$\frac{dy}{dt} + 4y(t) = 2x(t) \tag{1.11.1}$$

is linear.

Solution

To prove that the system is linear, we need to show that Equation 1.63 is satisfied. Let the inputs $x_1(t)$ and $x_2(t)$ cause the system responses $y_1(t)$ and $y_2(t)$, respectively, that is,

$$\frac{dy_1}{dt} + 4y_1(t) = 2x_1(t) \tag{1.11.2}$$

$$\frac{dy_2}{dt} + 4y_2(t) = 2x_2(t) \tag{1.11.3}$$

Multiplying Equation 1.11.2 by a constant A and Equation 1.11.3 by constant B and adding, we get

$$\frac{d}{dt}\left[Ay_1 + By_2\right] + 4\left[Ay_1 + By_2\right] = 2[Ax_1 + Bx_2]$$

This satisfies the system equation (1.11.1) with

$$x = Ax_1 + Bx_2$$
$$y = Ay_1 + By_2$$

(The time dependence is dropped for simplicity.) In other words, with the input as $x = Ax_1 + Bx_2$, the response is $y = Ay_1 + By_2$. Superposition holds and we conclude that the system is linear.

Practice Problem 1.11 Show that the system described by

$$y(t) = 2x(t) - 5$$

is nonlinear.

Answer: Proof: a part of the output signal does not depend on the input signal.

Example 1.12

A modulator with carrier frequency of ω_c gives an output

$$y(t) = x(t)\cos\omega_c t$$

Determine whether the system is (a) memoryless, (b) causal, or (c) time-invariant.

Solution

(a) The system is memoryless since the output $y(t)$ depends only on the present values of the input $x(t)$, that is, the output is a function of the input at only the present time.
(b) It is causal since the output $y(t)$ does not depend on the future values of the input $x(t)$. Since it is memoryless, it is also causal.
(c) To be time-invariant, an input $x(t - T)$ should produce an output $y(t - T)$. Let the input signal $x_o(t-t_o)$ produce an output signal y_o. Then

$$y_o(t) = x(t - t_o)\cos\omega_c t$$

But we know that

$$y(t - t_o) = x(t - t_o)\cos\omega_c(t - t_o) \neq y_o$$

The system fails to satisfy the time-invariance property of Equation 1.66. Hence it is not time-invariant. It is time-varying.

Practice Problem 1.12 Consider the system represented by $y(t) = x(t - 3)$. Discuss the properties of the system, that is, check whether it is memoryless, causal, and time-invariant.

Answer: Not memoryless, causal, and time-invariant.

1.8 APPLICATIONS

The information covered in this chapter is useful in many systems such as mass-spring-damper system, moving average filter, stock trading, electrical circuits, square-law device, and digital signal processing (DSP). We will briefly consider three simple applications related to what we have covered in this chapter. These applications deal with electric circuits, square-law device, and DSP system.

1.8.1 Electric Circuit

Electric circuits are mostly linear systems. Consider the one shown Figure 1.44. Our objective is to find the system equation for it. By applying Kirchhoff's voltage law, we obtain

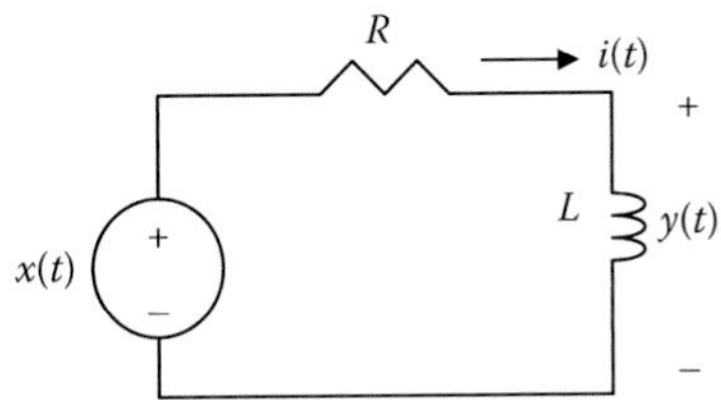

FIGURE 1.44 An electric circuit.

$$x(t) = Ri(t) + y(t) \tag{1.69}$$

But $i(t) = \dfrac{1}{L}\int_{-\infty}^{t} y(\lambda)\,d\lambda$. Substituting this in Equation 1.69,

$$y(t) = x(t) - \frac{R}{L}\int_{-\infty}^{t} y(\lambda)\,d\lambda \tag{1.70}$$

Or we may differentiate this to get

$$\frac{dy(t)}{dt} + \frac{R}{L}y(t) = \frac{dx}{dt} \tag{1.71}$$

This is an explicit input–output relationship between $x(t)$ (source voltage) and $y(t)$ (inductor voltage).

1.8.2 Square-Law Device

A square-law device is a device where either current or voltage depends on the square of the other. The input–output relationship is given by

$$y(t) = x^2(t) \tag{1.72}$$

Any system that is defined by Equation 1.72 is called a *square-law device*. To realize this system, we use a signal amplifier, as shown in Figure 1.45. For this reason, the square-law detector is also called a *signal multiplier*. It can be built using operational amplifiers and diodes. It is evident from Equation 1.72 that the system is nonlinear and memoryless. A square-law device is often found in the receiver front end of radar and communication devices.

1.8.3 DSP System

Since most signals are analog in nature, an analog signal can be processed directly in its analog form, as shown in Figure 1.46. DSP provides an alternative

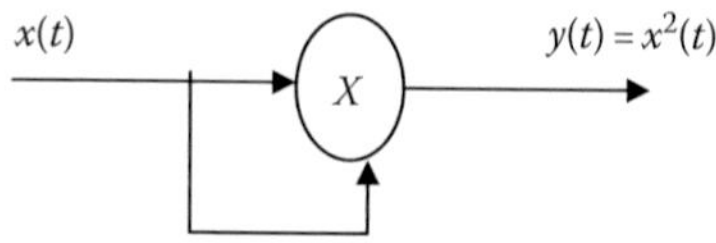

FIGURE 1.45 A square-law detector.

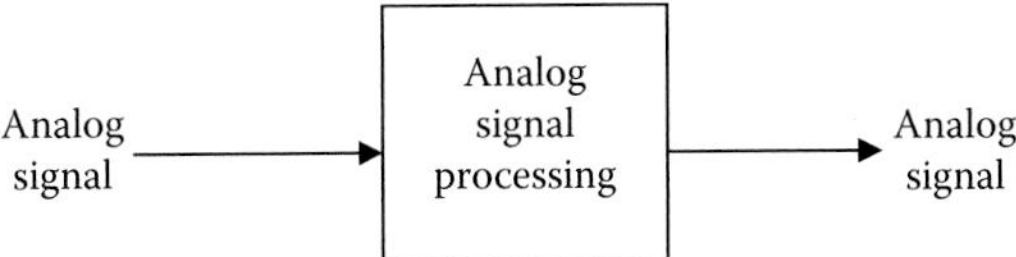

FIGURE 1.46 Analog signal processing.

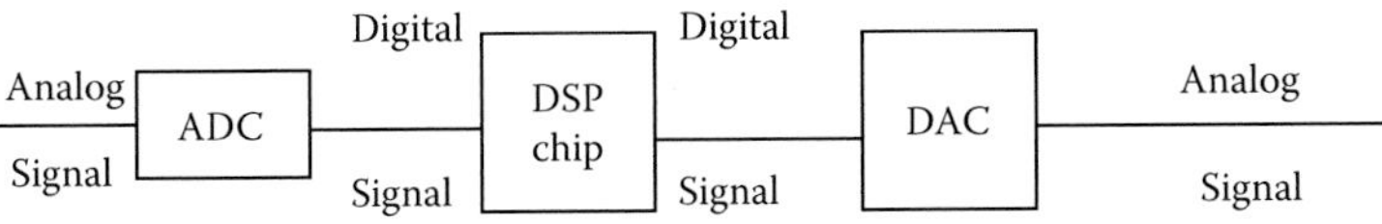

FIGURE 1.47 Digital signal processing.

means of processing the analog signal, as shown in Figure 1.47. DSP techniques are replacing analog signal processing in many fields such as telecommunications; radar and sonar; digital control; speech, audio, and video processing; and biomedicine. One main reason for preferring DSP is that, it allows programmability, which means that the same DSP hardware can be used for different applications.

A DSP is a special microprocessor designed for DSP. Most DSPs can be used to manipulate different kinds of information including sound, images, and video. The goal of a DSP is to process real-world analog signals. As illustrated in Figure 1.47, the first step is to convert the analog signal to digital signal using an ADC. The ADC quantizes the sampled analog signal and codes the quantized signal into an acceptable format. It is common for the sequence coming from the ADC to be converted back to analog to drive a motor, a stereo system, or whatever. The conversion of the digital sequence to an analog signal requires a digital-to-analog converter (DAC). Most PCs include ADCs and DACs in their sound card. In these modern days, DSP systems such as high-speed modems and cell phones are common.

1.9 COMPUTING WITH MATLAB®

A brief introduction to MATLAB is given in Appendix C. The reader is encouraged to read Appendix C before proceeding with this section.

MATLAB provides built-in functions for unit step function $u(t)$ and unit impulse function $\delta(t)$. The **unit step function** is called **Heaviside** or **stepfun**, while the impulse function is **Dirac**. **Heaviside**(t) is zero when $t < 0$, 1 for $t > 0$ and 0.5 for $t = 0$. stepfun(t,t0) returns a vector of the same length at t with zeros for $t < t0$ and ones for $t > t0$. **stepfun**(n,n0) works the same way for discrete signals. **Dirac**(t) is zero for all t, except $t = 0$, where it is infinite.

MATLAB treats continuous-time signals differently from discrete-time signals. We will treat them separately in the following two examples.

Example 1.13

Use MATLAB to plot

(a) The signal $x(t) = 10e^{-0.5t}\sin(4\pi t) + 20e^{-t}\sin(2\pi t)$
(b) The step function starting at $t = 1$ and going right

Solution

(a) The MATLAB code is shown in the following:

```
t = 0.1:2*pi/100:2*pi % t ranges from 0.1 to 2*pi
x =10* exp(-0.5*t).*sin(4*pi*t) + 20*exp(-t).*sin(2*pi*t); %
generates values for x
plot(t,x); %plot the signal
xlabel( 't in seconds')
ylabel('signal x(t)')
```

We let t vary from 0.1 to 2π s, with an increment of $2\pi/100$ s. Note the use of the .* operator. The terms 10*exp(−0.5*t), 20*exp(−t).* are all vectors. We use .* instead of * because we want the components of these vectors to be multiplied by their corresponding components of the other vectors. Thus, element-by-element operations require a dot before the operator. The plot is presented in Figure 1.48a.

(b) The MATLAB script for this case is presented as follows:

```
t = 0:0.1:5
t0 = 1;
u = stepfun(t,t0);
```

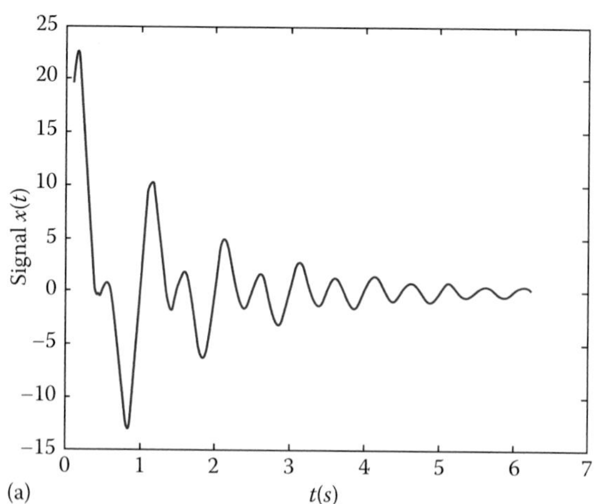

FIGURE 1.48 (a) For Example 1.13a and b for Example 1.13b. *(Continued)*

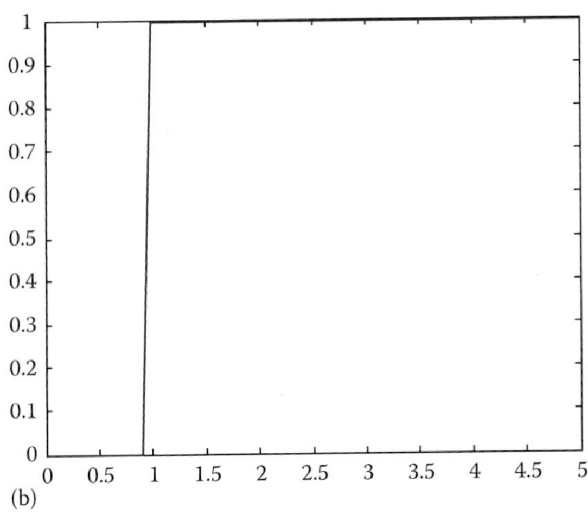

FIGURE 1.48 (*Continued*) (a) For Example 1.13a and b for Example 1.13b.

```
plot(t,u)
```

As shown in Figure 1.48b, the plot is a delayed unit step, that is, $u(t - 1)$.

Practice Problem 1.13 Using MATLAB, plot the signal

$$x(t) = e^{-0.2t}\cos(0.5t + 1.2),\ 0 < t < 20\text{ s}.$$

Answer: See Figure 1.49.

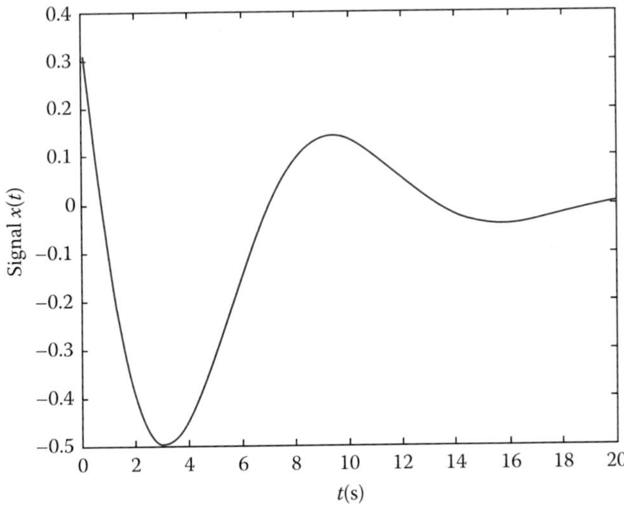

FIGURE 1.49 For Practice Problem 1.13.

Example 1.14

Using MATLAB, plot the following discrete signals:

(a) $$x[n] = \begin{cases} 1, & n = 0 \\ 3, & n = 1 \\ -2, & n = 3 \\ 2, & n = 4 \\ 0, & \text{otherwise} \end{cases}$$

(b) $f[n] = A(\alpha)^n$ with $A = 5$, $\alpha = -0.6$

Solution

(a) A plot of this discrete signal is generated by the following MATLAB commands:

```
n=-3:1:5; % n = -3, -4,…,4,5
x=[ 0 0 0 1 3 0 -2 2 0 ];
stem(n,x);
xlabel('n');
ylabel( 'x[n]');
       grid;
```

The plot is shown in Figure 1.50a.

(b) The MATLAB code for generating and plotting the discrete signal is as follows:

```
n=-5:1:5;
f=5*(-0.6).^n;
stem(n,f);
xlabel('n');
ylabel('f(n)');
```

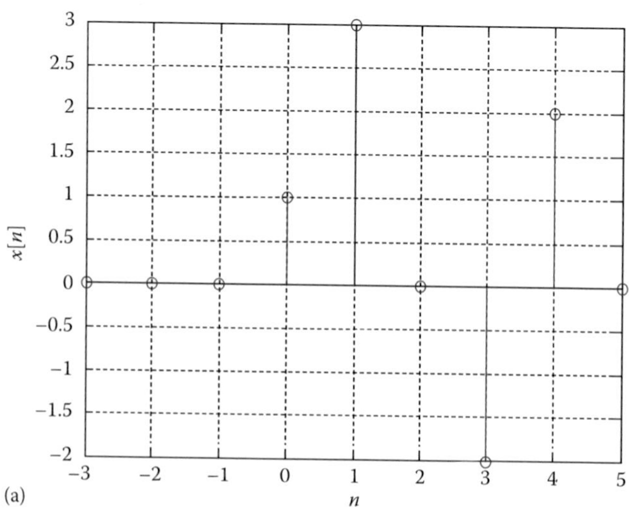

FIGURE 1.50 (a) For Example 1.13a. *(Continued)*

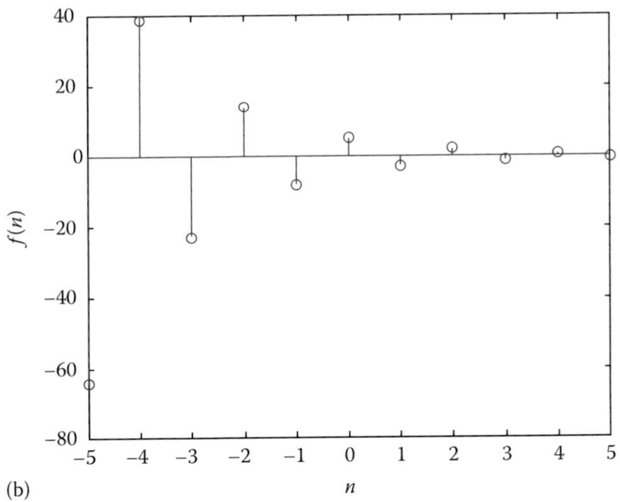

FIGURE 1.50 (*Continued*) (b) For Example 1.13b.

The plot is shown in Figure 1.50b. Notice that $f[n]$ is an alternating exponential sequence.

Practice Problem 1.14 Use MATLAB to plot the following discrete signals:

(a) $y[n] = 10\sin\left(\frac{2\pi n}{16}\right)$ for $n = 0$ to 32.

(b) The exponential sequence in Example 1.12b multiplied by a step sequence.

Answer: See Figure 1.51.

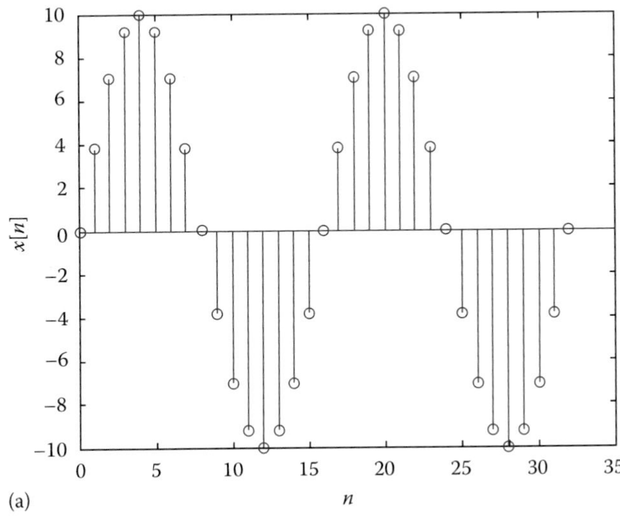

FIGURE 1.51 (a) For Practice Problem 1.14a. *(Continued)*

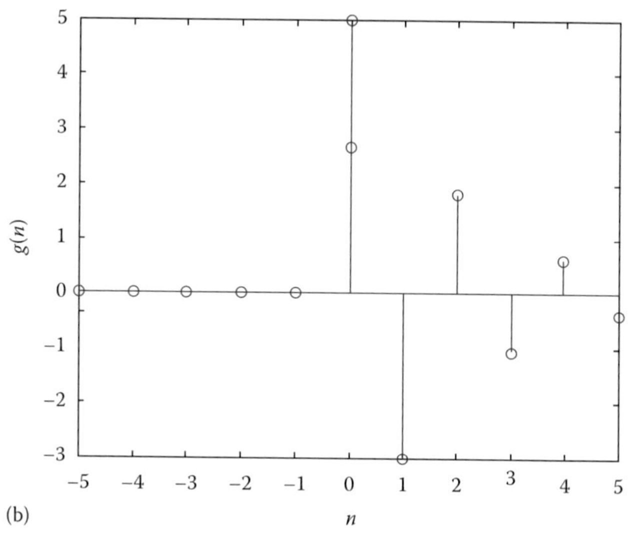

FIGURE 1.51 (*Continued*) (b) For Practice Problem 1.14b.

1.10 SUMMARY

1. A signal is a set of data or function that represents a variable of interest.
2. A system is a device, process, or algorithm that operates on input signal to yield an output signal.
3. A signal may be classified in many ways:
   a. A *continuous-time* signal $x(t)$ has a value specified at all time, whereas a *discrete-time* signal $x[n]$ is specified only at a finite set of time instants.
   b. A *periodic* signal repeats itself after period T, that is, $x(t + T) = x(t)$. An *aperiodic* signal is one that is not periodic.
   c. An *analog* signal is one that can take on any value over a continuum; a *digital* signal can only take a finite number of values.
   d. A signal $x(t)$ or $x[n]$ is an *energy* signal if and only if $0 < E < \infty$ and consequently $P = 0$, that is, it has finite energy. A signal $x(t)$ or $x[n]$ is a *power* signal if and only if $0 < P < \infty$ and consequently $E = \infty$; it has finite power.
   e. A signal is *even* if $x(t) = x(-t)$; it is *odd* if $x(t) = -x(-t)$.
4. Examples of standard signals are the singularity functions, which include the *unit impulse* $\delta(t)$ or $\delta[n]$, the *unit step* $u(t)$ or $u[n]$, and the *unit ramp* $r(t)$ or $r[n]$. The sifting property of the impulse function is

$$\int_a^b f(t)\delta(t-t_o)\,dt = f(t_o)$$

In addition to the singularity functions, we consider the unit rectangular function $\Pi(t/\tau)$, the unit triangular function $\Lambda(t/\tau)$, the sinusoidal function, and the exponential function.

5. The following operations can be performed on a signal:
   a. A signal $x(-t)$ is the time reversal of $x(t)$; it is obtained by reflecting $x(t)$ about $t = 0$.
   b. The signal $x(at)$ is compressed if $a > 1$ or expanded if $a < 1$.
   c. A signal $x(t - t_o)$ is delayed by t_o seconds, while the signal $x(t + t_o)$ is advanced by t_o seconds, assuming $t_o > 0$.
   d. Amplitude transformations take the general form of $y(t) = Ax(t) + B$.
6. A system may be classified in many ways:
   a. A system is *continuous-time* if the input and output signals are continuous-time; it is *discrete-time* if the input and output signals are discrete-time.
   b. A system is *causal* if its present response does not depend on the future values of the input; it is *noncausal* otherwise.
   c. A system is *linear* if it is characterized by the linearity property; it is *nonlinear* if the superposition does not hold.
   d. A system is *time-varying* if its parameters change with time; it is *time-invariant* if its parameters do not change with time.
   e. A *memoryless system* is one in which the current output depends only on the current input; a system with memory is one whose output depends on the present and past or future inputs.
7. Specific applications considered in this chapter are electric circuits, square-law device, and DSP system.
8. MATLAB can be used to process continuous-time and discrete-time signals.

REVIEW QUESTIONS

1.1 Which of the following is not a signal?
(a) Voltage, (b) blood, (c) torque, (d) pressure, (e) displacement

1.2 Any set of mathematical relationships between input and output variables constitutes a system.
(a) True, (b) false

1.3 Which of the following is not a system?
(a) Automobile speed, (b) a circuit, (c) a camera, (d) a computer program

1.4 The exponential signal is
(a) A power signal, (b) an energy signal, (c) neither a power nor an energy signal

1.5 The signals $\delta(t)$, $u(t)$, and $r(t)$ are known as
(a) Transformation functions, (b) expansion functions, (c) singularity functions, (d) basic functions

1.6 Which of these is not true?
(a) $t^2\delta(t) = 0$
(b) $\cos(t)\delta(t) = \delta(t)$
(c) $\sin(t)\delta(t-\pi) = \delta(t-\pi)$
(d) $\int_{-\infty}^{\infty} (2t^2 + 1)\delta(t)\, dt = 1$

1.7 Given the signal $x(t)$, which of the following is true for signal $x(2t + 8)$?
(a) $x(t)$ shifted to the left by eight units
(b) $x(t)$ is compressed by a factor of 2 and then shifted left by four units
(c) $x(t)$ is expanded by a factor of 2 and then shifted right by four units
(d) $x(t)$ is reflected about the vertical axis and then shifted

1.8 A system is described by $y(t) = \cos x(t) + 2 \sin x(t - 1)$. It is
(a) Causal, (b) noncausal, (c) memoryless, (d) with memory

1.9 Causal systems are also referred to as physically realizable systems.
(a) True, (b) false

1.10 The impulse function in MATLAB is
(a) Heaviside, (b) stepfun, (c) Dirac, (d) stem

Answers: 1.1(b), 1.2(a), 1.3(a), 1.4(c), 1.5(c), 1.6(c), 1.7(b), 1.8(a,d), 1.9(a), 1.10(c).

PROBLEMS

Section 1.3—Classifications of Signals

1.1 Show that the complex sinusoidal signal $x(t) = Ae^{j\omega_o t}$ is periodic.

1.2 Find the energy content of

$$x(t) = \begin{cases} e^{-3t}, & t \geq 0 \\ 0, & t < 0 \end{cases}$$

and determine whether it is a power or energy signal.

1.3 Let

$$x(t) = \begin{cases} e^{-2t} \sin t, & t \geq 0 \\ 0, & \text{otherwise} \end{cases}$$

calculate the energy content of the signal.

1.4 Find out if the following signals are power or energy signals or neither:
(a) $x(t) = 10 \sin 2\pi t$
(b) $y(t) = 2\left[u(t) - u(t-4)\right]$
(c) $z(t) = r(t) - r(t-2)$
(d) $h(t) = 10e^{2t}$

1.5 Repeat Problem 1.4 for the following signals:
(a) $x(t) = e^{-3t}u(t)$
(b) $x(t) = (e^{-2t} + 1)u(t)$
(c) $x(t) = t^{-1/2}u(t-2)$

Section 1.4—Basic Continuous-Time Signals

1.6 Express the signals in Figure 1.52 in terms of the unit step function.

1.7 Express the signals in Figure 1.53 in terms of the unit step and unit ramp functions.

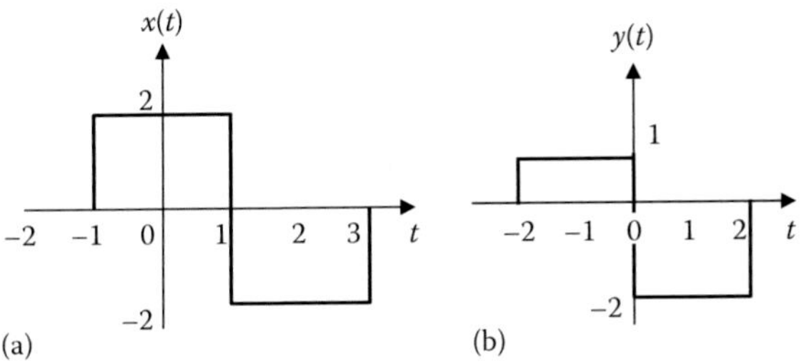

FIGURE 1.52 For Problem 1.6.

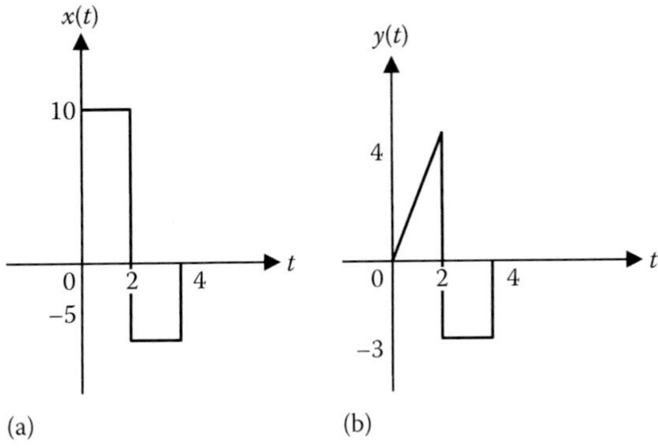

FIGURE 1.53 For Problem 1.7.

1.8 If $x(t) = \begin{cases} 0, & t < 0 \\ 8, & 0 < t < 2 \\ 2t + 6, & 2 < t < 6 \\ 0, & t > 6 \end{cases}$

Express $x(t)$ in terms of singularity functions.

1.9 (a) Express $x(t)$ in Figure 1.54 in terms of the unit step function.
(b) Sketch the derivative of $x(t)$.

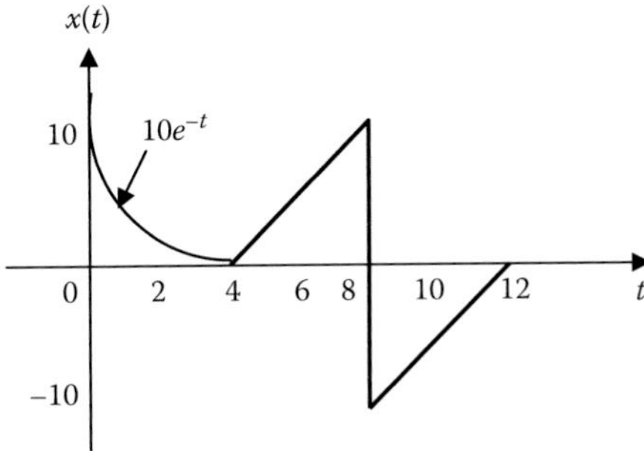

FIGURE 1.54 For Problem 1.9.

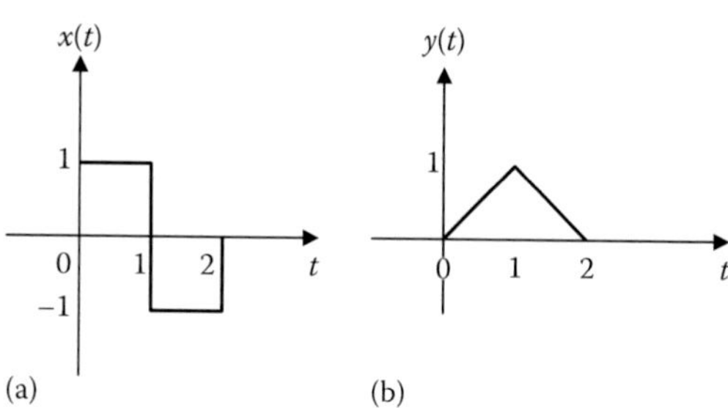

FIGURE 1.55 For Problem 1.10.

1.10 Express $x(t)$ and $y(t)$ in Figure 1.55 in terms of the step function.

1.11 Using Euler relation, prove the following trig identities:

(a) $\cos^2 t = ½(1 + \cos 2t)$

(b) $\sin(x + y) = \sin x \cos y + \cos x \sin y$

(c) $\sin x \sin y = ½\cos(x-y) - ½\cos(x + y)$

1.12 Evaluate the following:

(a) $\int_{-\infty}^{\infty} e^{-j\omega t}\delta(t-3)\,dt$

(b) $\int_{-\infty}^{\infty} \exp(t^2 - 2t + 1)\delta(t-2)\,dt$

(c) $\int_{-\infty}^{\infty} \left[5\delta(t) + e^{-t}\delta(t) + \cos 2\pi t\delta(t)\right]dt$

1.13 Evaluate the following integrals:

(a) $\int_{1}^{t} u(\lambda)\,d\lambda$

(b) $\int_{0}^{4} r(t-1)\,dt$

(c) $\int_{0}^{4} (t-6)^2\delta(t-2)\,dt$

(d) $\int_{-4}^{4} 2t^3\delta(t-4)\,dt$

1.14 Evaluate the following derivatives:

(a) $\frac{d}{dt}\left[u(t-1)u(t+1)\right]$

(b) $\frac{d}{dt}\left[r(t-6)u(t-2)\right]$

(c) $\frac{d}{dt}\left[\sin 4tu(t-3)\right]$

Section 1.5—Basic Discrete-Time Signals

1.15 Sketch the following discrete-time signals:

(a) $x[n] = 4(-0.9)^n,\ n \geq 0$

(b) $x[n] = 2(1.2)^n,\ n \geq 0$

(c) $x[k] = (0.8)^k \sin(2\pi k/8),\ k \geq 0$

1.16 Sketch $x[n] = \begin{cases} 10(0.6)^n, & -3 \leq n \geq 3 \\ 0, & \text{otherwise} \end{cases}$

1.17 Sketch the following discrete-time signals:

(a) $x[n] = \begin{cases} 1, & -3 \leq n \leq 3 \\ 0, & \text{otherwise} \end{cases}$

(b) $x(t) = \begin{cases} 1, & n = -1 \\ 2, & n = 0 \\ -1, & n = 2 \\ 2, & n = 3 \\ 0, & \text{otherwise} \end{cases}$

(c) $x[n] = 2\sin(n/4)\cos(n\pi/8)$

1.18 Sketch the following signals:

(a) $x[n] = u[n-2]$

(b) $y[n] = u[n+2]$

(c) $z[n] = u[n] - u[n-4]$

1.19 Sketch the following discrete-time signals:

(a) $x[n] = x[n] + u[-n]$

(b) $x[n] = u[3n]$

(c) $x[n] = 10\sin(2\pi n/20)$

1.20 Represent the signal in Figure 1.56 as a sum of impulses.

Section 1.6—Basic Operations on Signals

1.21 Let $x(t) = 10e^{4-3t}$. Evaluate and simplify each of the following:

(a) $x(2)$

(b) $x(1-t)$

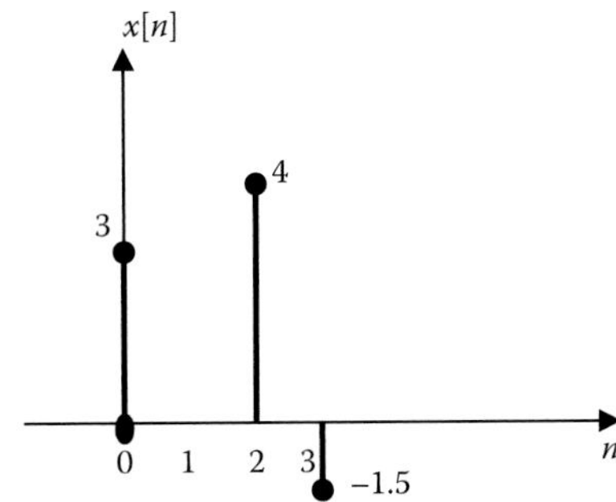

FIGURE 1.56 For Problem 1.20.

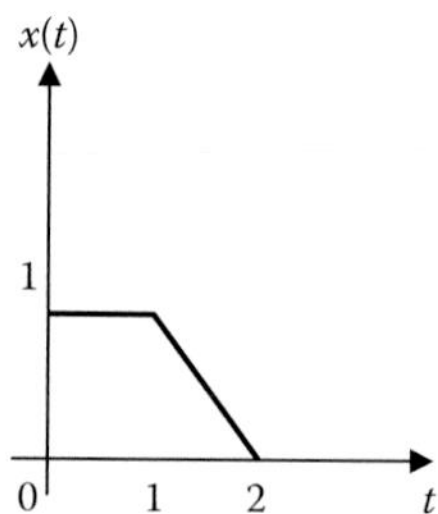

FIGURE 1.57 For Problem 1.22.

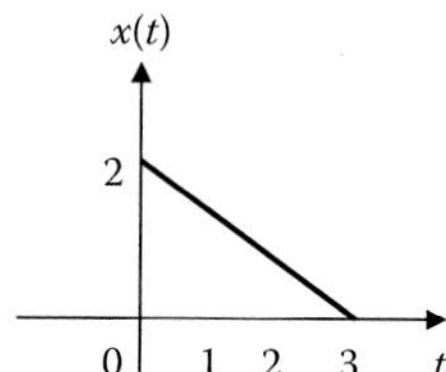

FIGURE 1.58 For Problem 1.24.

(c) $x(t/4 + 1)$

(d) $\frac{1}{2}[x(jt) + x(-jt)]$

1.22 Given the signal $x(t)$ shown in Figure 1.57, sketch (a) $x(t/2)$ and (b) $x(1 + t/2)$.

1.23 If $x(t) = \begin{cases} t+1, & 0 < t < 4 \\ 0, & \text{otherwise} \end{cases}$, sketch (a) $x(t)$ and (b) $y(t) = 1 + x(2t)$.

1.24 If $x(t)$ is the signal shown in Figure 1.58, sketch (a) $x(t - 2)$, (b) $x(3t)$, and (c) $y(t) = 1 + 2x(t)$.

1.25 Sketch each of the following continuous-time signals and their derivatives:

(a) $x(t) = u(t) - u(t-1)$ and dx/dt

(b) $y(t) = u(t + 1) - 2u(t) + u(t-1)$ and dy/dt

(c) $z(t) = t\left[u(t+1) - u(t-1)\right]$ and dz/dt

1.26 Given that $x(t) = u(t) - u(t-1)$, sketch

(a) $y(t) = x(-t)$

(b) $y(t) = x(3-t)$

(c) $y(t) = x(t-3)$

1.27 If $x(t) = \Pi(t + 2) + \Pi(t-2)$, sketch

(a) $y(t) = x(t-2)$

(b) $y(t) = x(t + 1) - x(t-1)$

1.28 Sketch these signals:

(a) $u\left[(t-1)/2\right]$

(b) $r[4-2t]$

(c) $\Pi(-2t + 1)$

(d) $\Lambda\left[(t-1)/3\right]$

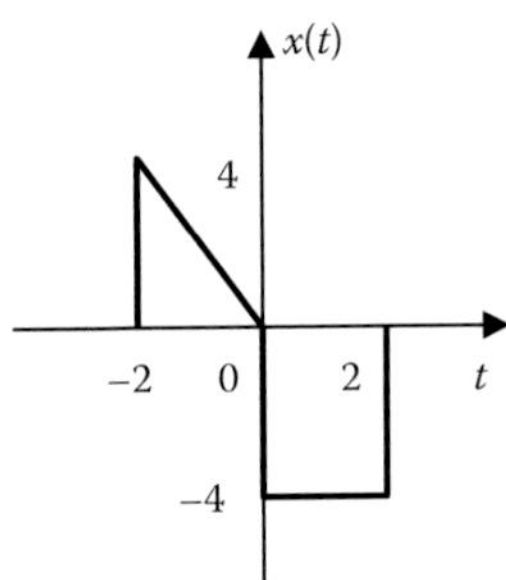

FIGURE 1.59 For Problem 1.29.

1.29 Given $x(t)$ in Figure 1.59, sketch
(a) $y(t) = -x(t-1)$
(b) $z(t) = 4x(t/2)$
(c) $h(t) = x(2-t)$

1.30 Given the two signals $x_1(t)$ and $x_2(t)$ in Figure 1.60, sketch the following:
(a) $y(t) = x_1(t) + x_2(t)$
(b) $z(t) = 0.5x_1(2t) + x_2(t)$
(c) $h(t) = x_1(t-1) - x_2(1-t)$

1.31 Given the discrete-time signal in Figure 1.61, sketch the following signals:
(a) $y[n] = x[n-3]$
(b) $z[n] = x[n] - x[n-1]$

1.32 Consider the discrete-time signal in Figure 1.62. Sketch the following signals:
(a) $x[n]u[2-n]$
(b) $x[n]\left[u[n+1]-u[n]\right]$
(c) $x[n]\delta[n-2]$

Section 1.7—Classifications of Systems

1.33 Show that the system described by the differential equation

$$\frac{dy(t)}{dt} + 2ty(t) = x(t)$$

is linear.

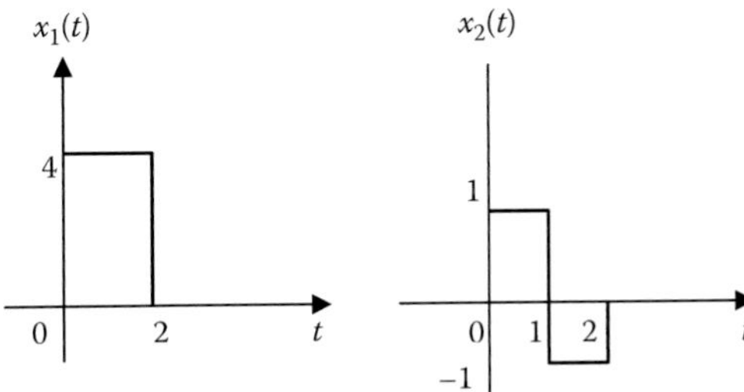

FIGURE 1.60 For Problem 1.30.

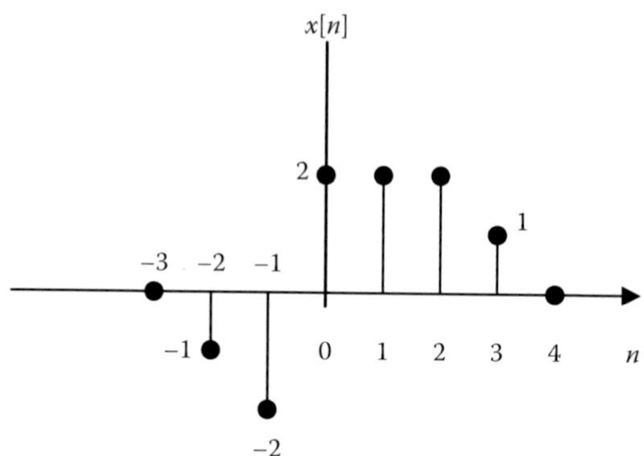

FIGURE 1.61 For Problem 1.31.

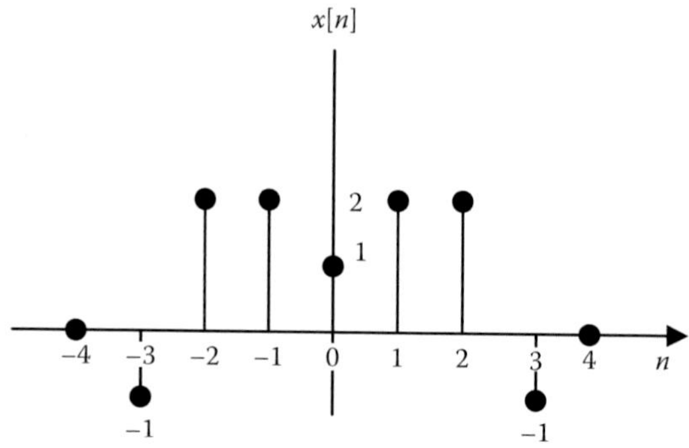

FIGURE 1.62 For Problem 1.32.

1.34 Show that the system described by the differential equation

$$\frac{dy(t)}{dt} + 4y(t) + 1 = x(t)$$

is nonlinear.

1.35 Which of the following systems can be classified as linear/nonlinear, time-varying/time-invariant?

(a) $\dfrac{d^2y(t)}{dt^2} + 2ty(t) = f(t)$

(b) $\dfrac{d^2y(t)}{dt^2} + y(t)\dfrac{dy(t)}{dt} + y(t) = 4f(t)$

(c) $\dfrac{d^4y(t)}{dt^4} - 5\dfrac{d^3y(t)}{dt^3} + 2\dfrac{dy(t)}{dt} - y(t) = \dfrac{df(t)}{dt} - 4f(t)$

(d) $t^2\dfrac{d^2y(t)}{dt^2} + y^2(t) = f(t)$

1.36 Determine which of the following systems is linear:

(a) $y(t) = \exp\left[x(t)\right]$

(b) $y(t) = \cos x(t)$

(c) $y(t) = t^2 x(t)$

1.37 Determine which of the following systems is linear or nonlinear:

(a) $\dfrac{dy}{dt} + y(t) = x^2(t)$

(b) $\left(\dfrac{dy}{dt}\right)^2 + y(t) = x(t)$

(c) $y(t) = \displaystyle\int_{-\infty}^{t} x(\lambda)\,d\lambda$

1.38 Determine whether the discrete-time system described by $y[n] = nx[n]$ is time-varying.

1.39 Determine whether the following systems are causal or noncausal, memoryless or with memory.

(a) $y(t) = e^{x(t)}\sin t$

(b) $y(t) = \displaystyle\int_0^t x(\tau)\tau\,d\tau$

1.40 Consider the integrator

$$y(t) = \int_0^t x(\lambda)\,d\lambda$$

Check for time-invariance.

1.41 Show whether the following systems are time-invariant:

(a) $y(t) = tx(t) + 5$

(b) $y[n] = x^2[n]$

Section 1.8—Applications

1.42 The unit delay element is shown in Figure 1.63a. Determine the input–output relation for the system in Figure 1.63b.

1.43 Write the differential equation for $i(t)$ in the circuit of Figure 1.64. Assume that the capacitor is initially uncharged and that the switch closes at $t = 0$.

1.44 Write the system equation for the electric circuit of Figure 1.65.

1.45 Consider the RC circuit in Figure 1.66. Obtain the relationship between the input $x(t)$ and the output $y(t)$.

Section 1.9—Computing with MATLAB®

1.46 Determine the numeric value of x in each of the following MATLAB instructions:

(a) t = pi/3; x = cos(t)

(b) t = 1:4; x = 2*sin(t)

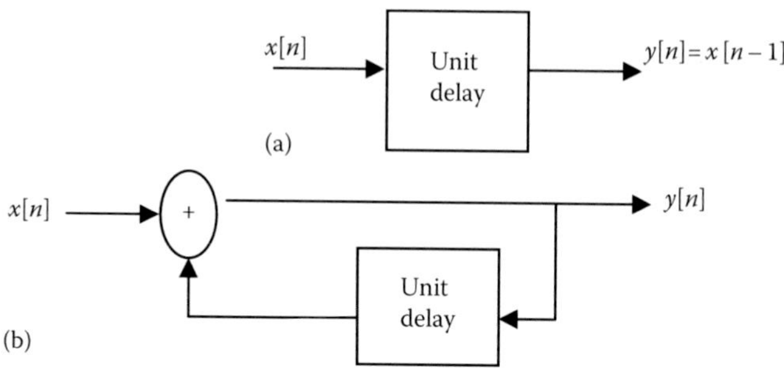

FIGURE 1.63 For Problem 1.42.

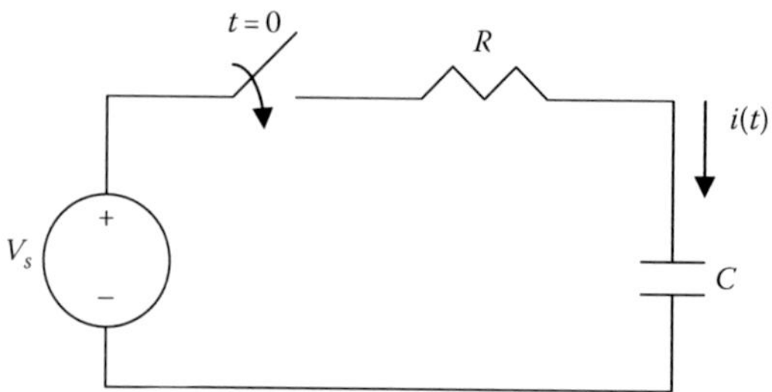

FIGURE 1.64 For Problem 1.43.

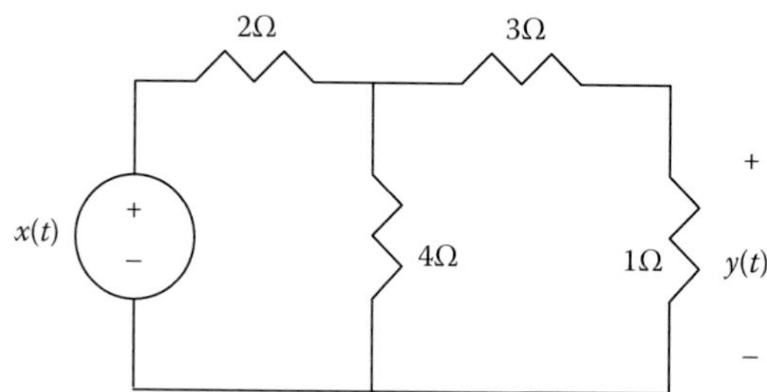

FIGURE 1.65 For Problem 1.44.

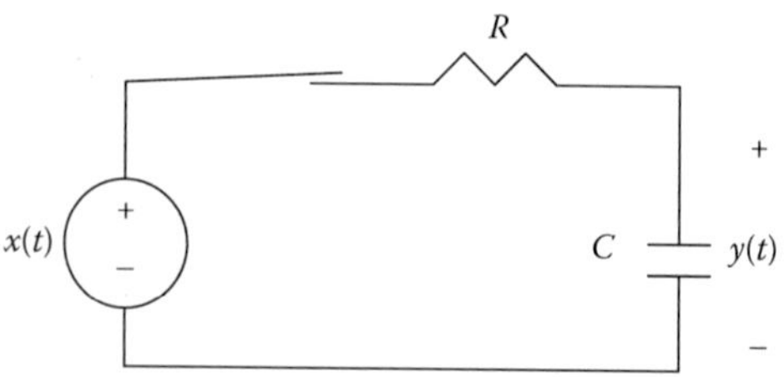

FIGURE 1.66 For Problem 1.45.

1.47 A signal is given by

$$x(t) = \begin{cases} \cos(3000\pi t), & 0 < t < 10\,\text{ms} \\ 0, & \text{otherwise} \end{cases}$$

Use MATLAB to plot the signal.

1.48 Use MATLAB to plot the signal

$$y(t) = \begin{cases} 2t, & 0 < t < 1 \\ \frac{1}{2}(t^2 - 4t + 3), & 1 < t < 3 \\ 0, & \text{otherwise} \end{cases}$$

1.49 Use MATLAB to plot these discrete-time signals:

(a) $x[n] = 10(0.7)^n$, $n \geq 0$

(b) $y[n] = 10(1.2)^n$, $n \geq 0$

1.50 Use MATLAB to plot the following signals over $-2 \leq t \leq 4$ s:

(a) $x(t) = 2\,r(t)$

(b) $y(t) = 5e^{-2t}\,u(t)$

(c) $z(t) = 4\cos 4t + 2\sin(2t - \pi/4)$

2 Convolution

The reason why we have two ears and only one mouth is that we may listen the more and talk the less.

—**Zeno**

ENHANCING YOUR COMMUNICATION SKILLS

Ability to communicate effectively is regarded by many as the most important step to an executive promotion.

Taking a class in signals and systems is a means to preparing yourself for a career in electrical engineering. Enhancing your **communication skills** while in school should also be part of that preparation because, as an engineer, a large part of your time will be spent in communicating. No matter how hard you try, you cannot avoid communication. Virtually everyone communicates and interacts at work. Communication skills will be essential to your success as an engineer.

ABET (Accreditation Board for Engineering and Technology) requires that graduates of engineering programs possess an ability to communicate effectively. It has been observed that graduating students are poorly prepared with respect to communication skills. It has also been observed that the ability to communicate is the most important factor to promotion, more important than motivation, education, and hard work.

Communication is the process of information transfer between the sender and the receiver through a medium. Oral, written, and graphical communication competence is an important professional skill for engineers. Professional engineers apply communication skills for many purposes, for example, emailing messages, describing solutions, pitching ideas, reporting experiments, discussing work with collaborators or customers, and presenting papers at conferences.

Effective communication is a learned skill. It takes times and effort to develop these skills and become an effective communicator. Developing the skills while in school is highly recommended. Look for continuing opportunities to develop and strengthen your reading, writing, listening, and speaking skills. You can do this through classroom presentations, team projects, active participation in student organizations, and enrolment in communication courses. Learn to write good laboratory reports in terms of grammar, spelling, accuracy, and coherence. The risks are less here than in the workplace. Effective communication can improve relationships at home, at workplace, and in social situations.

2.1 INTRODUCTION

As we saw in Chapter 1, a system may be regarded as a process that transforms an input signal into an output signal. The system must be capable of accepting input signals, operating on them, and producing output signals. The behavior of the system can be described mathematically either in the time domain, as is done in this chapter, or in the frequency domain, as in later chapters.

In this chapter, we will seek to understand a technique called *convolution*, which is a tool for time-domain analysis of systems. We will learn how to apply it in finding the response of LTI (linear, time-invariant) systems to input signals. The reason we limit our discussion to LTI systems is twofold. First, several physical systems can be modeled as LTI systems. Second, no general procedures exist for non-LTI systems. LTI systems can be analyzed in great detail because standard procedures are already available.

We will begin with applying convolution to continuous-time systems; we later will apply it to discrete-time systems. We use MATLAB® for evaluating the convolution of two signals. We will finally show how convolution can be applied in determining system stability and analyzing electric circuits.

2.2 IMPULSE RESPONSE

We recall from Chapter 1 that if $x(t)$ is the input signal and $y(t)$ is the output signal or response of a system, they are related through a transformation

$$y(t) = \mathbf{T}\,x(t) \tag{2.1}$$

where $\mathbf{T}$ is an operator transforming $x(t)$ into $y(t)$. The *impulse response* $h(t)$ is the response of the system when the input is the unit impulse function $\delta(t)$, that is,

$$h(t) = \mathbf{T}\,\delta(t) \tag{2.2}$$

The *impulse response* to an LTI system is the output of the system to a unit impulse function.

From Table 1.1, the input $x(t)$ can be expressed as

$$x(t) = \int_{-\infty}^{\infty} x(\tau)\delta(t-\tau)d\tau \tag{2.3}$$

where τ is a dummy variable. Equation 2.3 is the **sifting property** of the unit impulse.

The response $y(t)$ to the input $x(t)$ is obtained by combining Equations 2.1 and 2.3:

$$\begin{aligned} y(t) = \mathbf{T}x(t) &= \mathbf{T}\left\{\int_{-\infty}^{\infty} x(\tau)\delta(t-\tau)d\tau\right\} \\ &= \int_{-\infty}^{\infty} x(\tau)\mathbf{T}\{\delta(t-\tau)\}\,d\tau \end{aligned} \tag{2.4}$$

in which the interchange of operations is allowed by virtue of the system's linearity. Since we are assuming that the system is time-invariant, Equation 2.2 implies that

$$h(t-\tau) = \mathbf{T}\{\delta(t-\tau)\} \tag{2.5}$$

From Equations 2.4 and 2.5, we get

$$\boxed{y(t) = \int_{-\infty}^{\infty} x(\tau)h(t-\tau)d\tau} \tag{2.6}$$

This shows that an LTI system is characterized by its impulse response.

2.3 CONVOLUTION INTEGRAL

Equation 2.6 is known as the **convolution integral** or *superposition integral.* The convolution integral occurs frequently in science, mathematics, and engineering. The convolution of two signals $x(t)$ and $h(t)$ is usually written in terms of the operator * as

$$y(t) = x(t) * h(t) = \int_{-\infty}^{\infty} x(\tau)h(t-\tau)d\tau \tag{2.7}$$

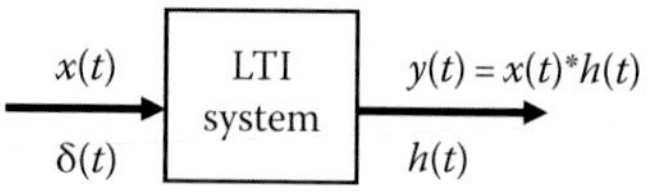

FIGURE 2.1 An illustration of convolution in an LTI system.

that is, $y(t)$ equals $x(t)$ convolved with $h(t)$. The asterisk denotes convolution in this chapter and should not be confused with complex conjugate. A block diagram illustration of convolution is given in Figure 2.1.

We can split the integral in Equation 2.7 into two parts:

$$y(t) = x(t) * h(t) = \underbrace{\int_{-\infty}^{t_o} x(\tau)h(t-\tau)d\tau}_{y_{zir}} + \underbrace{\int_{t_o}^{\infty} x(\tau)h(t-\tau)d\tau}_{y_{zsr}} == y_{zir}(t) + y_{zsr}(t) \quad (2.8)$$

where

$y_{zir}(t)$ is the zero-input response (or the natural response) of the system
$y_{zsr}(t)$ is the zero-state response (or the forced response) of the system
t_o is the initial time

Thus, the complete response of a physical system is divided into the zero-input response and the zero-state response.

The convolution integral in Equation 2.7 is a general one. It applies to any LTI system. However, the convolution integral can be simplified if we assume that a system has two properties. First, if $x(t) = 0$ for $t < 0$, then

$$y(t) = \int_{-\infty}^{\infty} x(\tau)h(t-\tau)d\tau = \int_{0}^{\infty} x(\tau)h(t-\tau)d\tau \quad (2.9)$$

Second, if we assume that the system is causal, $h(t) = 0$ for $t < 0$, then

$h(t - \tau) = 0$ for $t - \tau < 0$ or $\tau > t$, so that Equation 2.9 becomes

$$y(t) = x(t) * h(t) = \int_{0}^{t} x(\tau)h(t-\tau)d\tau \quad (2.10)$$

Some important properties of the convolution integral are listed in Table 2.1.

Property #1 states that the order in which two functions are convolved is unimportant. We will see shortly how to take advantage of this commutative property when performing graphical computation of the convolution integral.

Another important property of the convolution integral is the width property. If the durations of $x(t)$ and $h(t)$ are T_1 and T_2, respectively, then the duration of $y(t) = x(t)*h(t)$ is $T_1 + T_2$, as illustrated in Figure 2.2. If the areas under $x(t)$ and $h(t)$ are A_1 and A_2, respectively, then the area under $y(t) = x(t)*h(t)$ is A_1A_2. That is,

$$\text{Area under } y(t) = \text{Area under x}(t) \times \text{area under } h(t) \quad (2.11)$$

TABLE 2.1
Properties of the Convolution Integral

1. $x(t)*h(t)=h(t)*x(t)$ (Commutative)

2. $f(t)*[x(t)+y(t)]=f(t)*x(t)+f(t)*y(t)$ (Distributive)

3. $f(t)*[(x(t)*y(t)]=f[(t)*x(t)]*y(t)$ (Associative)

4. $f(t)*\delta(t)=\int_{-\infty}^{\infty} f(\tau)\delta(t-\tau)d\tau=f(t)$

5. $f(t)*\delta(t-t_o)=f(t-t_o)$

6. $f(t)*\delta'(t)=\int_{-\infty}^{\infty} f(\tau)\delta'(t-\tau)d\tau=f'(t)$

7. $f(t)*u(t)=\int_{-\infty}^{\infty} f(\tau)u(t-\tau)d\tau=\int_{-\infty}^{t} f(\tau)d\tau$

8. $u(t)*\delta'(t)=\int_{-\infty}^{\infty} u(\tau)\delta'(t-\tau)d\tau=u'(t)=\delta(t)$

9. If $x_1(t)*x_2(t)=y(t)$, then
   $x_1(t+t_1)*x_2(t+t_2)=y(t+t_1+t_2)$

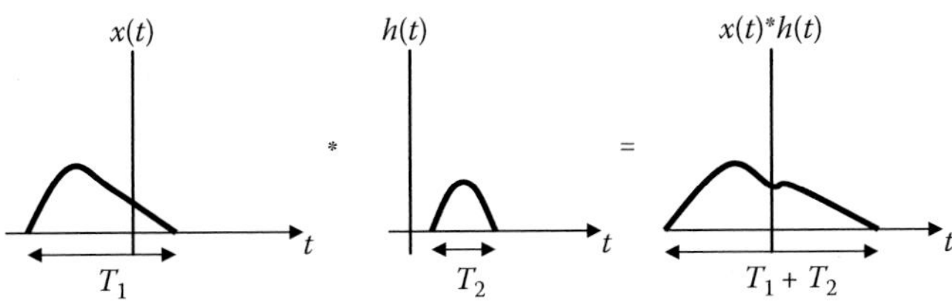

FIGURE 2.2 The width property of convolution.

Of course, this does not hold for signals extending to $\pm\infty$.
We can also add that if

$$x_1(t)*x_2(t)=y(t)$$

then

$$x_1(t+t_1)*x_2(t)=y(t+t_1) \tag{2.12}$$

$$x_1(t+t_1)*x_2(t+t_2)=y(t+t_1+t_2) \tag{2.13}$$

$$y(at)=ax_1(at)*x_2(at), \quad a>0 \text{ (time scaling)} \tag{2.14}$$

$$y(-t)=x_1(-t)*x_2(-t) \quad \text{(time reversal)} \tag{2.15}$$

TABLE 2.2
Abbreviated Convolution Table

No.	$x_1(t)$	$x_2(t)$	$x_1(t) * x_2(t)$
1.	$x(t)$	$\delta(t)$	$x(t)$
2.	$u(t)$	$u(t)$	$tu(t) = r(t)$
3.	$e^{at}\,u(t)$	$u(t)$	$\frac{e^{at}-1}{a}u(t)$
4.	$e^{-at}\,u(t)$	$e^{-bt}\,u(t)$	$\frac{1}{a-b}\left[e^{-bt}-e^{-at}\right]u(t)$
5.	$e^{at}\,u(t)$	$e^{bt}\,u(t)$	$\frac{1}{b-a}\left[e^{bt}-e^{at}\right]u(t)$
6.	$e^{at}\,u(t)$	$e^{at}\,u(t)$	$te^{at}\,u(t)$
7.	$te^{at}\,u(t)$	$e^{at}\,u(t)$	$\frac{1}{2}t^2e^{at}u(t)$

The convolution integral can be evaluated in three different ways:

1. Analytical method, which involves performing the integration by hand when $x(t)$ and $h(t)$ are specified analytically.
2. Graphical method, which is appropriate when $x(t)$ and $h(t)$ are provided in graphical form.
3. Numerical method, where we approximate $x(t)$ and $h(t)$ by numerical sequence and obtain $y(t)$ by discrete convolution using a digital computer.

We will apply the analytical method right away and apply graphical and numerical methods later. Table 2.2 shows some common signals and their convolution.

Example 2.1

Obtain the convolution of these two signals:

$$x(t) = e^{-t^2} \quad \text{and} \quad y(t) = 2t^2 \quad \text{for all } t.$$

Solution

Let $z(t) = x(t)*y(t)$. We first get $x(\tau) = e^{-\tau^2}$ and $y(t-\tau) = 2(t-\tau)^2$. Applying Equation 2.6, we obtain

$$z(t) = \int_{-\infty}^{\infty} x(\tau)y(t-\tau)d\tau = \int_{-\infty}^{\infty} e^{-\tau^2}[2(t-\tau)^2]d\tau$$

$$= \int_{-\infty}^{\infty} e^{-\tau^2}[2t^2 - 4t\tau + 2\tau^2]d\tau$$

$$= 2t^2\int_{-\infty}^{\infty} e^{-\tau^2}d\tau - 4t\int_{-\infty}^{\infty} \tau e^{-\tau^2}d\tau + 2\int_{-\infty}^{\infty} \tau^2 e^{-\tau^2}d\tau \tag{2.1.1}$$

The first integral on the left side is given by

$$\int_{-\infty}^{\infty} e^{-\tau^2} d\tau = \sqrt{\pi} \tag{2.1.2}$$

The second integral is zero because the integrand is an odd function of τ. The third function

$$\int_{-\infty}^{\infty} \tau^2 e^{-\tau^2} d\tau = 2\int_{0}^{\infty} \tau^2 e^{-\tau^2} d\tau = \frac{\sqrt{\pi}}{2} \tag{2.1.3}$$

substituting Equations 2.1.2 and 2.1.3 into Equation 2.1.1 gives

$$\begin{aligned} z(t) &= 2t^2\sqrt{\pi} - 0 + \sqrt{\pi} \\ &= 3.545t^2 + 1.772, \quad \text{for all } t \end{aligned}$$

Practice Problem 2.1 Determine the convolution of $f(t) = e^{-2t^2}$ and $g(t) = t$.

Answer: $0.4431t$.

Example 2.2

The input $x(t)$ and the impulse response $h(t)$ of an LTI system are given by $x(t) = u(t)$ and $h(t) = e^{-3t}u(t)$. Find the output response.

Solution

From Equation 2.6,

$$y(t) = \int_{-\infty}^{\infty} x(\tau)h(t-\tau)d\tau = \int_{-\infty}^{\infty} u(\tau)e^{-3(t-\tau)}u(t-\tau)d\tau \tag{2.2.1}$$

In this integral, t is a constant so that we can pull out the factor e^{-3t}.

By definition,

$$u(\tau) = \begin{cases} 1, & \tau > 0 \\ 0, & \tau < 0 \end{cases}$$

$$u(t-\tau) = \begin{cases} 1, & t-\tau > 0 \\ 0, & t-\tau < 0 \end{cases} = \begin{cases} 1, & \tau < t \\ 0, & \tau > t \end{cases}$$

Thus,

$$u(\tau)u(t-\tau) = \begin{cases} 1, & 0 < t < \tau \\ 0, & \text{otherwise} \end{cases}$$

The limits on the integral now becomes $0 < \tau < t$. Therefore, Equation 2.2.1 becomes

$$y(t) = e^{-3t}\int_0^t e^{3\tau}d\tau = \frac{e^{-3t}}{3}e^{3\tau}\Big|_0^t$$

$$= \frac{1}{3}(1 - e^{-3t}), \quad t > 0$$

or

$$y(t) = \frac{1}{3}(1 - e^{-3t})u(t)$$

Practice Problem 2.2 Consider a system in which the input $x(t) = r(t) = tu(t)$ and the impulse response $h(t) = u(t)$. Determine the output $y(t)$.

Answer: $y(t) = 0.5t^2u(t)$.

2.4 GRAPHICAL CONVOLUTION

We now consider graphical method of evaluating the convolution integral. This method usually involves four steps:

1. *Folding*: Take the mirror image of $h(\tau)$ about the ordinate (or vertical) axis to obtain $h(-\tau)$
2. *Shifting*: Displace or shift $h(-\tau)$ by t to obtain $h(t - \tau)$
3. *Multiplication*: Multiply $h(t - \tau)$ and $x(\tau)$ together
4. *Integration*: For a given t, integrate the product $h(t - \tau)x(\tau)$ over $0 < \tau < t$ to get $y(t)$ at t

The folding operation in step 1 is the reason for the term *convolution*. The function $h(t - \tau)$ scans or slides over $x(\tau)$.

To apply this procedure, we must be able to sketch $h(t - \tau)$. Sketching $h(t - \tau)$ is the key to the convolution process. It involves reflecting $h(\tau)$ about the vertical axis and shifting by t. This is illustrated in Figure 2.3. Analytically, we obtain $h(t - \tau)$ by replacing every t in $h(t)$ by $t - \tau$. Since convolution is commutative, it may be more convenient to apply steps 1 and 2 to $x(t)$ instead of $h(t)$.

Example 2.3

Obtain the convolution of the two signals in Figure 2.4.

Solution

Let $y(t) = x(t)*h(t)$, where $x(t)$ is a unit step and

$$h(t) = \begin{cases} 2 + t, & -2 < t < 0 \\ 1, & 0 < t < 2 \end{cases}$$

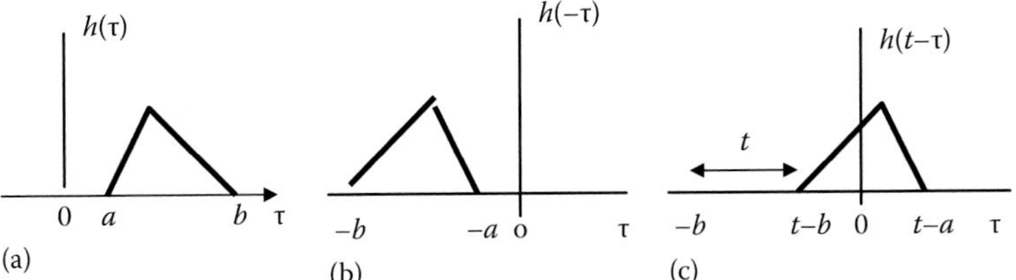

FIGURE 2.3 Folding and sliding the function $h(t)$.

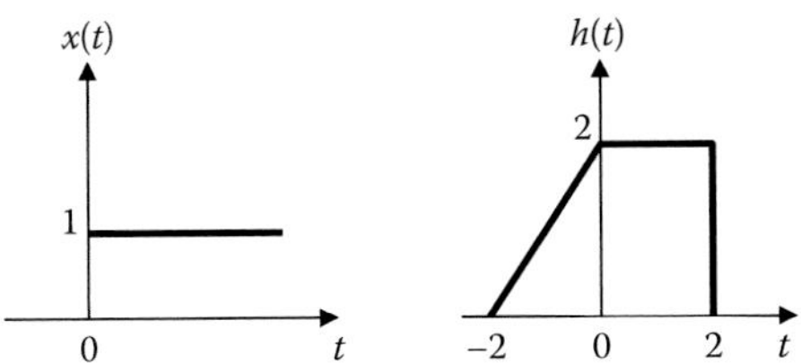

FIGURE 2.4 For Example 2.3.

In this case, it is easy to fold $x(t)$, the unit step function. Let

$$y(t) = x(t) * h(t) = \int x(t - \tau)h(\tau)d\tau$$

We follow the four steps to get $y(t)$. First, we fold $x(t)$ as shown in Figure 2.5a and shift it by t as shown in Figure 2.5b.

For $t < -2$, there is no overlap of the two signals, as shown in Figure 2.6a. Hence,

$$y(t) = x(t) * h(t) = 0, \quad t < -2 \tag{2.3.1}$$

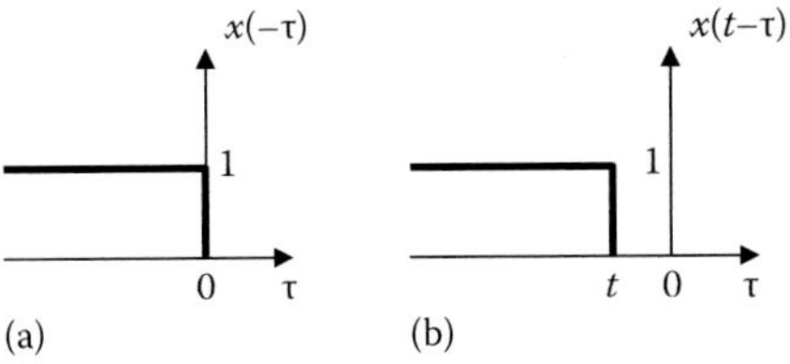

FIGURE 2.5 For Example 2.3: (a) Folding $x(t)$ and (b) shifting $x(-\tau)$ by t.

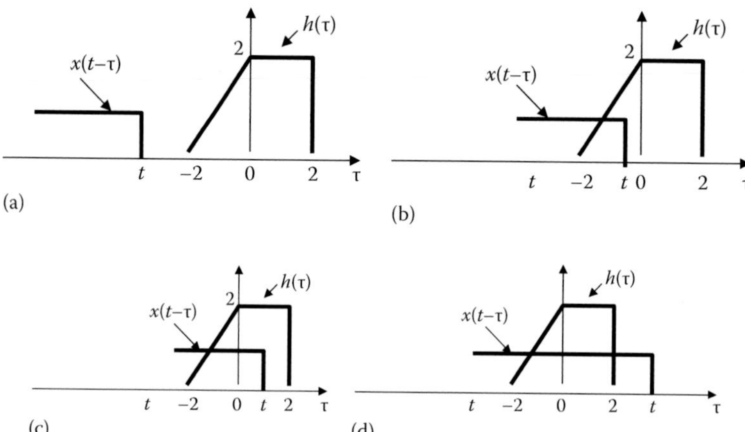

FIGURE 2.6 For Example 2.3: Overlapping of $x(t-\tau)$ and $h(\tau)$ for (a) $t<-2$, (b) $-2<t<0$, (c) $0<t<2$, and (d) $t>2$.

For $-2<t<0$, the two signals overlap between -2 and t, as shown in Figure 2.6b.

$$y(t)=\int_{-2}^{t} h(\tau)x(t-\tau)d\tau=(2+\tau)(1)d\tau\Big|_{-2}^{t}=2\tau+\frac{\tau^2}{2}\Big|_{-2}^{t}$$

$$=0.5t^2+2t+2,\quad -2<t<0 \tag{2.3.2}$$

For $0<t<2$, the two signals overlap between -2 and t, as shown in Figure 2.6c.

$$y(t)=\int_{-2}^{0}(2+\tau)(1)d\tau+\int_{0}^{t}(2)(1)d\tau$$

$$=\left(2\tau+\frac{\tau^2}{2}\right)\Big|_{-2}^{0}+2\tau\Big|_{0}^{t}=4-2+2t=2(t+1),\quad 0<t<2 \tag{2.3.3}$$

For $t>2$, the two signals overlap between -2 and 2, as shown in Figure 2.6d.

$$y(t)=\int_{-2}^{0}(2+\tau)(1)d\tau+\int_{0}^{2}(2)(1)d\tau$$

$$=\left(2\tau+\frac{\tau^2}{2}\right)\Big|_{-2}^{0}+2\tau\Big|_{0}^{2}=4-2+4=6,\quad t>2 \tag{2.3.4}$$

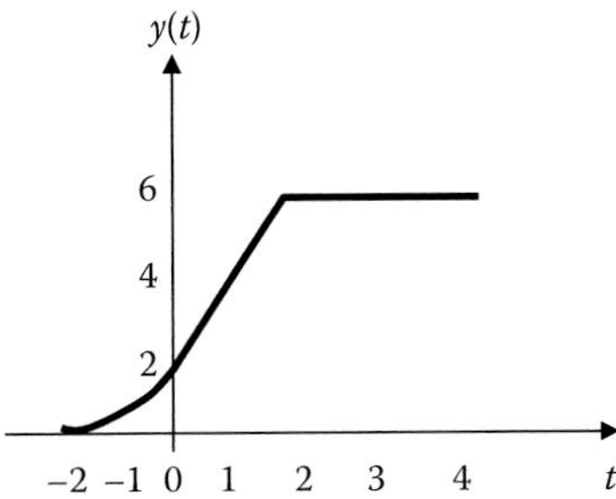

FIGURE 2.7 For Example 2.3: Convolution of signals $x(t)$ and $h(t)$.

Combining the results in Equations 2.3.1 through 2.3.4, we obtain

$$y(t) = \begin{cases} 0.5t^2 + 2t + 2, & -2 \le t \le 0 \\ 2(t+1), & 0 \le t \le 2 \\ 6, & t > 2 \\ 0, & \text{otherwise} \end{cases} \tag{2.3.5}$$

which is sketched in Figure 2.7. Notice that $y(t)$ in this equation is continuous; that is, it has no discontinuities. This fact can be used to check the results as we move from one range of t to another.

Practice Problem 2.3 Find the convolution of the two signals in Figure 2.8.

Answer: $z(t) = x(t) * y(t) = \begin{cases} \frac{1}{2}t^2, & 0 \le t \le 1 \\ \frac{1}{2}(2t-1), & 1 < t < 2 \\ \frac{1}{2}\left(3 + 2t - t^2\right), & 2 \le t \le 3 \\ 0, & \text{otherwise} \end{cases}$

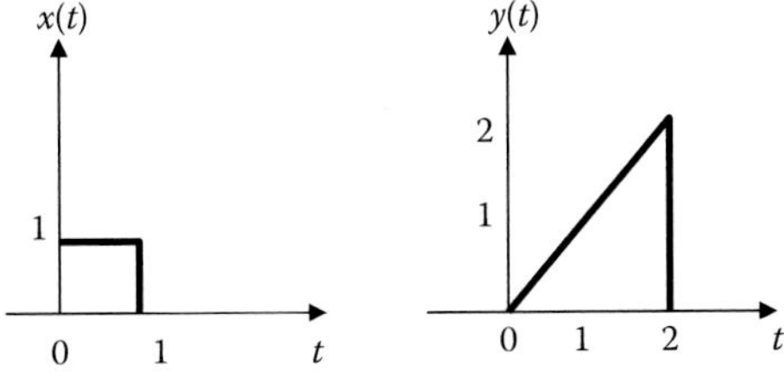

FIGURE 2.8 For Practice Problem 2.3.

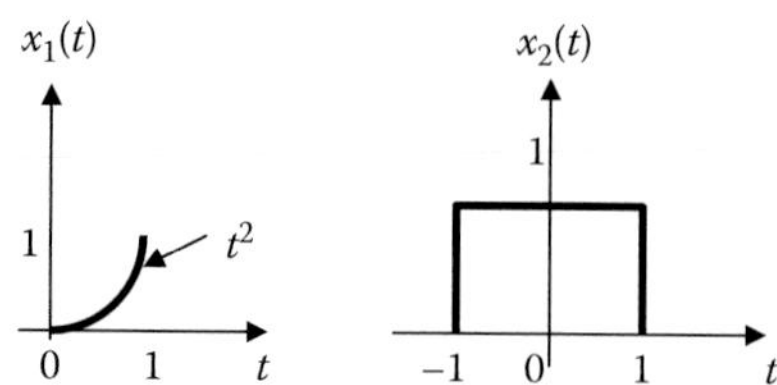

FIGURE 2.9 For Example 2.4.

Example 2.4

Given $x_1(t)$ and $x_2(t)$ in Figure 2.9, find $y(t) = x_1(t) * x_2(t)$.

Solution

We fold $x_1(t)$ so that

$$y(t) = x_1(t) * x_2(t) = \int x_1(t-\tau)x_2(\tau)d\tau$$

For $t < -1$, there is no overlapping, as shown in Figure 2.10a. Hence, $y(t) = 0$.

For $-1 < t < 0$, there is overlapping between −1 and t, as in Figure 2.10b,

$$y(t) = \int_{-1}^{t} (t-\tau)^2(1)d\tau = \left.\frac{(t-\tau)^3}{-3}\right|_{-1}^{t} = \frac{(t+1)^3}{3}, \quad -1 < t < 0$$

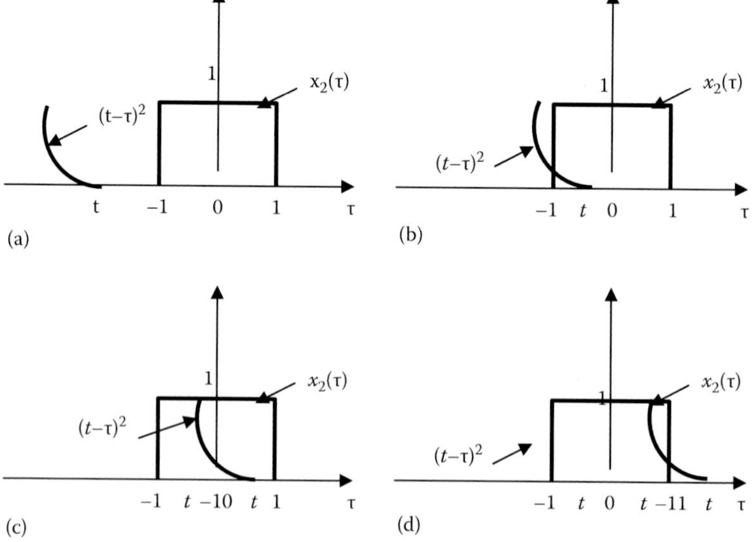

FIGURE 2.10 For Example 2.4: (a) $t < -1$, (b) $-1 < t < 0$, (c) $0 < t < 1$, and (d) $t > 1$.

For $0 < t < 1$, the two signals overlap between $t - 1$ and t, as shown in Figure 2.10c.

$$y(t) = \int_{t-1}^{t} (t-\tau)^2 (1) d\tau = \left. \frac{(t-\tau)^3}{-3} \right|_{t-1}^{t} = \frac{1}{3}, \quad 0 < t < 1$$

For $1 < t < 2$, they overlap between $t - 1$ and 1, as depicted in Figure 2.10d.

$$y(t) = \int_{t-1}^{t} (t-\tau)^2 (1) d\tau = \left. \frac{(t-\tau)^3}{-3} \right|_{t-1}^{1} = \frac{1}{3} - \frac{1}{3}(t-1)^3, \quad 1 < t < 2$$

For $t > 2$, there is no overlapping so that $y(t) = 0$. Thus,

$$y(t) = \begin{cases} \frac{1}{3}(t+1)^3, & -1 < t < 0 \\ \frac{1}{3}, & 0 < t < 1 \\ \frac{1}{3} - \frac{1}{3}(t-1)^3, & 1 < t < 2 \\ 0, & \text{otherwise} \end{cases}$$

Practice Problem 2.4 Graphically convolve $x(t)$ and $y(t)$ shown in Figure 2.11.

Answer: $z(t) = x(t) * y(t) = \begin{cases} \frac{1}{\pi}(1 - \cos \pi t), & 0 < t < 1 \\ \frac{1}{\pi}(\cos \pi t - 1), & 1 < t < 2 \\ 0, & \text{otherwise} \end{cases}$

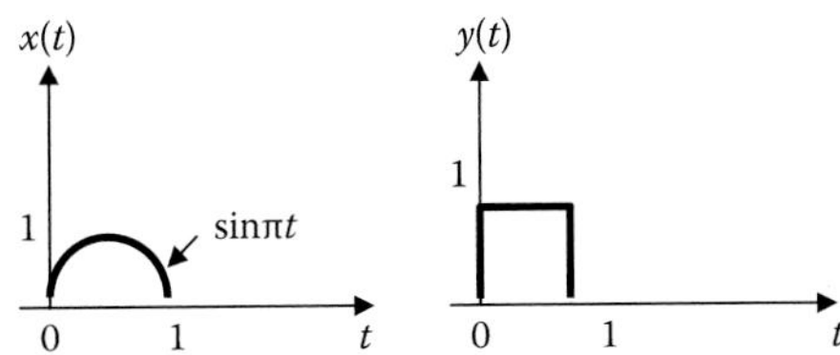

FIGURE 2.11 For Practice Problem 2.4.

2.5 BLOCK DIAGRAM REPRESENTATION

The commutative, associative, and distributive properties of the convolution integral, listed in Table 2.1, can be illustrated in the block diagrams shown in Figure 2.12. Notice that the commutative property, illustrated in Figure 2.12a, shows that $x(t)$ and $h(t)$ are interchangeable. The associative property, illustrated in Figure 2.12b, states that a series or cascaded arrangement of LTI systems can be replaced by a single system using convolution. The distributive property, illustrated in Figure 2.12c, indicates that a parallel arrangement of LTI systems can be combined into a single system through addition.

Example 2.5

Find the impulse response of the system shown in Figure 2.13. Let

$$h_1(t) = 3\delta(t)$$

$$h_2(t) = 2e^{-t}u(t)$$

$$h_3(t) = 4e^{-2t}u(t)$$

$$h_4(t) = e^{-3t}u(t)$$

Solution

Comparing Figure 2.13 with Figure 2.12, we notice that

$$h(t) = h_1(t) * h_2(t) + h_3(t) * h_4(t) \tag{2.5.1}$$

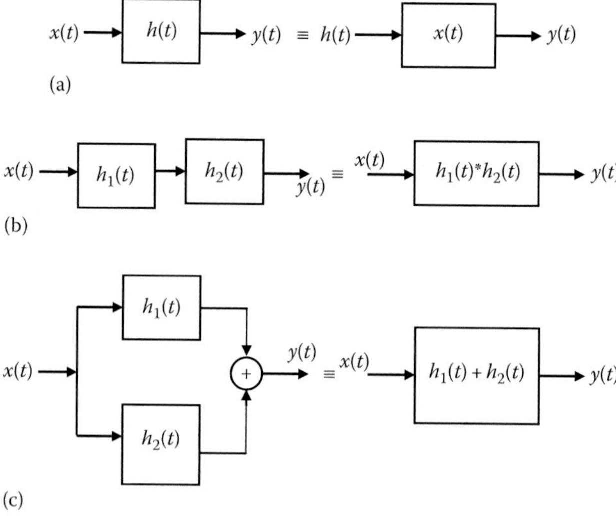

FIGURE 2.12 Block diagrams illustrating properties of convolution.

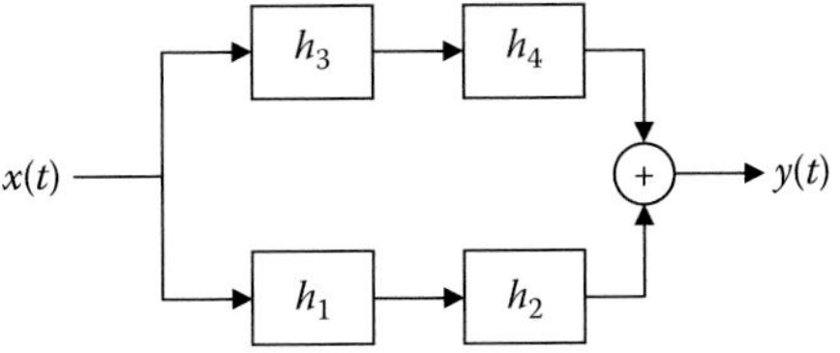

FIGURE 2.13 For Example 2.5.

Using the fact that $x(t)*\delta(t) = x(t)$, we obtain the first convolution on the right-hand side as

$$h_1(t) * h_2(t) = 3\delta(t) * \left\{2e^{-t}u(t)\right\} = 6e^{-t}u(t) \tag{2.5.2}$$

For the second convolution,

$$h_3(t) * h_4(t) = \int_{-\infty}^{\infty} 4e^{-2\tau}u(\tau)e^{-3(t-\tau)}u(t-\tau)d\tau$$

but

$$u(\tau)u(t-\tau) = \begin{cases} 1, & 0 < t < \tau \\ 0, & \text{otherwise} \end{cases}$$

Therefore,

$$h_3(t) * h_4(t) = \int_0^t 4e^{-2\tau}e^{-3(t-\tau)}d\tau = 4e^{-3t}\int_0^t e^{(3-2)\tau}d\tau$$

$$= 4e^{-3t}e^{\tau}\Big|_0^t = 4(e^{-2t} - e^{-3t}), \quad t > 0$$

$$= 4(e^{-2t} - e^{-3t})u(t) \tag{2.5.3}$$

Substituting Equations 2.5.2 and 2.5.3 into Equation 2.5.1 gives

$$h(t) = 6e^{-t}u(t) + 4(e^{-2t} - e^{-3t})u(t)$$

Practice Problem 2.5 For the system shown in Figure 2.14,

$$h_1(t) = 2e^{-t}u(t)$$

$$h_2(t) = 4e^{-2t}u(t)$$

Determine the overall impulse response $h(t)$.

Answer: $h(t) = 8(e^{-t} - e^{-2t})u(t)$.

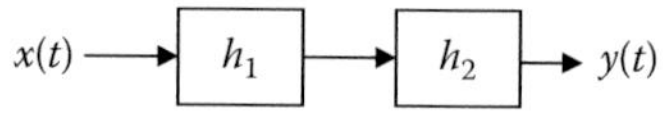

FIGURE 2.14 For Practice Problem 2.5.

2.6 DISCRETE-TIME CONVOLUTION

When an input is applied to a discrete-time system, the response or output sequence can be determined in a way similar to using the impulse response and the convolution integral for continuous-time systems. To refresh our mind, we will redefine the unit step sequence $u[n]$ as

$$u[n] = \begin{cases} 0, & n < 0 \\ 1, & n \geq 0 \end{cases} \tag{2.16}$$

The unit impulse sequence is also redefined as

$$\delta[n] = \begin{cases} 0, & n \neq 0 \\ 1, & n = 0 \end{cases} \tag{2.17}$$

The signal only has value at $n = 0$. The displaced delta function is

$$\delta[n-k] = \begin{cases} 0, & n \neq k \\ 1, & n = k \end{cases} \tag{2.18}$$

We said in Chapter 1 that an alternative way of expressing any discrete signal $x[n]$ is

$$x[n] = \sum_{k=-\infty}^{\infty} x[k]\delta[n-k] \tag{2.19}$$

that is, we can represent $x[n]$ as a weighted sum of delayed impulses. The multiplication property of the impulse function is

$$\boxed{\delta[n]x[n-k] = x[-k]\delta[n]} \tag{2.20}$$

$$\delta[n-k]x[n] = x[k]\delta[n-k] \tag{2.21}$$

The impulse response $h[n]$ of a discrete-time LTI system is the response of the system when the input is $\delta[t]$, that is,

$$h[n] = \mathbf{T}\{\delta[t]\} \tag{2.22}$$

This is illustrated in Figure 2.15.

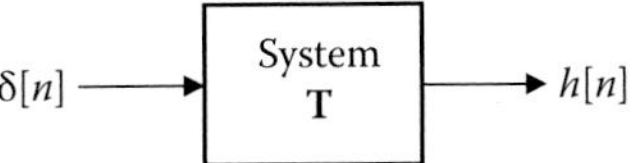

FIGURE 2.15 Block diagram illustrating impulse response.

The convolution of the discrete input signal $x[n]$ and the impulse response $h[n]$ is

$$y[n] = x[n] * h[n] \tag{2.23}$$

and is defined as

$$y[n] = \sum_{k=-\infty}^{\infty} x[k]h[n-k] \tag{2.24}$$

Thus,

$$\boxed{y[n] = x[n] * h[n] = \sum_{k=-\infty}^{\infty} x[k]h[n-k]} \tag{2.25}$$

This is known as the **convolution sum** or *superposition sum* for the system response. Notice that as in continuous-time convolution, one of the signals is time-inverted, shifted, and then multiplied by the other. By the change of variables $m = n - k$ or $k = n - m$, we have

$$y[n] = \sum_{m=-\infty}^{\infty} h[m]x[n-m] = h[n] * x[n] \tag{2.26}$$

This shows that the order of summation is immaterial; that is, discrete convolution is commutative. This and other properties of the convolution operation are listed in Table 2.3. It is evident from the convolution sum in Equation 2.23 that if we know $h[n]$, we can find the system response $y[n]$ to any input $x[n]$ (Figure 2.15).

TABLE 2.3
Properties of the Convolution Sum

1. $x[n] * h[n] = h[n] * x[n]$ (Commutative)
2. $f[n] * \left[x[n] + y[n]\right] = f[n] * x[n] + f[n] * y[n]$ (Distributive)
3. $f[n] * \left[x[n] * y[n]\right] = \left[f[n] * x[n]\right] * y[n]$ (Associative)
4. $x[n-m] * h[n-k] = y[n-m-k]$ (Shifting)
5. $x[n] * \delta[n] = x[n]$

If both $x[n]$ and $h[n]$ are causal, that is, $x[n]$ and $h[n]$ are zero for all integers $n < 0$, the summation in Equation 2.25 becomes

$$y[n] = \sum_{k=0}^{n} h[k]x[n-k], \quad n \geq 0 \tag{2.27}$$

where $y[n] = 0$ for $n < 0$.

The convolution of an M-point sequence with an N-point sequence produces an $(M + N - 1)$-point sequence.

Just as the continuous-time convolution involves some steps, evaluating the convolution sum requires the following steps:

1. The signal $h[k]$ is time-reversed to get $h[-k]$ and then shifted by n to form $h[n - k]$ or $h[-(k - n)]$, which should be regarded as a function of k with parameter n.
2. For a fixed value of n, multiply $x[k]$ and $h[n - k]$ for all values of k.
3. The product $x[k]h[n - k]$ is summed over all k to produce a single value of $y[n]$.
4. Repeat steps 1–3 for various values of n to produce the entire output $y[n]$.

Table 2.4 presents a list from which convolution sums can be determined directly for a variety of signal pairs.

Example 2.6

Let $r[n]$ be the convolution of two unit step sequences, that is,

$$r[n] = u[n] * u[n]$$

TABLE 2.4
Abbreviated Table for Convolution Sums

No.	$x_1[n]$	$x_2[n]$	$x_1[n] * x_2[n]$
1.	$x[n]$	$\delta[n-k]$	$x[n-k]$
2.	$u[n]$	$u[n]$	$(n+1)u[n] = r[n]$
3.	$a^n u[n]$	$u[n]$	$\left[\dfrac{1-a^{n+1}}{1-a}\right]u[n]$
4.	$a^n u[n]$	$b^n u[n]$	$\left[\dfrac{a^{n+1}-b^{n+1}}{a-b}\right]u[n]$
5.	$u[n]$	$nu[n]$	$\tfrac{1}{2}(n)(n+2)u[u]$
6.	$nu[n]$	$nu[n]$	$\dfrac{1}{6}n(n+1)(n-1)u[n]$
7.	$a^n u[n]$	$nu[n]$	$\left[\dfrac{a(a^n-1)+n(1-a)}{(1-a)^2}\right]u[n]$
8.	$a^n u[n]$	$a^n u[n]$	$(n+1)a^n u[n]$

Find $r[n]$.

Solution

$$r[n] = u[n] * u[n] = \sum_{k=\infty}^{\infty} u[k]u[n-k]$$

The functions $u[k]$, $u[-k]$, and $u[n-k]$ are shown in Figure 2.16. The convolution takes place when we multiply the sequences in Figure 2.16a with Figure 2.16c and d. It is good to show $u[n-k]$ for $n < 0$ and for $n > 0$ as in Figure 2.16c and d, respectively. For $n < 0$, the nonzero values of $u[k]$ and $u[n-k]$ do not overlap so that $u[k]u[n-k] = 0$ for all values of k. This implies that $r[n] = 0$ for $n < 0$. For $n \geq 0$, the nonzero values of $u[k]$ and $u[n-k]$ overlap. This overlap begins with $u[k]$ at $k = 0$ and ends with $u[n-k]$ at $k = n$. Hence,

$$r[n] = \sum_{k=0}^{n} u[k]u[n-k] = \sum_{k=0}^{n} (1) = n+1$$

Therefore,

$$r[n] = (n+1)u[n]$$

This is the unit ramp sequence and is shown in Figure 2.16e.

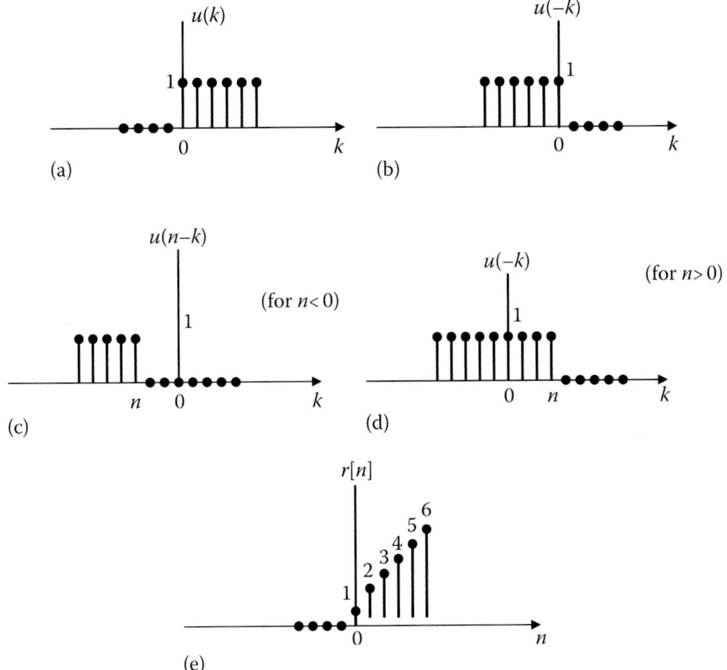

FIGURE 2.16 For Example 2.6.

Practice Problem 2.6 Let the input $x[n]$ and impulse response of a system be given by

$$x[n] = u[n], \quad h[n] = a^n u[n], \quad 0 < a < 1$$

find the output $y[n] = x[n] * h[n]$.

Answer: $y[n] = \frac{1 - a^{n+1}}{1 - a} u[n]$

Example 2.7

Consider $x[n]$ and $h[n]$ as shown in Figure 2.17. (The signals are all zero outside the ranges indicated.) Find $y[n] = x[n]*h[n]$:

(a) Analytically and (b) graphically.

Solution

(a) Using Equation 2.19, we can write

$$x[n] = \delta[n-1] + \delta[n-2] + \delta[n-3]$$

$$h[n] = \delta[n] + \delta[n-1] + \delta[n-2]$$

$$\begin{aligned} y[n] = x[n] * h[n] &= x[n] * \{\delta[n] + \delta[n-1] + \delta[n-1]\} \\ &= x[n] + x[n-1] + x[n-2] \\ &= \delta[n-1] + \delta[n-2] + \delta[n-3] \\ &\quad + \delta[n-2] + \delta[n-3] + \delta[n-4] \\ &\quad + \delta[n-3] + \delta[n-4] + \delta[n-5] \\ &= \delta[n-1] + 2\delta[n-2] + 3\delta[n-3] + 2\delta[n-4] + \delta[n-5] \end{aligned}$$

This is illustrated in Figure 2.18.

(b) The graphical convolution is shown in Figure 2.19. We do the convolution for $n = 0, 1, 2, 3$, and 4. Looking at the trend, we can similarly conclude that for $n = 5$, $y = 1$, and for $n > 5$, $y = 0$. Thus, $y[n]$ is as shown in Figure 2.18, that is,

$$y[n] = [1, 2, 3, 2, 1]$$

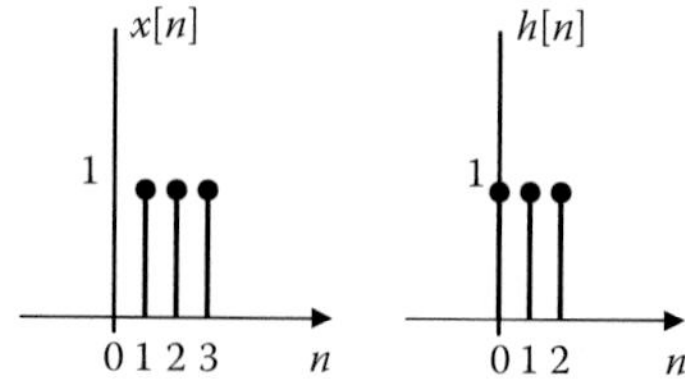

FIGURE 2.17 For Example 2.7.

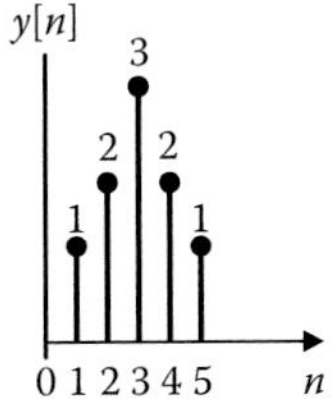

FIGURE 2.18 For Example 2.7; $y[n] = x[n]*h[n]$.

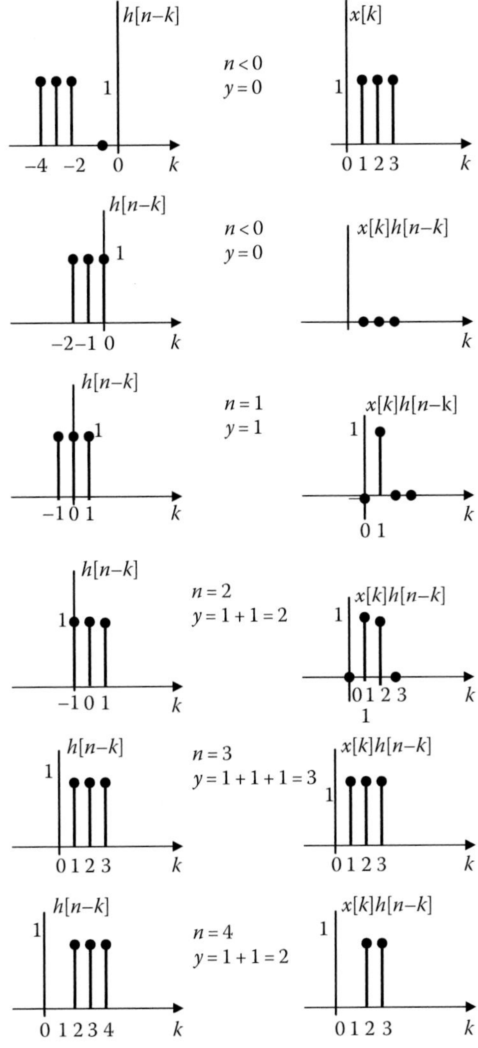

FIGURE 2.19 For Example 2.7(b); graphical convolution.

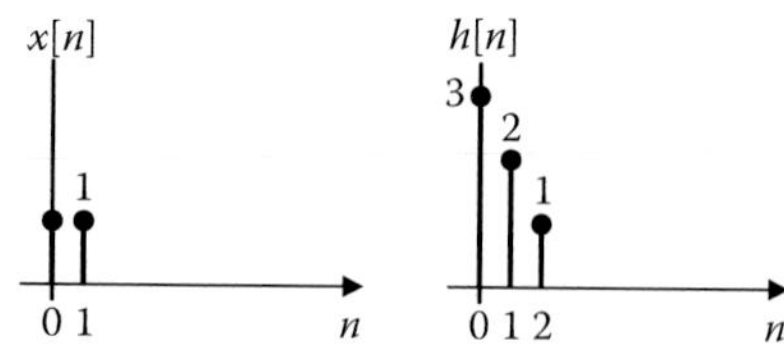

FIGURE 2.20 For Practice Problem 2.7.

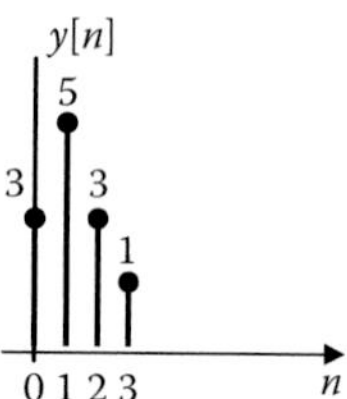

FIGURE 2.21 Solution for Practice Problem 2.7.

Practice Problem 2.7 Find the convolution of the discrete signals shown in Figure 2.20. The signals are all zero outside the range indicated.

Answer: See Figure 2.21.

Example 2.8

Consider a discrete-time LTI system for which the input and impulse functions are given by

$$x[n] = (0.7)^n u[n], \quad h[n] = (0.2)^n u[n]$$

Find the response $y[n]$.

Solution

$$y[n] = x[n] * h[n] = \sum_{k=-\infty}^{\infty} x[k]h[n-k]$$

But $x[k] = (0.7)^k u[k]$ and $h[n-k] = (0.2)^{n-k} u[n-k]$
Because $x[n]$ and $h[n]$ are causal, we can use Equation 2.27.

$$y[n] = x[n] * h[n] = \sum_{k=0}^{n} (0.7)^k (0.2)^{n-k}, \quad n = 0, 1, 2, \ldots$$

$$= (0.2)^n \sum_{k=0}^{n} \left(\frac{0.7}{0.2}\right)^k$$

This is a geometric series. We notice from Appendix A,

$$\sum_{k=0}^{n}\left(\frac{a}{b}\right)^k = \frac{1-(a/b)^{n-1}}{1-(a/b)}$$

Taking $a = 0.7$ and $b = 0.2$

$$y[n] = 0.2^n \frac{1-(0.7/0.2)^{n+1}}{1-0.7/0.2} = -2\left[(0.2)^{n+1} - (0.7)^{n+1}\right]$$

Therefore,

$$y[n] = 2\left[(0.7)^{n+1} - (0.2)^{n+1}\right]u[n]$$

Practice Problem 2.8 Suppose for a discrete-time LTI system, the input $x[n]$ and the impulse response $h[n]$ are

$$x[n] = 4^n u[n] \text{ and } h[n] = u[n]$$

Find the output signal $y[n]$.

Answer: $y[n] = \begin{cases} 4^{n+1}, & n < 0 \\ \frac{4}{3}, & n \geq 0 \end{cases}$

2.7 BLOCK DIAGRAM REALIZATION

Just as with continuous-time case, we can find the impulse response of systems connected in series (cascade) or in parallel. Figure 2.22a illustrates the commutative property, that is, $x[n]$ and $h[n]$ are interchangeable. The associative property, illustrated in Figure 2.22b, shows that when two systems (with impulse responses $h_1[n]$ and $h_2[n]$) are connected in series, the overall impulse response is $h_1[n]*h_2[n]$. The distributive property, illustrated in Figure 2.22c, shows that when two systems are connected in parallel, the impulse response of the composite system is $h_1[n] + h_2[n]$.

2.8 DECONVOLUTION

We know that if the impulse response $h[n]$ of a system is known, we can find the response $y[n]$ to an input $x[n]$ as simply the convolution of $x[n]$ and $h[n]$. If we know $x[n]$ and $y[n]$, how do we get $h[n]$? The process of getting $h[n]$, given $x[n]$ and $y[n]$, is known as *deconvolution*. The process is also known as inverse filtering or system identification.

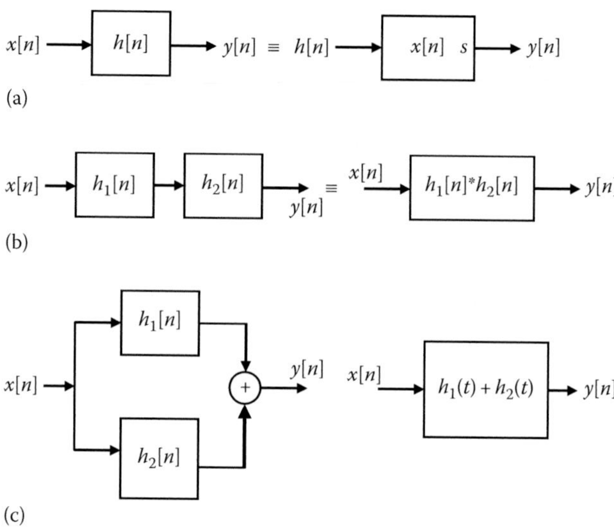

FIGURE 2.22 Block diagrams illustrating interconnected systems.

Deconvolution is the process of obtaining one of the constituent signals in the convolution sum.

Deconvolution is often encountered in practice when trying to measure the response of the communication channel. It has no direct mathematical definition in the continuous-time domain. Discrete convolution can be done in two ways: polynomial division and recursive algorithm. Both methods will be shown in the following example. We will also use a MATLAB command for performing discrete deconvolution.

Example 2.9

Let $x[n] = \{3, 0, 2, 6\}$ and $y[n] = \{6, 10, 25, 20, 38, 42\}$. Find $h[n]$.

Solution

We will obtain the corresponding $h[n]$ in two ways.

Method 1 (Long division):

We regard the given sequences $x[n]$ and $y[n]$ as coefficients of the following polynomials in descending order.

$$x(z) = 3z^3 + 0z^2 + 2z + 6, \quad y(z) = 6z^5 + 12z^4 + 25z^3 + 20z^2 + 38z + 42$$

We now perform a long division $y(z)/x(z)$.

$$\begin{array}{r} 2z^2+4z+7 \\ 3z^3+0z^2+2z+6\overline{\big)6z^5+12z^4+25z^3+20z^2+38z+42} \\ \underline{6z^5+0z^4+4z^3+12z^2} \\ 12z^4+21z^3+8z^2+38z+42 \\ \underline{12z^4+0z^3+8z^2+24z} \\ 21z^3+0z^2+14z+42 \\ \underline{21z^3+0z^2+14z+42} \\ 0 \end{array}$$

From this, we obtain the polynomial $h(z) = 2z^2 + 4z + 7$ or $h[n] = \{2, 4, 7\}$.

Method 2 (recursive algorithm):
By definition,

$$y[n] = x[n] * h[n] = \sum_{k=0}^{n} h[k]x[n-k] \tag{2.9.1}$$

This convolution sum can be cast as a recursive algorithm. For $n = 0$,

$$y[0] = x[0]h[0] \rightarrow h[0] = y[0]/x[0] = 6/3 = 2 \tag{2.9.2}$$

We separate the term containing $x[0]$ in Equation 2.9.1.

$$y[n] = h[n]x[0] + \sum_{k=0}^{n-1} h[k]x[n-k] \tag{2.9.3}$$

From this, we obtain for $n > 0$,

$$h[n] = \frac{1}{x[0]}\left[y[n] - \sum_{k=0}^{n-1} h[k]x[n-k]\right] \tag{2.9.4}$$

We will need to evaluate this at $N_y - N_x + 1$ points, where N_y and N_x are the lengths of $y[n]$ and $x[n]$, respectively. In this example,

$$N_y = 6, N_x = 4 \text{ so that } N_y - N_x + 1 = 6 - 4 + 1 = 3$$

that is, we need three evaluations of $h[n]$. From Equation 2.9.2, we already got $h[0] = 2$. We now apply Equation 2.9.4 twice. For $n = 1$,

$$h[1] = \frac{1}{x[0]}\left[y[1] - \sum_{k=0}^{0} h[k]x[1-k]\right] = \frac{1}{x[0]}\big[y[1] - h[0]x[1]\big] = \frac{1}{3}[12 - 2\times 0] = 4$$

For $n = 2$,

$$h[2] = \frac{1}{x[0]}\left[y[2] - \sum_{k=0}^{1} h[k]x[2-k]\right] = \frac{1}{x[0]}\left[y[2] - h[0]x[2] - h[1]x[1]\right]$$

$$= \frac{1}{3}[25 - 2\times 2 - 4\times 0] = 7$$

thus, $h = \{2, 4, 7\}$ as we obtained previously.

Practice Problem 2.9 Given that $x[n] = \{1, 8, 4, 5\}$ and $y[n] = \{4, 38, 66, 61, 46, 14, 5\}$, find $h[n]$.

Answer: $h[n] = \{4, 6, 2, 1\}$.

2.9 COMPUTING WITH MATLAB®

MATLAB provides a simple way of performing convolution. Given two signals x and h, we can use MATLAB to do the convolution with the built-in function **conv**(x,h). The function **conv** returns a vector of length $L_x + L_h - 1$, where L_x is the length of x and L_h is the length of h. This produces an output that will be longer than the input. Often the additional data can be discarded.

It does not matter when x and h are continuous-time or discrete-time signals. We use **conv** function for both cases. In fact, the function **conv** is actually designed for discrete-time convolution. To use it for continuous-time convolution, we need to evaluate the convolution integral numerically. Consider the convolution integral

$$y(t) = \int_0^t x(\tau)h(t-\tau)d\tau \tag{2.28}$$

Let T, the step size or sampling period, be small and let $t = kT$ and $\tau = nT$, the convolution integral becomes a convolution summation which can be expressed as

$$y(kT) \approx T\sum_{n=0}^{kT} x(nT)h((k-n)T) = T\sum_{n=0}^{k} x[n]h[k-n] \tag{2.29}$$

This approximates a rectangular rule integration. Equation 2.29 can be written as

$$y(k) \approx T\sum_{n=0}^{k} x[n]h[k-n] \tag{2.30}$$

which looks like discrete-time convolution except for multiplying by T.

For the purpose of illustration, we will solve one example for discrete-time signals and one example for continuous-time signals.

The MATLAB command **deconv**(y,h) can be used to find the discrete deconvolution of $y[n]$ and $x[n]$, producing $h[n]$. While **conv** can be used to multiply polynomials, **deconv** can be used to divide polynomials.

Example 2.10

Use MATLAB to find the convolution of the sequences:

$$x_1[n] = \{0.2, 1.4, 2.6, 5.1, 3.4, 8.4\}$$

$$x_2[n] = \{1.0, 4.2, 3.7, 0.8, 3.9\}$$

Solution

The MATLAB to perform the convolution is as follows:

$$x_1 = \left[0.2,\ 1.4,\ 2.6,\ 5.1,\ 3.4,\ 8.4\right]$$

$$x_2 = \left[1.0,\ 4.2,\ 3.7,\ 0.8,\ 3.9\right]$$

$$y = \text{conv}(x_1, x_2)$$

The result is

$$y = 0.2000\ 2.2400\ 9.2200\ 21.3600\ 36.3400\ 49.0900\ 62.0800\ 53.6900\ 19.9800\ 32.7600$$

Thus,

$$x_1[n] * x_2[n] = \{0.2000\ 2.2400\ 9.2200\ 21.3600\ 36.3400\ 49.0900\ 62.0800\ 53.6900\ 19.9800\ 32.7600\}$$

Practice Problem 2.10 Using MATLAB, find the convolution of these sequences:

$$x_1[n] = \{2, 5, 4, 3, 5, 1, 0\},\ x_2[n] = \{1, 6, 3, 9, 4, 7, 2\}$$

Answer: $y[n]$ = [2 17 40 60 88 110 103 98 58 45 17 2 0]

Example 2.11

A system is represented by its impulse response:

$$h(t) = \frac{1}{4}\left(e^{-2t} - e^{-t}\right)$$

Find and plot the response when the input is $x(t) = \cos(t)\, u(t)$.

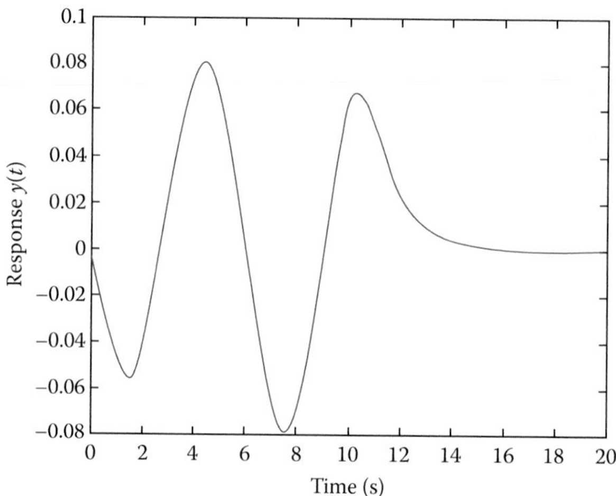

FIGURE 2.23 For Example 2.11.

Solution

To use the function **conv** for continuous-time convolution, we need to mutilply **conv** with a sampling period or step size, as shown in Equation 2.30. The MATLAB code is as follows, while the plot is shown in Figure 2.23.

```
T = 0.1; % sampling period
t = 0:T:10;
x = cos(t); % calculates x(t)
h = 0.25*(exp(-2*t) - exp(-t)); % calculates h(t)
y = T*conv(x,h); %this contains Lx + Lh - 1
t0 = (0:200)*T
plot(t0,y) % or use this plot(t,y(1:101))
xlabel('Time (s)')
ylabel('Response y(t)')
```

Practice Problem 2.11 Repeat Example 2.11 when

$$h(t) = \frac{1}{2}\left(e^{-t} - e^{-4t}\right)u(t), \qquad x(t) = e^{-t}\sin(t)u(t)$$

Answer: See Figure 2.24.

Example 2.12

Repeat Example 2.9 using MATLAB.

Solution

In MATLAB, we enter the following lines:

```
x = [3 0 2 6]
y = [6 10 25 20 38 42]
h = deconv(y,x)
```

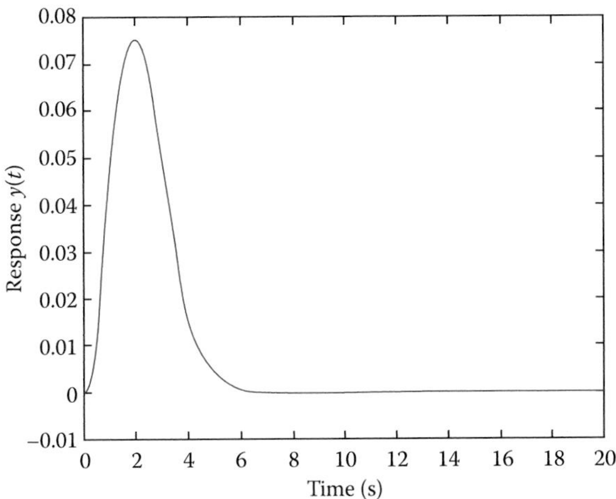

FIGURE 2.24 For Practice Problem 2.11.

This produces

```
h = [2 4 7]
```

Practice Problem 2.12 Repeat Practice Problem 2.9 using MATLAB.

Answer: $h = [4\ 6\ 2\ 1]$

2.10 APPLICATIONS

There are several applications of the convolution technique developed in this chapter. Convolution is used in circuit analysis, statistics, probability theory, optics, acoustics, digital image processing, and computational fluid dynamics. We will employ convolution to determine the stability of systems. We will also consider an application in circuit analysis.

2.10.1 BIBO Stability of Continuous-Time Systems

A system is bounded-input bounded-output (BIBO) stable if every bounded-input signal produces a bounded-output. We can determine the condition for stability by applying the convolution integral. Suppose we are given an LTI system and we apply a bounded-input $x(t)$. The boundedness of $x(t)$ can be expressed as $|x(t)| < K$, where K is a constant. The output $y(t)$ can be bounded as follows. By definition,

$$y(t) = x(t) * h(t) = \int_{-\infty}^{\infty} x(t-\tau)h(\tau)d\tau$$

Therefore,

$$|y(t)| = \left| \int_{-\infty}^{\infty} x(t-\tau)h(\tau)d\tau \right| \le \int_{-\infty}^{\infty} |x(t-\tau)||h(\tau)|d\tau \le \int_{-\infty}^{\infty} K|h(\tau)|d\tau$$

$$= K \int_{-\infty}^{\infty} |h(\tau)| d\tau \qquad (2.31)$$

This shows that $|y(t)|$ is finite if $\int_{-\infty}^{\infty} K|h(\tau)|d\tau$ is finite. This reduces to $\int_{-\infty}^{\infty} |h(\tau)|d\tau < \infty$ since K is finite. For an LTI system, the condition for stability reduces to

$$\int_{-\infty}^{\infty} |h(\tau)|d\tau < \infty \qquad (2.32)$$

A continuous-time system is **BIBO stable** if its impulse response $h(t)$ is absolutely integrable.

Example 2.13

An LTI system has the impulse function $h(t) = e^{-2t}u(t)$. Determine whether the system is stable or not.

Solution

Using Equation 2.32,

$$\int_{-\infty}^{\infty} |h(\tau)|d\tau = \int_{0}^{\infty} e^{-2\tau}d\tau = \frac{e^{-2\tau}}{-2}\Big|_{0}^{\infty} = \frac{1}{2} < \infty$$

Since $h(t)$ is integrable, we conclude that the system is stable.

Practice Problem 2.13 Repeat Example 2.13 with $h(t) = u(t)$. This is the impulse response for a continuous-time integrator.

Answer: Unstable.

2.10.2 BIBO Stability of Discrete-Time Systems

The same analysis applies to discrete-time systems. Suppose we have an LTI system and we apply a bounded input $x(t)$, with $|x[n]| < K$, where K is a constant. We can bound the output as follows. We recall that

$$y[n] = x[n] * h[n] = \sum_{k=-\infty}^{\infty} x[n-k]h[k]$$

Therefore,

$$\left|y[n]\right| = \left|\sum_{k=-\infty}^{\infty} x[n-k]h[k]\right| \leq \sum_{k=-\infty}^{\infty} |\, x[n-k]\,||\, h[k]\,| \leq \sum_{k=-\infty}^{\infty} K\,|\, h[k]\,|$$

$$= K\sum_{k=-\infty}^{\infty} |\, h[k]\,| \tag{2.33}$$

This shows that $|y[n]|$ is finite if $\sum_{k=-\infty}^{\infty} |\, h[k]\,|$ is finite. Thus, for stability ($|y[n]|<\infty$)

$$\sum_{k=-\infty}^{\infty} |\, h[k]\,| < \infty \tag{2.34}$$

A discrete-time system is *BIBO stable* if its impulse response $h[n]$ is absolutely summable.

Example 2.14

A discrete-time system has an impulse response $h[n] = \frac{1}{n}u[n]$. Determine if the system is stable.

Solution

From Equation 2.34

$$\sum_{k=-\infty}^{\infty} |\, h[k]\,| = \sum_{k=0}^{\infty} \frac{1}{n} = \lim_{N\to\infty} \sum_{k=0}^{N} \frac{1}{n}$$

which does not converge as $N\to\infty$. Thus, $h[n]$ is not absolutely summable and we conclude that the system is unstable.

Practice Problem 2.14 A filter is described by $h[n] = (-0.6)^n$. Is the system stable?

Answer: Stable.

2.10.3 Circuit Analysis

We now apply convolution to analyze electric circuits. The following example will illustrate the procedure.

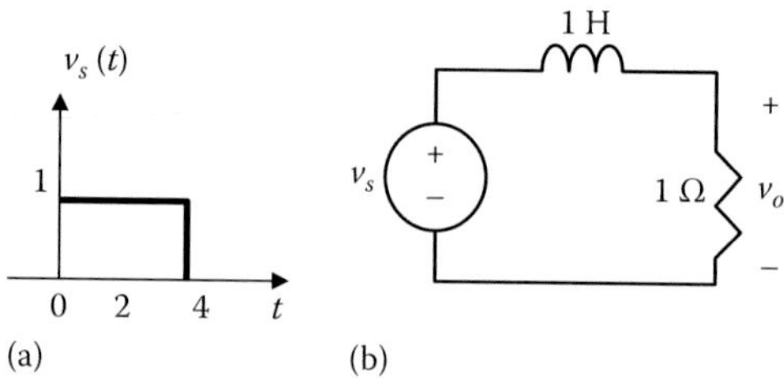

FIGURE 2.25 For Example 2.15.

Example 2.15

The pulse $v_s(t)$ in Figure 2.25a is applied the RL circuit in Figure 2.25b. Use the convolution integral to find the response $v_o(t)$.

Solution

This problem requires that we know the impulse response of the circuit. Throughout this chapter, we assume that the impulse response $h(t)$ is given. We will learn how to find it in the next chapter. For now, let us assume that the impulse response of the circuit is

$$h(t) = e^{-t}\, u(t)$$

From Figure 2.25a, $v_s(t) = u(t) - u(t-4)$. Once we know impulse response $h(t)$ and input signal $v_s(t)$, we can find the response $v_o(t)$:

$$v_o(t) = h(t) * v_s(t) = \int_0^t v_s(\tau)h(t-\tau)d\tau$$

$$= \int_0^t \left[u(\tau) - u[\tau - 4]\right] e^{-(t-\tau)} d\tau \qquad (2.15.1)$$

Since $u(\tau - 4) = 0$ for $0 < \tau < 4$, the integrand involving $u(\tau)$ is nonzero for all τ, whereas the integrant involving $u(\tau - 4)$ is nonzero only for $\tau > 4$. The best way to handle the integral is to evaluate the two parts separately. For $0 < t < 4$,

$$v_o' = \int_0^t (1)e^{-(t-\tau)} d\tau = e^{-t}\int_0^t (1)e^{\tau} d\tau$$

$$= e^{-t}(e^t - 1) = 1 - e^{-t}, \quad 0 < t < 4 \qquad (2.15.2)$$

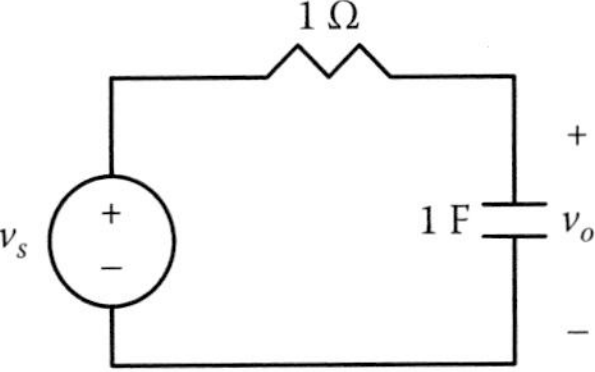

FIGURE 2.26 For Practice Problem 2.15.

For $t > 4$,

$$v_o'' = \int_4^t (1)e^{-(t-\tau)}d\tau = e^{-t}\int_4^t (1)e^{\tau}d\tau$$

$$= e^{-t}(e^t - e^4) = 1 - e^4 e^{-t}, \quad t > 4 \tag{2.15.3}$$

Substituting Equations 2.15.2 and 2.15.3 into Equation 2.15.1 gives

$$v_o(t) = v_o'(t) - v_o''(t)$$

$$= (1 - e^{-t})\left[u(t-4) - u(t)\right] - (1 - e^4 e^{-t})u(t-4)$$

$$= \begin{cases} 1 - e^{-t}, & 0 < t < 4 \\ e^4 e^{-t} - 1, & t > 4 \end{cases} \tag{2.15.4}$$

Practice Problem 2.15 Find the response $v_o(t)$ of the RC circuit of Figure 2.26. Assume that the impulse response of the circuit is $h(t) = e^{-t}u(t)$.

Answer: $5\ (e^{-t} - e^{-2t})u(t)\ V$

2.11 SUMMARY

1. This chapter discusses time-domain analyses of linear, time-invariant (LTI) continuous-time and discrete-time systems.
2. Every LTI system is characterized by its impulse response $h(t)$ or $h[n]$.
3. The analysis of an LTI system can be reduced to finding the response of the system to input signals. Given the impulse response $h(t)$, the system's response $y(t)$ to an input signal $x(t)$ is given by the convolution integral:

$$y(t) = x(t) * h(t) = \int_{-\infty}^{\infty} x(\tau)h(t-\tau)d\tau$$

   Thus, convolution is a time-domain technique for determining the response of a system to an input.
4. Convolution is obtained analytically or graphically by performing four operations: (1) folding or time reversal to get $h(-\tau)$, (2) shifting to get $h(t-\tau)$, (3) multiplying $h(t-\tau)$ and $x(\tau)$, and (4) integrating $h(t-\tau)x(\tau)$.

5. The composite impulse response $h(t)$ of two systems (with impulses $h_1(t)$ and $h_2(t)$) in series is the convolution of the individual impulses, that is, $h(t) = h_1(t)*h_2(t)$.
6. The composite impulse response $h(t)$ of two systems (with impulses $h_1(t)$ and $h_2(t)$) in parallel is the sum of the individual impulses, that is, $h(t) = h_1(t) + h_2(t)$.
7. Block diagram representation helps us understand the dynamic behavior of systems.
8. For discrete-time system, the impulse response $h[n]$, the input $x[n]$, and the output $y[n]$ are related by the convolution sum:

$$y[n] = x[n] * h[n] = \sum_{k=-\infty}^{\infty} x[k]h[n-k]$$

9. Convolution (both in continuous-time and discrete-time) is commutative, associative, and distributive.
10. Deconvolution is the process of finding the impulse response $h[n]$ of a system given its input $x[n]$ and its output $y[n]$.
11. The function **conv**(x,h) is used in MATLAB to perform both continuous-time and discrete-time convolution. The MATLAB function **deconv**(y,x) performs discrete deconvolution.
12. A continuous-time system is BIBO stable if its impulse response is absolutely integrable. A discrete-time system is BIBO stable if its impulse response is absolutely summable.
13. A simple application of convolution is circuit analysis. If we know the impulse response of a circuit, we can determine the response to any input with the use of convolution.

REVIEW QUESTIONS

2.1 It is improper to talk about an impulse response for a non-LTI system.
(a) True, (b) false

2.2 Which of these is not true?
(a) $x(t)*h(t) = h(-t)*x(t)$
(b) $x(t)*\delta(t) = x(t)$
(c) $x(t)*\delta'(t) = x'(t)$
(d) $u(t)*\delta'(t) = \delta(t)$

2.3 The methods introduced in this chapter are known as time-domain method because they involve functions of time.
(a) True, (b) false

2.4 When two systems with impulse responses $h_1(t)$ and $h_2(t)$ are connected in series, the impulse response of the composite system is
(a) $h_1(t) + h_2(t)$, (b) $h_1(t)*h_2(t)$, (c) $h_1(t)/h_2(t)$, (d) none of the above

2.5 Consider the two signals in Figure 2.27. They are convolved to get $y(t)$. The duration of $y(t)$ is
(a) $-1 < t < 2$, (b) $-2 < t < 2$, (c) $-1 < t < 3$, (d) none of the above

2.6 From the previous question, the area of $y(t)$ is
(a) 2, (b) 4, (c) 6, (d) none of the above

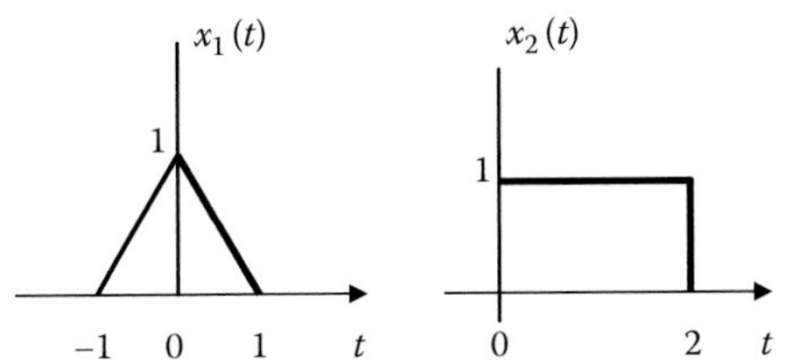

FIGURE 2.27 For Review Questions 2.5 and 2.6.

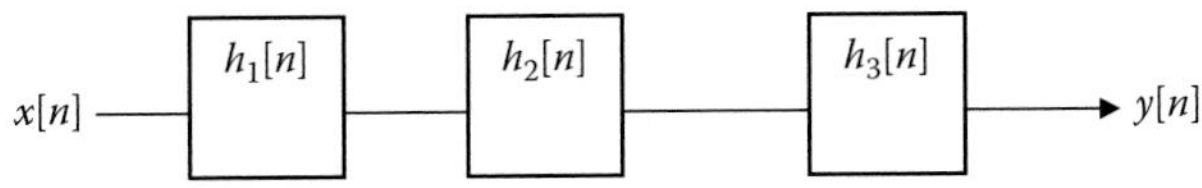

FIGURE 2.28 For Review Question 2.9.

2.7 Which of the following is not true?
(a) $\delta[n] = u[n] - u[n-1]$
(b) $u[n] = \sum_{k=-\infty}^{n} \delta[k]$
(c) $x[n] * \delta[n] = x[n]$
(d) $x[n] * h[n] = -h[n] * x[n]$

2.8 If two sequences of length 8 and 10 are convolved to give $y[n]$, the length of $y[n]$ is
(a) 8, (b) 10, (c) 17, (d) 18, (e) none of the above

2.9 Consider the interconnection of three LTI system in Figure 2.28. The overall impulse response is
(a) $h_1[n] + h_2[n] + h_3[n]$
(b) $h_1[n] * h_2[n] * h_3[n]$
(c) None of the above

2.10 The MATLAB function **conv** is used for
(a) Continuous-time convolution
(b) Discrete-time convolution
(c) Both (a) and (b)

Answers: 2.1(a), 2.2(a), 2.3(a), 2.4(b), 2.5(c), 2.6(a), 2.7(d), 2.8(c), 2.9(b), 2.10(c)

PROBLEMS

Section 2.3—Convolution Integral

2.1 Show that $x_1(t) * x_2(t) = x_2(t) * x_1(t)$.

2.2 Show that
(a) $f(t) * \delta(t) = f(t)$
(b) $f(t) * \delta(t - t_o) = f(t - t_o)$
(c) $f(t) * u(t) = \int_{-\infty}^{t} f(\tau) d\tau$

2.3 Show $x(t) * \frac{d}{dt}\delta(t) = \frac{d}{dt}x(t)$.

2.4 Given the following signals

$$x(t) = 2\delta(t), \quad y(t) = 4u(t), \quad z(t) = e^{-2t}u(t),$$

Evaluate the following operations.
(a) $x(t)*y(t)$
(b) $x(t)*z(t)$
(c) $y(t)*z(t)$
(d) $y(t)*[y(t) + z(t)]$

2.5 Use the convolution integral to find
(a) $t*e^{at}\ u(t)$
(b) $\cos(t)*\cos(t)u(t)$

2.6 Write the expression for $y(t) = x(t)*\delta(t-2)$ when
(a) $x(t) = t^2$
(b) $x(t) = t(t + 0.5)^2$
(c) $x(t) = 4e^{-2t}\cos(2\pi t - \pi/2)$

2.7 Sketch these functions
(a) $x(t) = \Pi(t/2)*\ [\delta(t+1) + \delta(t+2)]$
(b) $y(t) = [\Lambda(t+1) - \Lambda(t-1)]*\delta(t)$
(c) $z(t) = \Pi(t/2)*\ \Lambda(t)$

2.8 Show that
(a) $u(t)*\ u(t) = r(t)$
(b) $u(t-1)*\ u(t-3) = r(t-4)$
(c) $\Pi(t/\tau)*\ \Pi(t/\tau) = \tau\Lambda(t/\tau)$

2.9 For an LTI system, the impulse response is $h(t) = e^{-2t}u(t)$. If the input to the system is $x(t) = 10e^{-2t}u(t)$, find the output $y(t)$.

2.10 The input to an LTI system is $x(t) = \cos(2t)$, while the impulse response of the system is

$$h(t) = e^{-|t|} = \begin{cases} e^{t}, & t < 0 \\ e^{-t} & t \geq 0 \end{cases}$$

Determine the system's output $y(t)$.

2.11 The impulse response of a filter is $h(t) = e^{-2t}u(t) - \delta(t)$. Find the response of the filter to the input $e^{-t}u(t)$.

2.12 An LTI system has its impulse response $h(t) = u(t)$. Find the response corresponding to the input $x(t) = e^{-2t}u(t)$.

2.13 Given that the impulse response of a system is $h(t) = e^{-2t}u(t)$, determine the system's response $y(t)$ if the input is
(a) $u(t)$
(b) $e^{-t}u(t)$
(c) $\cos(2t)u(t)$

2.14 The impulse response of a low-pass filter is $h(t) = e^{-t}u(t)$. Determine its step response, that is, the output when the input is a unit step.

2.15 If $h(t)$ is the impulse response of a system and $x(t)$ is the input, what kinds of systems are described by: (a) $h(t) = \delta(t)$, (b) $h(t) = (d/dt)\delta(t)$, (c) $h(t) = u(t)$.

2.16 Given that $h(t) = 4e^{-2t}u(t)$ and $x(t) = \delta(t) - 2e^{-2t}u(t)$, find $y(t) = x(t)*h(t)$.

Section 2.4—Graphical Convolution

2.17 Verify the area property of convolution in Example 2.3; that is, if A_{x1} and A_{x2} are the areas under $x_1(t)$ and $x_2(t)$ respectively, the area A_y under $y(t) = x_1(t)*x_2(t)$ is $A_y = A_{x1}A_{x2}$.

2.18 Perform a graphical convolution of the two signals in Figure 2.29.

2.19 Find the convolution of $x(t)$ and $h(t)$ in Figure 2.30.

2.20 Determine the convolution $x_1(t)*x_2(t)$ for each pair of signals in Figure 2.31.

2.21 Suppose $x(t) = u(t) - u(t-2)$, determine $y(t) = x(t)*x(t)$.

2.22 Obtain the convolution of the two signals in Figure 2.32.

Section 2.5—Block Diagram Representation

2.23 Two systems are connected in parallel as in Figure 2.33. If the impulse responses of the systems are given by

$$h_1(t) = 4e^{-t}u(t) \quad \text{and} \quad h_1(t) = e^{-2t}u(t)$$

Find the impulse response of the overall system.

2.24 Determine the overall impulse response for the system shown in Figure 2.34.

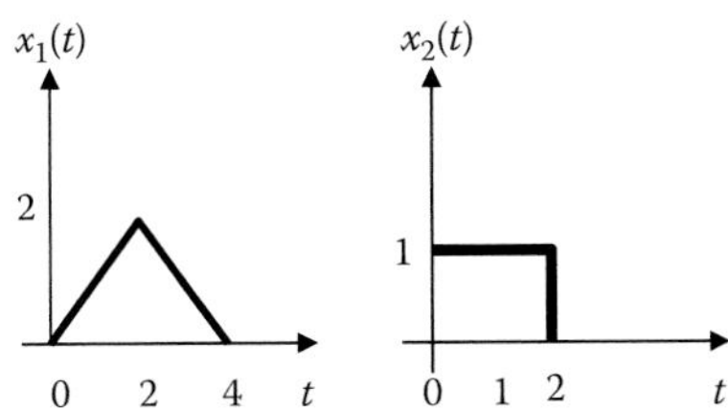

FIGURE 2.29 For Problem 2.18.

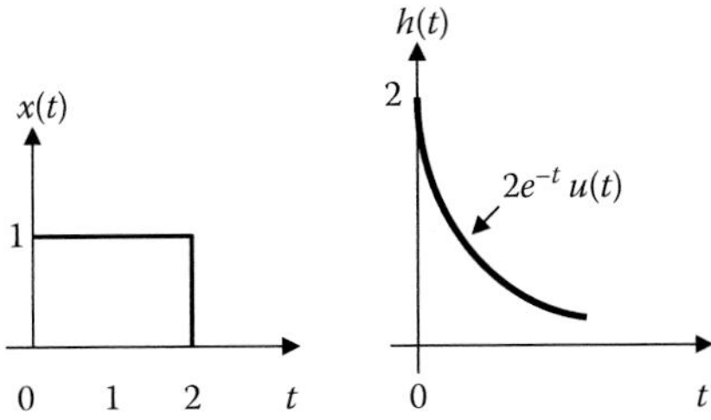

FIGURE 2.30 For Problem 2.19.

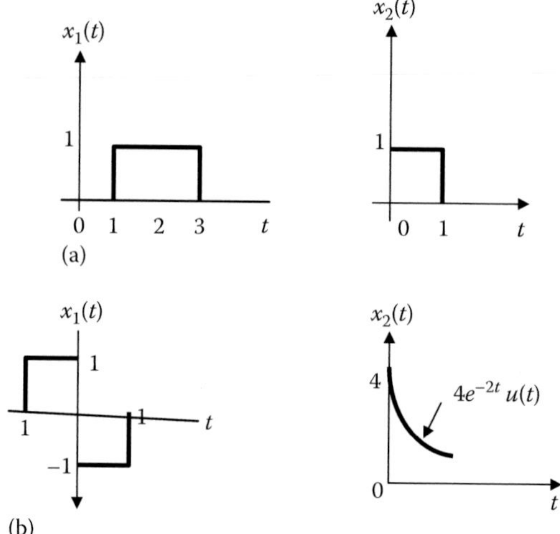

FIGURE 2.31 For Problem 2.20.

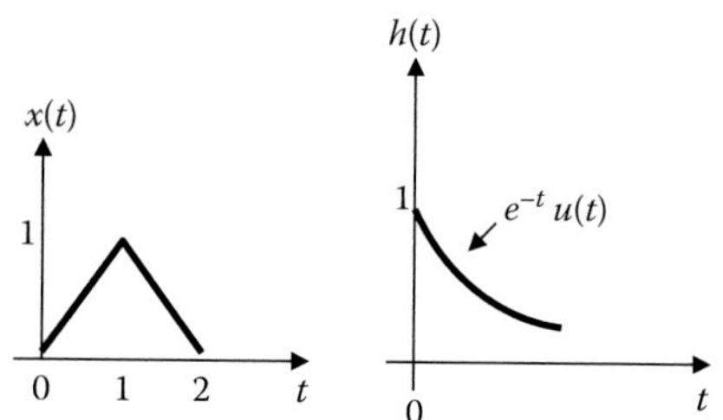

FIGURE 2.32 For Problem 2.22.

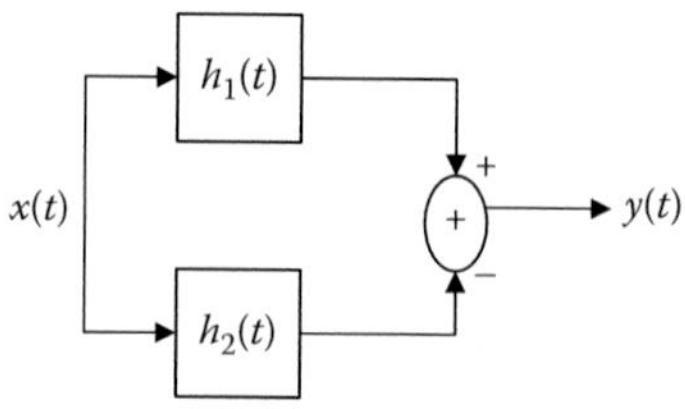

FIGURE 2.33 For Problem 2.23.

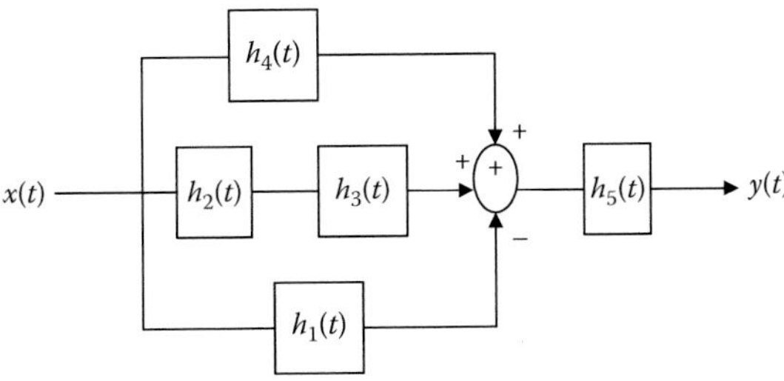

FIGURE 2.34 For Problem 2.24.

Section 2.6—Discrete-Time Convolution

2.25 The following equalities are useful in our studies. Prove each of them.

(a) $$\sum_{k=0}^{N-1} a^k = \begin{cases} N, & a = 1 \\ \dfrac{1-a^N}{1-a}, & a \neq 1 \end{cases}$$

(b) $$\sum_{k=0}^{\infty} a^k = \frac{1}{1-a}, \quad |a| < 1$$

(c) $$\sum_{k=0}^{\infty} ka^k = \frac{a}{(1-a)^2}, \quad |a| < 1$$

2.26 Consider a discrete-time system with impulse response

$$h[n] = (0.4)^n u[n]$$

If $y[n]$ is the output of the system due to the input

$$x[n] = \delta[n-1] + 3\delta[n]$$

find $y[2]$ and $y[5]$.

2.27 Determine $y[n] = x[n]*h[n]$ for the following pairs of signals:

(a) $x[n] = u[n], \quad h[n] = 4^n u[n]$

(b) $x[n] = h[n] = 2^n u[n]$

(c) $x[n] = (0.3)^n u[n], \quad h[n] = 2^n u[n]$

2.28 Find the convolution of the signals $x_1[n]$ and $x_2[n]$ given as

$$x_1[n] = 2(3)^n u[n], \quad x_2[n] = 3(2)^n u[n]$$

2.29 Obtain the convolution of $x[n]$ and $h[n]$ shown in Figure 2.35.

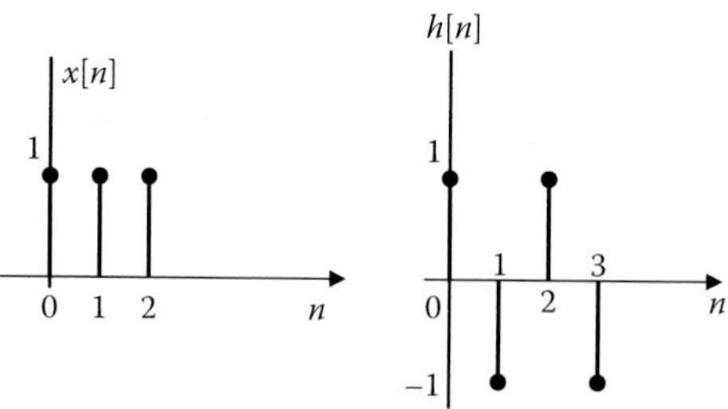

FIGURE 2.35 For Problem 2.29.

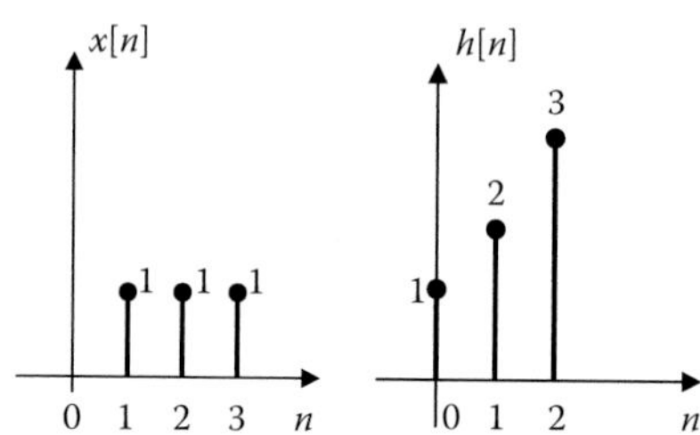

FIGURE 2.36 For Problem 2.31.

2.30 Given that $x[n]=\begin{cases}1, & n=0\\ -1, & n=1\\ 0, & \text{otherwise}\end{cases}$, $h[n]=\begin{cases}1, & n=0\\ 3, & n=1\\ 2, & n=3\\ 0, & \text{otherwise}\end{cases}$

(a) Sketch $x[n]$ and $h[n]$.
(b) Find $x[n]*h[n]$.

2.31 Find the discrete convolution of $x[n]$ and $h[n]$ shown in Figure 2.36.

2.32 Given that $x[n] = u[n]$ and $h[n] = (0.4)^n + (0.5)^{n+1}$, $n \geq 0$. Find $y[n] = x[n]*h[n]$.

Hint: $\sum_{k=0}^{n} a^k = \frac{1-a^{n+1}}{1-a}$.

Section 2.7—Block Diagram Realization

2.33 Two systems are described by

$$h_1[n]=(0.4)^n u[n], \qquad h_2[n]=\delta[n]+0.5\delta[n-1]$$

Determine the response to the input $x[n] = (0.4)^n u[n]$ if
(a) The two systems are connected in parallel
(b) The two systems are connected in cascade

2.34 For the interconnection of LTI systems in Figure 2.37, find the overall impulse response.

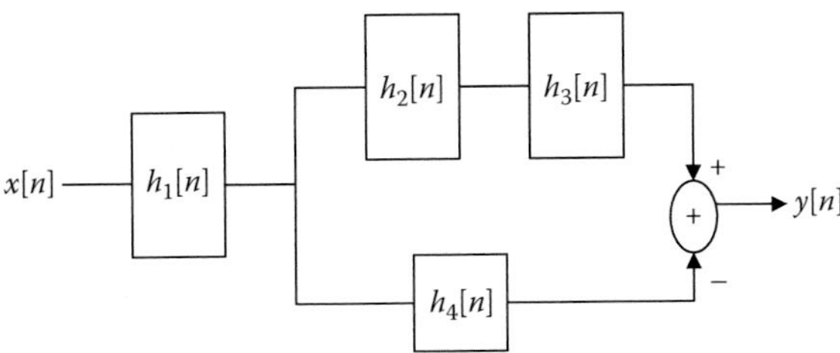

FIGURE 2.37 For Problem 2.34.

Section 2.8—Deconvolution

2.35 Find the impulse response of a system which yields $y[n]$ = [2 3 −2 4] when excited by the input $x[n]$ = [1 0 2].

2.36 The input $x[n]$ = [1 −1] to a system produces the output $y[n]$ = [4 2 5 1]. Determine the impulse response.

Section 2.9—Computing with MATLAB®

2.37 (a) Using MATLAB, write a program to convolve $x(t) = e^{-t}u(t)$ and $h(t) = u(t) - u(t-2)$.
(b) Do the convolution analytically and compare this with MATLAB result in part (a).

2.38 A filter has the impulse response as

$$h[n] = \{1,1,1,1,1,-1,-1,-1,-1,-1\}$$

Use MATLAB to determine the output due to $x[n] = \cos(\pi n/2)$.

2.39 An LTI discrete system has the impulse response $h[n] = (0.6)^n\ u[n]$. Use MATLAB to calculate the response of the system to input $x[n] = u[n]$ and plot it.

2.40 Repeat the previous problem for $x[n] = \cos(n\pi/6)u[n]$.

2.41 Given that $x[n]$ = [1 −1 2 4] and $y[n]$ = [2 6 4 0 8 5 12], use MATLAB to find $h[n]$.

Section 2.10—Applications

2.42 Determine the stability of the LTI system with the following impulse responses.
(a) $h(t) = e^{2t}u(t)$
(b) $h(t) = \sin 2t\ u(t)$
(c) $h(t) = e^{-t}\cos 2t\ u(t)$

2.43 A system has an impulse response $h(t) = u(t+1) - u(t-1)$. Is the system stable?

2.44 State which of the systems represented by the following impulse responses are stable or unstable.
(a) $h[n] = \delta[n]$
(b) $h[n] = (-0.5)^n u[n]$
(c) $h[n] = u[n] - u[n-10]$

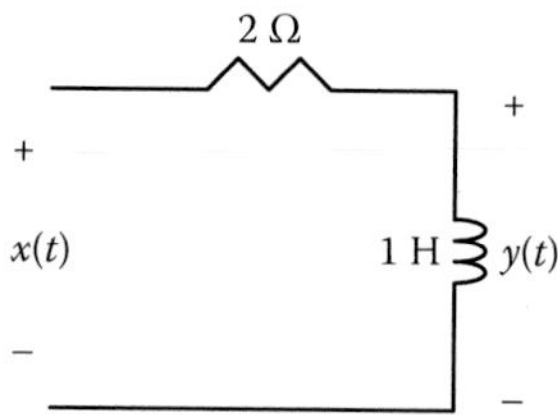

FIGURE 2.38 For Problem 2.48.

2.45 An accumulator has impulse response $h[n] = u[n]$. Check if an accumulator is BIBO stable.

2.46 The input to the circuit discussed in Example 2.15 is $v_s(t) = u(t - 1) + \delta(t - 2)$. Use convolution to determine the output.

2.47 Determine the output of the circuit discussed in Practice Problem 2.15 if $v_s(t) = u(t)$.

2.48 If the filter in Figure 2.38 has the impulse response $h(t) = \delta(t) - 2e^{-2t}u(t)$, find the step response.

3 The Laplace Transform

The brave man is the man who faces or fears the right thing for the right purpose in the right manner at the right moment.

—**Aristotle**

HISTORICAL PROFILE

Pierre Simon Laplace

Pierre Simon Laplace (1749–1827) was a French astronomer and mathematician, whose work was important to the development of mathematical astronomy and statistics. He also put the theory of mathematical probability on a sound footing. He first discovered Laplace's equation and Laplace transform, which is discussed in this chapter. To Laplace, the universe is nothing but a giant problem in calculus.

Laplace was born in Beaumont-en-Auge, Normandy, France on March 23, 1749, and died in Paris on March 5, 1827. Laplace became a professor of mathematics at the age of 20. He is remembered as one of the greatest scientists of all time. He is sometimes referred to as the "Newton of France." He was widely known for his five-volume work, *Celestial Mechanics*, which supplemented the work of Newton on astronomy. After the publication of this work, Laplace continued to apply his ideas of physics to other problems such as capillary action, double refraction, the velocity of sound, the theory of heat, rotation of the cooling earth, and elastic fluids. He was born and died a Catholic.

3.1 INTRODUCTION

We saw in Chapter 2 that system analysis in the time-domain involves the evaluation of the convolution integral or sum since the response of the system is given by the convolution of the input and the impulse response. This chapter introduces the *Laplace transform*, a very powerful, alternative tool for analyzing systems. The Laplace transform is named after Pierre Simon Laplace (1749–1827), the French astronomer and mathematician. In contrast to the time-domain models studied in the previous chapter, the Laplace transform is a frequency-domain representation that makes analysis and design of linear systems simpler.

The Laplace transform is a well-established tool in analyzing continuous-time, linear systems. It is important for a number of reasons. First, it is applicable to a wider variety of inputs. The Laplace transform of an unbounded signal can be found. Second, Laplace transform is powerful for providing us in one single operation the complete response, that is, the steady-state plus transient or homogeneous, and particular. Third, it automatically includes the initial conditions in the system analysis. It allows us to convert ordinary differential equations into algebraic equations, which are easier to manipulate and solve. It converts convolution into a simple multiplication. Fourth, we can apply Laplace transform to generate the transfer function representation of a continuous-time LTI system.

The chapter begins with the definition of the Laplace transform and uses that definition to derive the transform of some basic, important functions. We will consider some properties of Laplace transform which are helpful in obtaining the Laplace transform of other functions. We then consider the inverse Laplace transform. We apply all these to solving integro-differential equations and system analysis, especially in circuit and control systems. We finally demonstrate how MATLAB® can be used to do most of what we cover in this chapter.

3.2 DEFINITION OF THE LAPLACE TRANSFORM

There are two types of Laplace transforms:

1. The two-sided (or bilateral) Laplace transform:

   The two-sided Laplace transform allows time functions to be nonzero for negative time.

$$\mathcal{L}[x(t)] = X(s) = \int_{-\infty}^{\infty} x(t)e^{-st}dt \tag{3.1}$$

   where s is the complex frequency given by

$$s = \sigma + j\omega \tag{3.2}$$

   where σ (in Np/s) and ω (in rad/s) are the real and imaginary parts of s, respectively. The advantage of the bilateral Laplace transform is that it can handle both causal and noncausal signals over $-\infty$ to ∞.

2. The one-sided (or unilateral) form:
   A one-sided Laplace transform is zero for negative time, that is, $t < 0$. It is found from Equation 3.1 by setting the lower limit of the integral equal to zero.

$$\mathcal{L}[x(t)] = X(s) = \int_{0^-}^{\infty} x(t)e^{-st}\,dt \tag{3.3}$$

   where 0^- denotes a time just before 0.

   The one-sided Laplace transform is a more commonly used transform than the two-sided Laplace transform, because we encounter only positive-time signals in practice and it is defined for positive-time signals.

   According to Equation 3.3, the t-domain signal $x(t)$ is changed into the s-domain function $X(s)$.

The **Laplace transform** of a signal $x(t)$ is the integration of the product of $x(t)$ and e^{-st} over the interval from 0 to $+\infty$.

In inverse Laplace transform the s-domain $X(s)$ is changed back to t-domain $x(t)$ and is given as

$$\mathcal{L}^{-1}[X(s)] = x(t) = \frac{1}{2\pi j}\int_{\sigma_1 - j\infty}^{\sigma_1 + j\infty} X(s)e^{st}\,ds \tag{3.4}$$

where the integration is performed along a straight line ($\sigma_1 + j\omega$, $-\infty < \omega < \infty$) as shown in Figure 3.1. The Laplace transform pair $x(t)$ and $X(s)$ are represented as

$$x(t) \Leftrightarrow X(s) \tag{3.5}$$

because there is one-to-one correspondence between $x(t)$ and $X(s)$.

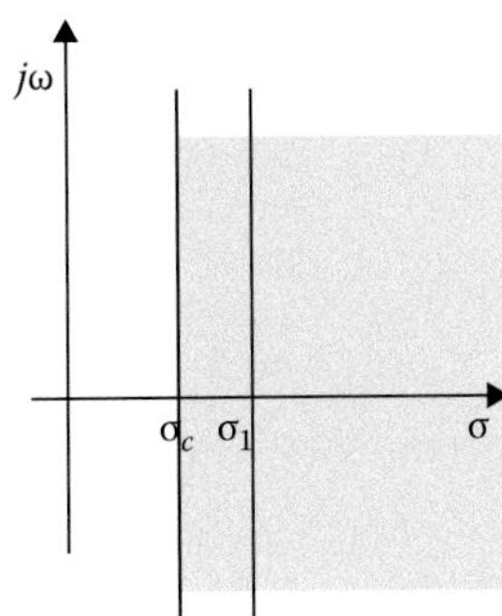

FIGURE 3.1 Region of convergence for the Laplace transform.

A signal $x(t)$ is Laplace transformable if the integral in Equation 3.3 exists; that is, in order for $x(t)$ to have a Laplace transform, the integral in Equation 3.3 must converge. The integral converges when

$$\left|\int_{0^-}^{\infty} x(t)e^{-(\sigma+j\omega)t}dt\right| < \infty \tag{3.6}$$

This can be simplified as

$$\left|\int_{0^-}^{\infty} x(t)e^{-(\sigma+j\omega)t}dt\right| \le \int_{0^-}^{\infty} \left|x(t)e^{-(\sigma+j\omega)t}dt\right|$$

$$\le \int_{0^-}^{\infty} |x(t)|\left|e^{-(\sigma+j\omega)t}dt\right| < \infty \tag{3.7}$$

Since $|e^{j\omega t}| = 1$ for any value of t,

$$\int_{0^-}^{\infty} |x(t)|e^{-\sigma t}dt < \infty \tag{3.8}$$

for some real value of $\sigma = \sigma_c$. The inequality in Equation 3.8 states that the Laplace transform exists. Hence, the region of convergence (ROC) for Laplace transform is $\text{Re}(s) = \sigma > \sigma_c$, as shown in Figure 3.1, the ROC means, the range of s for which the Laplace transform converges. We cannot define $X(s)$ outside the region of convergence. The portion of the s plane in which the integral converges is found by the value of σ. All signals that we come across within the system analysis fulfill the convergence criterion in Equation 3.8 and have their corresponding Laplace transforms.

Example 3.1

Find the Laplace transform of the following functions and establish the ROC for each case.

(a) $e^{-5t}u(t)$
(b) $\delta(t)$

Solution

(a) For the exponential function, shown in Figure 3.2a, the Laplace transform is

$$\mathcal{L}[e^{-5t}u(t)] = \int_{0^-}^{\infty} e^{-5t}e^{-st}dt = -\frac{1}{s+5}e^{-(s+5)t}\Big|_{0^-}^{\infty} = \frac{1}{s+5} \tag{3.1.1}$$

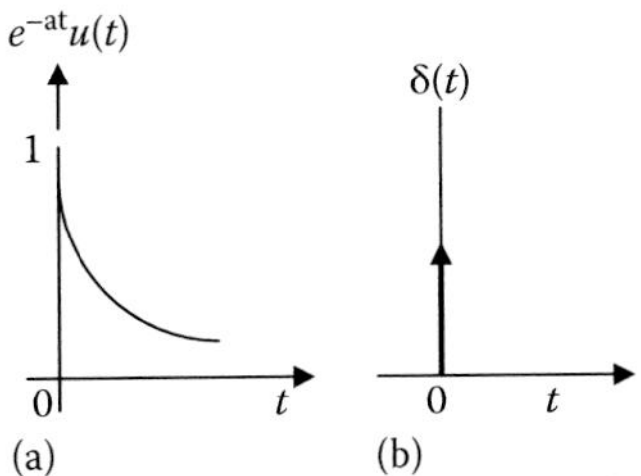

FIGURE 3.2 For Example 3.1: (a) Exponential function, (b) unit impulse function.

Notice that $\lim_{t\to\infty} e^{-(s+5)t} = 0$ only if Re(s + 5) > 0 or Re(s) > −5. Alternatively, the ROC is obtained from

$$\left|e^{-(s+5)t}\right| = \left|e^{-(\sigma+5)t}\right|\left|e^{-j\omega t}\right| = \left|e^{-(\sigma+5)t}\right| < \infty$$

which is valid when $\sigma + 5 > 0$ or $\sigma > -5$. The result in Equation 3.1.1 is so important that it deserves boxing. In general,

$$\boxed{\mathcal{L}[e^{-at}u(t)] = \frac{1}{s+a}} \tag{3.1.2}$$

A special case of this is when $a = 0$. $e^{-at}u(t)$ becomes $u(t)$ and we have

$$\mathcal{L}[u(t)] = \frac{1}{s} \tag{3.1.3}$$

(b) For the unit impulse function, shown in Figure 3.2b,

$$\mathcal{L}[\delta(t)] = \int_{0^-}^{\infty} \delta(t)e^{-st}dt = e^{-0} = 1 \quad \text{for all } s \tag{3.1.4}$$

The sifting property in Equation 1.28 has been applied in Equation 3.1.4. In this case, the transform does not depend on s and hence the region of convergence is the entire s plane.

Practice Problem 3.1 Determine the Laplace transform of these functions and specify the region of convergence for each:

(a) $h(t) = \frac{t^2}{2}u(t)$

(b) $e^{-j\omega t}u(t)$

Answers: (a) $1/s^3$, Re(s) > 0, (b) $1/(s + j\omega)$, Re(s) > 0.

Example 3.2

Find the Laplace transform of $x(t) = \sin \omega t\, u(t)$.

Solution

The Laplace transform of the cosine function is

$$X(s) = \mathcal{L}[\sin \omega t] = \int_0^\infty (\sin \omega t) e^{-st} dt = \int_0^\infty \left(\frac{e^{j\omega t} - e^{-j\omega t}}{2j} \right) e^{-st} dt$$

$$= \frac{1}{2j} \int_0^\infty \left(e^{-(s-j\omega)t} - e^{-(s+j\omega)t} \right) dt$$

$$= \frac{1}{2j} \left(\frac{1}{s - j\omega} - \frac{1}{s + j\omega} \right) = \frac{\omega}{s^2 + \omega^2}$$

Practice Problem 3.2 Find the Laplace transform of $x(t) = 10 \cos \omega t$.

Answer: $\dfrac{10s}{s^2 + \omega^2}$

3.3 PROPERTIES OF THE LAPLACE TRANSFORM

The properties of the Laplace transform are summarized below, which are useful to obtain the transform of many signals. Some of the properties are linearity, scaling, time shifting, frequency shifting, time differentiation, time convolution, time integration, frequency differentiation, time periodicity, modulation, and initial and final values.

3.3.1 LINEARITY

Let the Laplace transform of two signals $x_1(t)$ and $x_2(t)$ be $X_1(s)$ and $X_2(s)$, respectively. Then the linearity property states that the Laplace transform of the sum of two functions of time is equal to the sum of the transforms of each function and is given as

$$\boxed{\mathcal{L}\left[a_1 x_1(t) + a_2 x_2(t) \right] = a_1 X_1(s) + a_2 X_2(s)} \tag{3.9}$$

where a_1 and a_2 are constants.

The linearity property is explained as follows:

$$\mathcal{L}[a_1 x_1(t) + a_2 x_2(t)] = \int_0^\infty [a_1 x_1(t) + a_2 x_2(t)] e^{-at} dt$$

$$= a_1 \int_0^\infty x_1(t) e^{-at} dt + a_2 \int_0^\infty x_2(t) e^{-at} dt$$

$$= a_1 X_1(s) + a_2 X_2(s) \tag{3.10}$$

For example, consider

$$x(t) = 3e^{4t} + 5e^{-6t}. \tag{3.11}$$

By using linearity property, $X(s) = \mathcal{L}[3e^{4t} + 5e^{-6t}] = L[3e^{4t}] + L[5e^{-6t}]$

$$X(s) = \frac{3}{s-4} + \frac{5}{s+6} \tag{3.12}$$

3.3.2 Scaling

Let $X(s)$ be the Laplace transform of the signal $x(t)$. Let a be any positive real constant, then Laplace transform of $x(at)$ is given as

$$\mathcal{L}[x(at)] = \int_0^{\infty} x(at)e^{-st}dt \tag{3.13}$$

Let $\lambda = at$, $d\lambda = a\ dt$, we obtain

$$\mathcal{L}[x(at)] = \int_0^{\infty} x(\lambda)e^{-\lambda}e^{-\lambda\left(\frac{s}{a}\right)}\frac{d\lambda}{a} = \frac{1}{a}\int_0^{\infty} x(\lambda)e^{-\lambda\left(\frac{s}{a}\right)}d\lambda \tag{3.14}$$

Thus, the scaling property is obtained as

$$\boxed{\mathcal{L}[x(at)] = \frac{1}{a}X\left(\frac{s}{a}\right), \quad a > 0} \tag{3.15}$$

The unilateral Laplace transform is valid only for causal signals and not valid for noncausal signal (when a is negative i.e., $a < 0$).

For example, consider

$$\mathcal{L}[r(t)] = \frac{1}{s^2} \tag{3.16}$$

By applying the scaling property,

$$\mathcal{L}[r(2t)] = \frac{1}{2}\frac{1}{(s/2)^2} = \frac{2}{s^2} \tag{3.17}$$

3.3.3 Time Shifting

Often we can see that there is a signal delay in control systems due to transmission delays or other effects. Let $X(s)$ be the Laplace transform of signal $x(t)$. Then time shifting is given as

$$\boxed{\mathcal{L}\left[x(t-a)u(t-a)\right]=e^{-\alpha s}X(s)} \tag{3.18}$$

If a function is delayed in time by a, then the Laplace transform of the function is multiplied by $e^{-\alpha s}$. The time shifting property can be proved as follows:

$$\mathcal{L}[x(t-a)u(t-a)]=\int_0^\infty x(t-a)u(t-a)e^{-st}dt, \quad a\geq 0 \tag{3.19}$$

If $u(t-a)=0$ for $t<a$ and $u(t-a)=1$ for $t>a$
then,

$$\mathcal{L}[x(t-a)u(t-a)]=\int_0^\infty 0e^{-st}dt+\int_a^\infty x(t-a)(1)e^{-st}dt \tag{3.20}$$

by substituting $\lambda=t-a$, $d\lambda=dt$, and $t=\lambda+a$. As $t\to a$, $\lambda\to 0$, and as $t\to\infty$, $\lambda\to\infty$
thus,

$$\mathcal{L}[x(t-a)u(t-a)]=\int_0^\infty x(\lambda)e^{-s(\lambda+a)d\lambda}$$

$$=e^{-as}\int_0^\infty x(\lambda)e^{-s\lambda}d\lambda$$

$$=e^{-as}X(s) \tag{3.21}$$

For example, consider

$$g(t)=u(t-4)-u(t-6). \tag{3.22}$$

We know that $\mathcal{L}[u(t)]=\frac{1}{s}$

Using linearity property and the time shifting property,

$$G(s)=\mathcal{L}[u(t-4)-u(t-6)]=\frac{1}{s}e^{-4s}-\frac{1}{s}e^{-6s}=\frac{1}{s}\left(e^{-4s}-e^{-6s}\right) \tag{3.23}$$

3.3.4 Frequency Shifting

Let $X(s)$ be the Laplace transform of the signal $x(t)$. Then the frequency shifting is given as

$$\mathcal{L}\left[e^{-at}x(t)u(t)\right] = \int_0^{\infty} e^{-at}x(t)e^{-st}dt = \int_0^{\infty} x(t)e^{-(s+a)t}dt = X(s+a)$$

$$\boxed{\mathcal{L}\left[e^{-at}x(t)u(t)\right] = X(s+a)} \tag{3.24}$$

The Laplace transform turns multiplication by e^{-at} in the time-domain into $s + a$ in the transform $X(s)$.

For example, consider

$$\cos\omega t u(t) \Leftrightarrow \frac{s}{s^2+\omega^2} \tag{3.25}$$

By applying the frequency property and replacing every s by $s + a$, the Laplace transform of the cosine function is given by

$$\mathcal{L}[e^{-at}\cos\omega t u(t)] = \frac{s+a}{(s+a)^2+\omega^2} \tag{3.26}$$

3.3.5 Time Differentiation

Time differentiation property is important for solving differential equations. Let $X(s)$ be the Laplace transform of the signal $x(t)$. The Laplace transform of the derivative of $x(t)$ is given as

$$\mathcal{L}\left[\frac{dx}{dt}u(t)\right] = \int_{0^-} \frac{dx}{dt}e^{(-st)}dt \tag{3.27}$$

By differentiation by parts, $U = e^{-st}$, $dU = -se^{-st}dt$, and $dV = \left(\frac{dx}{dt}\right)dt = dx$, $V = x(t)$,

then

$$\mathcal{L}\left[\frac{dx}{dt}u(t)\right] = x(t)e^{-st}\Big|_{0^-}^{\infty} - \int_0^{\infty} x(t)\left[-se^{-st}\right]dt$$

$$= 0 - x(0^-) + s\int_{0^-}^{\infty} x(t)e^{-st}dt$$

$$= sX(s) - x(0^-) \tag{3.28}$$

$$\boxed{\mathcal{L}\left[x'(t)\right] = sX(s) - x(0^-)}$$

By differentiating the above equation, we get

$$\mathcal{L}\left[\frac{d^2x}{dt^2}\right] = s\mathcal{L}\left[x'(t)\right] - x'(0^-) = s\left[sX(s) - x(0^-)\right] - x'(0^-)$$

$$= s^2X(s) - sx(0^-) - x'(0^-) \tag{3.29}$$

$$\boxed{\mathcal{L}\left[x''(t)\right] = s^2X(s) - sx(0^-) - x'(0^-)}$$

In the same way, for the nth derivative, we get

$$\mathcal{L}\left[\frac{d^nx}{dt^n}\right] = s^nX(s) - s^{n-1}x(0^-) - s^{n-2}x'(0^{-1}) - \cdots - s^0x^{(n-1)}(0^-) \tag{3.30}$$

For example, consider $x(t) = \cos\omega t\, u(t)$. Then at $t = 0$, we have $x(0) = 1$. By applying time differentiation, the Laplace transform of the derivative of $x(t)$ is

$$\mathcal{L}\left[\frac{d}{dt}\cos\omega t u(t)\right] = s\left(\frac{s}{s^2+\omega^2}\right) - \cos(0^-)u(0^-)$$

$$= \frac{s^2}{s^2+\omega^2} \tag{3.31}$$

3.3.6 Time Convolution

Let $X(s)$ and $H(s)$ be the Laplace transform of the signal $x(t)$ and $h(t)$. Then the convolution property is given as

$$\boxed{\mathcal{L}[x(t) * h(t)] = X(s)H(s)} \tag{3.32}$$

In time-domain the convolution property,

$$x(t) * h(t) = \int_0^\infty x(t-\tau)h(\tau)\,d\tau$$

Considering the Laplace transform on both sides

$$\mathcal{L}\left[x(t) * h(t)\right] = \int_0^\infty \left[\int_0^\infty h(\tau)x(t-\tau)d\tau\right] e^{-st}dt$$

$$\mathcal{L}\left[x(t) * h(t)\right] = \int_0^\infty h(\tau)\left[\int_0^\infty x(t-\tau)e^{-st}dt\right]d\tau \tag{3.33}$$

Let $\lambda = t - \lambda$ so that $t = \tau + \lambda$ and $dt = d\lambda$

$$\mathcal{L}\left[x(t) * h(t)\right] = \int_0^\infty h(\tau)\left[\int_0^\infty x(\lambda)e^{-s(t+\lambda)}d\lambda\right]d\tau$$

$$= \int_0^\infty h(\tau)e^{-st}d\tau \int_0^\infty x(\lambda)e^{-s\lambda}d\lambda = H(s)X(s) \tag{3.34}$$

For example, consider

$$g(t) = u(t-3) - u(t-4).$$

By applying the Laplace transform on both sides

$$G(s) = \frac{1}{s}\left(e^{-3s} - e^{-4s}\right)$$

Considering the convolution of $g(t)*g(t)$, which is $G^2(s)$,

$$G^2(s) = \frac{1}{s^2}\left(e^{-3s} - e^{-4s}\right)^2 = \frac{1}{s^2}\left(e^{-6s} - 2e^{-7s} + e^{-8s}\right) \tag{3.35}$$

By applying inverse Laplace transform, we get

$$x(t) = g(t) * g(t) = (t-6)u(t-6) - 2(t-7)u(t-7) + (t-8)u(t-8)$$

$$= r(t-6) - 2r(t-7) + r(t-8) \tag{3.36}$$

3.3.7 Time Integration

Let $X(s)$ be the Laplace transform of the signal $x(t)$. Then Laplace transform of the integral is given as

$$\mathcal{L}\left[\int_0^t x(t)dt\right] = \frac{1}{s}X(s)$$

From the convolution property, we get

$$\mathcal{L}\left[\int_0^t x(\lambda)d\lambda\right] = \mathcal{L}\left[\int_0^\infty x(\lambda)u(t-\lambda)d\lambda\right] = \mathcal{L}\left[u(t) * x(t)\right] = \frac{1}{s}X(s)$$

$$\mathcal{L}\left[\int_0^t x(t)dt\right] = \frac{1}{s}X(s) \tag{3.37}$$

For example, let $x(t) = u(t)$. We know that $X(s) = 1/s$. By applying the time integration property,

$$\mathcal{L}\left[\int_0^t u(t)dt\right] = L[t] = \frac{1}{s}\frac{1}{s} = \frac{1}{s^2}$$

3.3.8 Frequency Differentiation

Let $X(s)$ be the Laplace transform of the signal $x(t)$. Then Laplace transform of the frequency differentiation is given as

$$\boxed{\mathcal{L}\left[tx(t)\right] = -\frac{dX(s)}{ds}}$$

$X(s)$ is given as

$$X(s) = \int_0^\infty x(t)e^{-st}dt$$

Differentiating both sides, we get

$$\frac{dX(s)}{ds} = \int_0^\infty x(t)(-te^{-st})dt = \int_{0^-}^\infty \left(-tx(t)\right)e^{-st}dt = \mathcal{L}\left[-tx(t)\right]$$

The frequency differentiation property is given as

$$\mathcal{L}\left[tx(t)\right] = -\frac{dX(s)}{ds} \tag{3.38}$$

Similarly, the general form of frequency differentiation property is given as

$$\mathcal{L}\left[t^n x(t)\right] = (-1)^n \frac{d^n X(s)}{ds^n} \tag{3.39}$$

For example,

$$\mathcal{L}\left(t\cos\omega t\right) = -\frac{d}{ds}\left[\frac{s}{s^2+\omega^2}\right] = \frac{s^2-\omega^2}{(s^2+\omega^2)^2} \tag{3.40}$$

$$\mathcal{L}\left(t\sin\omega t\right) = -\frac{d}{ds}\left[\frac{\omega}{s^2+\omega^2}\right] = \frac{2\omega s}{(s^2+\omega^2)^2} \tag{3.41}$$

3.3.9 Time Periodicity

Consider a periodic signal $x(t)$ and it can be expressed as a sum of time-shifted functions as (Figures 3.3 and 3.4)

$$\begin{aligned} x(t) &= x_1(t) + x_2(t) + x_3(t) + \cdots \\ &= x_1(t) + x_1(t-T)u(t-T) + x_1(t-2T)u(t-2T) + \cdots \end{aligned} \tag{3.42}$$

where $0 < t < T$. By applying the Laplace transform and the time-shifting property, we get

$$\begin{aligned} X(s) &= X_1(s) + X_1(s)e^{-Ts} + X_1(s)e^{-2Ts} + X_1(s)e^{-3Ts} + \cdots \\ &= X_1(s)\left[1 + e^{-Ts} + e^{-2Ts} + e^{-3Ts} + \cdots\right] \end{aligned} \tag{3.43}$$

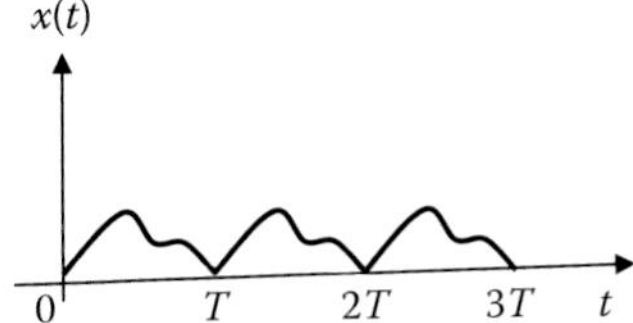

FIGURE 3.3 A periodic function.

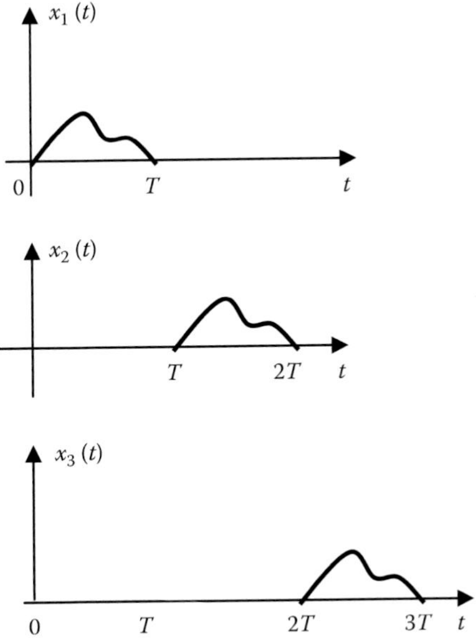

FIGURE 3.4 Decomposition of the periodic function in Figure 3.3.

But

$$1 + a + a^2 + a^3 + \cdots = \frac{1}{1-a} \tag{3.44}$$

if $|a| < 1$.

$$\boxed{X(s) = \frac{X_1(s)}{1 - e^{-Ts}}} \tag{3.45}$$

For example, consider

$$x_1(t) = \begin{cases} \sin \pi t, & 0 < t < 1 \\ 0, & 1 < t < 2 \end{cases}$$

$$X_1(s) = \int_0^{\infty} x_1(t) e^{-st} dt = \int_0^1 \sin \pi t \, e^{-st} \, dt \tag{3.46}$$

$$\int e^{at} \sin bt \; dt = \frac{e^{at}}{a^2 + b^2}(a \sin bt - b \cos bt)$$

where

$a = -s$
$b = \pi$

$$X_1(s) = \frac{e^{-st}}{s^2 + \pi^2}\left[-s \sin \pi t - \pi \cos \pi t\right]\Big|_0^1 = \frac{\pi(1 + e^{-s})}{s^2 + \pi^2}, \quad T = 2 \tag{3.47}$$

$$X(s) = \frac{X_1(s)}{1 - e^{-Ts}}$$

$$= \frac{\pi(1 + e^{-s})}{(s^2 + \pi^2)(1 - e^{-2s})} \tag{3.48}$$

3.3.10 Modulation

Let $X(s)$ be the Laplace transform of the signal $x(t)$, then for any real number ω,

$$\boxed{\begin{aligned} \mathcal{L}\left[x(t) \cos \omega t\right] &= \frac{1}{2}\left[X(s + j\omega) + X(s - j\omega)\right] \\ \mathcal{L}\left[x(t) \sin \omega t\right] &= \frac{j}{2}[X(s + j\omega) - X(s - j\omega)] \end{aligned}} \tag{3.49}$$

The signals $x(t)$ cos ωt and $x(t)$ sin ωt can be expressed as

$$
\begin{aligned}
x(t)\cos\omega t &= \frac{1}{2}x(t)\left[e^{-j\omega t}+e^{j\omega t}\right] \\
x(t)\sin\omega t &= \frac{1}{2j}x(t)\left[e^{-j\omega t}-e^{j\omega t}\right]
\end{aligned} \tag{3.50}
$$

From Equation 3.24,

$$\mathcal{L}\left[e^{\pm j\omega t}x(t)\right]=X(s\pm j\omega) \tag{3.51}$$

For example, we know that

$$u(t)\Leftrightarrow\frac{1}{s} \tag{3.52}$$

$$
\begin{aligned}
u(t)\cos(\omega_o t) &\Leftrightarrow \frac{1}{2}\left[\frac{1}{s+j\omega_o}+\frac{1}{s-j\omega_o}\right]=\frac{s}{s^2+\omega_o^2} \\
u(t)\sin(\omega_o t) &\Leftrightarrow \frac{1}{2j}\left[\frac{1}{s+j\omega_o}-\frac{1}{s-j\omega_o}\right]=\frac{\omega_o}{s^2+\omega_o^2}
\end{aligned} \tag{3.53}
$$

3.3.11 Initial and Final Values

The initial value $x(0)$ and final value $x(\infty)$ are used to relate frequency domain expressions to the time-domain as time approaches zero and infinity, respectively. These properties can be derived by using the differentiation property as

$$sX(s)-x(0^-)=\mathcal{L}\left[\frac{dx}{dt}\right]=\int_0^\infty\frac{dx}{dt}e^{-st}dt \tag{3.54}$$

By taking limits, we get

$$\lim_{s\to\infty}\left[sX(s)-x(0)\right]=0$$

Since $x(0)$ is not dependent on s,

$$\boxed{x(0)=\lim_{s\to\infty}sX(s)} \tag{3.55}$$

This is called the initial-value theorem. $x(0)$ can be found using $X(s)$, without using the inverse transform of $x(t)$.

If we consider $s \to 0$, then

$$sX(s) - x(0^-) = \int_{0^-}^{\infty} \frac{dx}{dt} e^{-0t} dt = \int_{0^-}^{\infty} dx = x(\infty) - x(0^-)$$

$$\boxed{x(\infty) = \lim_{s \to 0} sX(s)} \tag{3.56}$$

This is known as the final-value theorem. In the final-value theorem, all poles of $X(s)$ must be located in the left half of the s-plane.

For example, consider

$$x(t) = 4 + e^{-2t} \cos 3t \; u(t) \Leftrightarrow X(s) = \frac{4}{s} + \frac{s+2}{(s+2)^2 + 3^2} \tag{3.57}$$

By applying the initial-value theorem,

$$\begin{aligned} x(0) &= \lim_{s \to \infty} sX(s) = \lim_{s \to \infty} 4 + \frac{s^2 + 2s}{s^2 + 4s + 13} \\ &= \lim_{s \to \infty} 4 + \frac{1 + \frac{2}{s}}{1 + \frac{4}{s} + \frac{13}{s^2}} = 4 + \frac{1+0}{1+0+0} \\ x(0) &= 5 \end{aligned} \tag{3.58}$$

As another example, consider

$$x(t) = (5 + e^{-2t}) \; u(t) \Leftrightarrow X(s) = \frac{5}{s} + \frac{1}{s+2} \tag{3.59}$$

Using the final-value theorem,

$$x(\infty) = \lim_{s \to 0} sX(s) = \lim_{s \to 0} \left(5 + \frac{s}{s+2} \right) = 5 + 0 = 5 \tag{3.60}$$

The final-value theorem cannot be used to find the value of $x(t) = \cos 2t$, because $X(s)$ has poles at $s = \pm j\sqrt{2}$, which are not in the left half of the s plane.

For convenience, Table 3.1 provides the derived list of properties of Laplace transform, while Table 3.2 gives the Laplace transform of some common functions. For signals in Table 3.2, we have omitted the factor $u(t)$ except where it is necessary because we are only interested in functions that exist for $t > 0$.

TABLE 3.1
Properties of the Laplace Transform

Property	$x(t)$	$X(s)$
Linearity	$a_1x_1(t) + a_1x_2(t)$	$a_1X_1(s) + a_2X_2(s)$
Scaling	$x(at)$	$\frac{1}{a}X\left(\frac{s}{a}\right), \quad a > 0$
Time shifting	$x(t-a)\,u(t-a)$	$e^{-as}X(s)$
Frequency shifting	$e^{-at}x(t)$	$X(s+a)$
Time differentiation	$\frac{dx}{dt}$	$sX(s) - x(0^-)$
Second derivative	$\frac{d^2x}{dt^2}$	$s^2X(s) - sx(0^-) - x'(0^-)$
Third derivative	$\frac{d^3x}{dt^3}$	$s^3X(s) - s^2x(0^-) - sx'(0^-) - x''(0^-)$
nth derivative	$\frac{d^nx}{dt^n}$	$s^nX(s) - s^{n-1}x(0^-) - s^{n-2}x'(0^-) - \ldots - x^{(n-1)}(0^-)$
Time convolution	$x(t)*h(t)$	$X(s)H(s)$
Frequency convolution	$x_1(t)x_2(t)$	$\frac{1}{2\pi j}X_1(s)*X_2(s)$
Time integration	$\int_0^t x(t)dt$	$\frac{1}{s}X(s)$
Frequency integration	$\frac{x(t)}{t}$	$\int_s^\infty X(\lambda)d\lambda$
Frequency		
Differentiation	$tx(t)$	$-\frac{d}{ds}X(s)$
Multiplication by	$t^n \quad t^nx(t)$	$(-1)^n\frac{d^n}{dx^n}X(s)$
Time periodicity	$x(t) = x(t+nT)$	$\frac{X_1(s)}{1-e^{-sT}}$
Modulation		
Multiplication by cos ωt	$x(t)\cos\omega t$	$\frac{1}{2}\left[X(s+j\omega) + X(s-j\omega)\right]$
Multiplication by sin ωt	$x(t)\sin\omega t$	$\frac{j}{2}\left[X(s+j\omega) - X(s-j\omega)\right]$
Initial value	$x(0)$	$\lim_{s\to\infty} sX(s)$
Final value	$x(\infty)$	$\lim_{s\to 0} sX(s)$

TABLE 3.2
Laplace Transform Pairs[a]

Entry No. $x(t)$	$X(s)$
1. $\delta(t)$	1
2. $u(t)$	$1/s$
3. $u(t) - u(t-a)$	$\dfrac{1-e^{-as}}{s}$
4. e^{-at}	$\dfrac{1}{s+a}$
5. t	$\dfrac{1}{s^2}$
6. t^n	$\dfrac{n!}{s^{n+1}}$
7. te^{-at}	$\dfrac{1}{(s+a)^2}$
8. $t^n e^{-at}$	$\dfrac{n!}{(s+a)^{n+1}}$
9. $\sin \omega t$	$\dfrac{\omega}{s^2+\omega^2}$
10. $\cos \omega t$	$\dfrac{s}{s^2+\omega^2}$
11. $\sin(\omega t + \theta)$	$\dfrac{s\sin\theta + \omega\cos\theta}{s^2+\omega^2}$
12. $\cos(\omega t + \theta)$	$\dfrac{s\cos\theta - \omega\sin\theta}{s^2+\omega^2}$
13. $e^{-at}\sin\omega t$	$\dfrac{\omega}{(s+a)^2+\omega^2}$
14. $e^{-at}\cos\omega t$	$\dfrac{s+a}{(s+a)^2+\omega^2}$
15. $t \sin \omega t$	$\dfrac{2\omega s}{(s^2+\omega^2)^2}$
16. $t \cos \omega t$	$\dfrac{s^2-\omega^2}{(s^2+\omega^2)^2}$
17. $\sinh \omega t$	$\dfrac{\omega}{s^2-\omega^2}$
18. $\cosh \omega t$	$\dfrac{s}{s^2-\omega^2}$

[a] Defined for $t \ge 0$; $x(t) = 0$ for $t < 0$.

Example 3.3

Obtain the Laplace transform of

$$x(t) = \delta(t) - 3u(t) + 5e^{-2t}u(t).$$

Solution

By the linearity property,

$$X(s) = \mathcal{L}[\delta(t)] - 3\mathcal{L}[u(t)] + 5\mathcal{L}[e^{-2t}u(t)]$$
$$= 1 - 3\frac{1}{s} + 5\frac{1}{s+2} = \frac{s^3 + 4s - 6}{s(s+2)}$$

Practice Problem 3.3 Find the Laplace transform of $x(t) = [\cos 3t + e^{-5t}]u(t)$

Answer: $X(s) = \dfrac{2s^2 + 5s + 9}{(s+5)(s^2+9)}$

Example 3.4

Determine the Laplace transform of $x(t) = t^2u(t-1)$.

Solution

We can rewrite $x(t)$ as

$$x(t) = (t-1+1)^2 u(t-1) = (t-1)^2 u(t-1) - 2(t-1)u(t-1) + u(t-1)$$

Using the time-shifting property, we obtain

$$X(s) = e^{-s}\left(\frac{2}{s^3} - \frac{2}{s^2} + \frac{1}{s}\right)$$

Practice Problem 3.4 Find the Laplace transform of $x(t) = e^{-3t}\cos 2t u(t)$

Answer: $\dfrac{s+3}{(s+3)^2 + 4}$

Example 3.5

Determine the transform of the gate function in Figure 3.5a and the triangular pulse in Figure 3.5b.

Solution

(a) The gate function in Figure 3.5b can be expressed as

$$g(t) = u(t) - u(t-1)$$

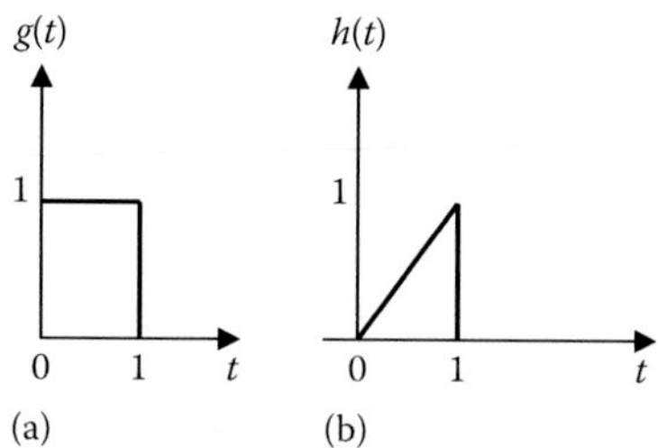

FIGURE 3.5 For Example 3.5.

We know that the Laplace transform of $u(t)$ is $1/s$. We apply the time shift property to the second term and obtain

$$G(s) = \left(\frac{1}{s} - \frac{e^{-s}}{s}\right) = \frac{1}{s}\left(1 - e^{-s}\right)$$

(b) The signal may be described as $h(t) = tg(t)$. Using the frequency differentiation property,

$$H(x) = -\frac{dG(s)}{ds} = -\frac{d}{ds}\left(\frac{1-e^{-s}}{s}\right) = \frac{1}{s^2} - \frac{e^{-s}}{s^2} - \frac{e^{-s}}{s}$$

Alternatively, we can express $h(t) = r(t) - r(t-1) - u(t-1)$. Taking the Laplace transform for each term gives the same result.

Practice Problem 3.5 Find the Laplace transform of the signals $x(t)$ and $y(t)$ in Figure 3.6.

Answers: (a) $X(s) = \dfrac{1-e^{-(s+2)}}{s+2}$, (b) $Y(s) = \dfrac{\pi(1+e^{-s})}{s^2+\pi^2}$

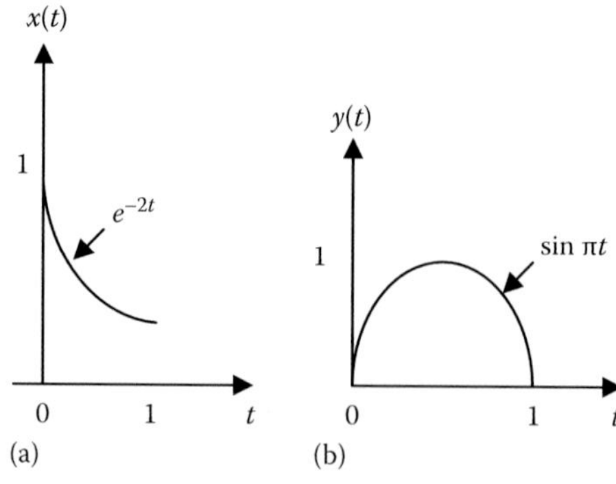

FIGURE 3.6 For Practice Problem 3.5.

Example 3.6

Let the Laplace transform of a signal $x(t)$ be

$$X(s) = \frac{s+2}{s^2+3s+5}$$

Determine the Laplace transform of the following signals:

(a) $y(t) = 3x(t/2)$
(b) $z(t) = tx(t-1)$
(c) $h(t) = 3\dfrac{dx}{dt}$
(d) $g(t) = \displaystyle\int_0^t x(\tau)d\tau$

Solution

(a) We apply the scaling property with $a = 1/2$.

$$Y(s) = 3\big(2X(2s)\big) = \frac{6(2s+2)}{(2s)^2+3(2s)+5} = \frac{12(s+1)}{4s^2+6s+5}$$

(b) $z(t) = tx(t-1) = (t-1)x(t-1) + x(t-1)$
Taking the Laplace transform of each term and using the shifting and differentiation properties, we obtain

$$Z(s) = -e^{-s}\frac{d}{ds}X(s) + e^{-s}X(s) = -e^{-s}\frac{\left[(s^2+3s+5)(1)-(s+2)(2s+3)\right]}{(s^2+3s+5)^2} + e^{-s}X(s)$$

$$= \frac{(s^2+4s+1)e^{-s}}{(s^2+3s+5)^2} + \frac{(s+2)e^{-s}}{(s^2+3s+5)}$$

(c) Applying the differentiation property,

$$H(s) = 3[sX(s) - x(0)]$$

The initial-value theorem gives

$$x(0) = \lim_{s\to\infty} sX(s) = \lim_{s\to\infty}\frac{s^2+2s}{s^2+3s+5} = \lim_{s\to\infty}\frac{1+2/s}{1+3/s+5/s^2} = 1$$

$$H(s) = 3\left[\frac{s^2+2s}{s^2+3s+5} - 1\right] = \frac{-3(s+5)}{s^2+3s+5}$$

(d) Using the integration property,

$$G(s) = \frac{1}{s}X(s) = \frac{s+2}{s(s^2+3s+5)}$$

Practice Problem 3.6 Let $H(s) = \mathcal{L}[h(t)] = \dfrac{s+2}{s(s+5)}$. Find the Laplace transform of:

(a) $h(t-2)$, (b) $h(4t)$, (c) $\dfrac{d}{dt}h(t)$, (d) $h(t)*h(t)$

Answers: (a) $\dfrac{(s+2)e^{-2s}}{s(s+5)}$, (b) $\dfrac{s+8}{s(s+20)}$, (c) $\dfrac{-3}{s+5}$, (d) $\dfrac{(s+2)^2}{s^2(s+5)^2}$

Example 3.7

Determine the initial and final values of $x(t)$, if they exist, given that

$$X(s) = \frac{s^2-2s+1}{(s-2)(s^2+2s+4)}$$

Solution

$$x(0) = \lim_{s\to\infty} sX(s) = \lim_{s\to\infty} \frac{s^3-2s^2+s}{(s-2)(s^2+2s+4)} = \lim_{s\to\infty} \frac{1-\dfrac{2}{s}+\dfrac{1}{s^2}}{\left(1-\dfrac{2}{s}\right)\left(1+\dfrac{2}{s}+\dfrac{4}{s^2}\right)} = 1$$

One pole ($s = 2$) is not in the left-half plane. $x(\infty)$ does not exist.

Practice Problem 3.7 Find the initial and final values of

$$X(s) = \frac{s^2+3}{s^3+4s^2+6}$$

Answer: $x(0) = 1$; $x(\infty)$ does not exist.

3.4 THE INVERSE LAPLACE TRANSFORM

For a given $X(s)$, we can obtain the corresponding $x(t)$ by using Equation 3.4, but this requires contour integration, which is beyond the scope of this text. We will use an easier approach by employing the partial fraction expansion of $X(s)$ and matching entries in Table 3.2.

In most cases, we can express $X(s)$ in the standard form of

$$X(s) = \frac{N(s)}{D(s)} \tag{3.61}$$

where

$N(s)$ is the numerator polynomial
$D(s)$ is the denominator polynomial

Equation 3.61 can be expressed as

$$X(s)=\frac{N(s)}{D(s)}=\frac{b_m s^m+b_{m-1}s^{m-1}+b_{m-2}s^{m-2}+\cdots+b_1 s+b_0}{a_n s^n+a_{n-1}s^{n-1}+a_{n-2}s^{n-2}+\cdots+a_1 s+a_0} \tag{3.62}$$

We can express $X(s)$ in terms of factors of $N(s)$ and $D(s)$ as

$$X(s)=\frac{N(s)}{D(s)}=k\frac{(s+z_1)(s+z_2)\cdots(s+z_m)}{(s+p_1)(s+p_2)\cdots(s+p_n)} \tag{3.63}$$

where $k = b_m/a_n$. The roots of $N(s) = 0$ are called the *zeros* of $X(s)$, that is, the values $-z_1,-z_2,\ldots,z_m$ are the zeros of $X(s)$.

Similarly, the roots of $D(s) = 0$ are the *poles* of $X(s)$, that is, the values $-p_1,-p_2,\ldots,p_n$ are the poles of $X(s)$. We assume that $X(s)$ is proper ($m < n$), that is, the degree of $N(s)$ is less than the degree of $D(s)$. (In case $X(s)$ is not proper ($m > n$), we use long division to reduce it proper fraction.) We use partial fraction to break $X(s)$ down into simple terms whose inverse transform is obtained by matching entries in Table 3.2. Partial fraction expansions are useful in finding the inverse Laplace, inverse Fourier, or inverse z-transforms. Thus, to find the inverse Laplace transform of $X(s)$ requires two steps:

1. Use partial fraction expansion to break $X(s)$ into simple terms.
2. Refer to Table 3.2 to obtain the inverse Laplace transform of each term.

We will now consider the three common forms $X(s)$ may take and how we apply the two steps to each form.

3.4.1 Simple Poles

A simple pole is a first-order pole. If the poles $s = -p_1,-p_2,\ldots,-p_n$ are simple and distinct ($p_i \neq p_j$ for all $i \neq j$), then $D(s)$ in Equation 3.63 becomes a product of factors, so that

$$X(s)=\frac{N(s)}{(s+p_1)(s+p_2)\cdots(s+p_n)} \tag{3.64}$$

Using partial fraction expansion, we decompose $X(s)$ in Equation 3.64 as

$$X(s)=\frac{k_1}{s+p_1}+\frac{k_2}{s+p_2}+\cdots+\frac{k_n}{s+p_n} \tag{3.65}$$

To get the expansion coefficients $k_1, k_2,\ldots, k_n$ (known as the *residues* of $X(s)$), we multiply both sides of Equation 3.65 by $(s + p_i)$ for each pole. For example, if we multiply by $(s + p_1)$, we obtain

$$(s+p_1)X(s) = k_1 + \frac{(s+p_1)k_2}{s+p_2} + \cdots + \frac{(s+p_1)k_n}{s+p_n} \tag{3.66}$$

Setting $s = -p_1$ in Equation 3.66 leaves only k_1 on the right-hand side of Equation 3.66. By isolating k_1 this way, we obtain

$$(s+p_1)X(s)\Big|_{s=-p_1} = k_1 \tag{3.67}$$

In general,

$$\boxed{k_i = (s+p_i)X(s)\Big|_{s=-p_i}} \tag{3.68}$$

This is referred to as *Heaviside's formula*. From Table 3.2, the inverse transform of each term in Equation 3.65 is

$$\mathcal{L}^{-1}[k/(s+a)] = ke^{-at}u(t) \tag{3.69}$$

Applying Equation 3.69 to each term in Equation 3.65 leads to the inverse Laplace transform of $X(s)$ as

$$\boxed{x(t) = \left(k_1e^{-p_1t} + k_2e^{-p_2t} + \cdots + k_ne^{-p_nt}\right)u(t)} \tag{3.70}$$

3.4.2 Repeated Poles

If $X(s)$ has repeated poles, we employ a different strategy. Let $X(s)$ have n repeated poles at $s = -p$, then we may represent $X(s)$ as

$$X(s) = \frac{k_n}{(s+p)^n} + \frac{k_{n-1}}{(s+p)^{n-1}} + \cdots + \frac{k_2}{(s+p)^2} + \frac{k_1}{s+p} + X_1(s) \tag{3.71}$$

where $X_1(s)$ stands for the remaining part of $X(s)$ that does not contain the repeated poles. We find the coefficient k_n as

$$k_n = (s+p)^n X(s)\Big|_{s=-p} \tag{3.72}$$

as in Equation 3.68 To determine k_{n-1}, we multiply each term in Equation 3.71 by $(s + p)^n$, differentiate, and then set $s = -p$ to get rid of the other coefficients except k_{n-1}. Thus, we obtain

$$k_{n-1} = \frac{d}{ds}\left[(s+p)^n X(s)\right]\Big|_{s=-p} \tag{3.73}$$

We repeat the same process to get

$$k_{n-2} = \frac{1}{2!}\frac{d^2}{ds^2}\left[(s+p)^n X(s)\right]\Big|_{s=-p} \tag{3.74}$$

The mth repetition gives

$$k_{n-m} = \frac{1}{m!}\frac{d^m}{ds^m}\left[(s+p)^n X(s)\right]\Big|_{s=-p} \tag{3.75}$$

where $m = 1, 2, 3,\ldots, n - 1$. From Table 3.2,

$$\mathcal{L}^{-1}\left[\frac{1}{(s+a)^n}\right] = \frac{t^{n-1}e^{-at}}{(n-1)!} \tag{3.76}$$

We apply this to each term in Equation 3.71 and obtain

$$\boxed{x(t) = \left(k_1 e^{-pt} + k_2 t e^{-pt} + \frac{k_3}{2!}t^3 e^{-pt} + \cdots + \frac{k_n}{(n-1)!}t^{n-1}e^{-pt}\right)u(t) + x_1(t)} \tag{3.77}$$

3.4.3 Complex Poles

We now consider when $X(s)$ has complex poles. Complex poles occur in complex conjugate pairs so that the number of poles is even. A pair of complex poles is simple if it is not repeated. Simple complex poles may be handled the same way as simple real poles, but the procedure involves complex algebra, which can be intimidating. An easier approach involves *completing the squares.* The idea is to express each complex pole pair (or quadratic term) in $D(s)$ as a complete square such as $(s + \alpha)^2 + \beta^2$ and then use Table 3.2 to find the inverse of the term.

For the sake of simplicity, we will assume that $X(s)$ has only one complex pole pair. If $X(s)$ has complex roots, its denominator $D(s)$ will contain a quadratic term and $X(s)$ may have the form

$$X(s) = \frac{A_1 s + A_2}{s^2 + as + b} + X_1(s) \tag{3.78}$$

where $X_1(s)$ stands for the remaining part of $X(s)$ that does not contain this pair of complex poles. We complete the square by letting

$$s^2 + as + b = s^2 + 2\alpha s + \alpha^2 + \beta^2 = (s+\alpha)^2 + \beta^2 \tag{3.79}$$

and we also let

$$A_1 s + A_2 = A_1(s+\alpha) + B_1\beta$$

Thus, Equation 3.78 becomes

$$X(s) = \frac{A_1(s+\alpha)}{(s+\alpha)^2+\beta^2} + \frac{B_1\beta}{(s+\alpha)^2+\beta^2} + X_1(s) \tag{3.80}$$

Using entries in Table 3.2, we obtain the inverse Laplace transform of each term in Equation 3.80 so that

$$\boxed{x(t) = \left(A_1 e^{-\alpha t}\cos\beta t + B_1 e^{-\alpha t}\sin\beta t\right)u(t) + x_1(t)} \tag{3.81}$$

The sine and cosine terms can be combined if desired. We use

$$A\cos\theta + B\sin\theta = R\cos(\theta-\phi) \tag{3.82}$$

where

$$R = \sqrt{A^2+B^2}$$

$$\phi = \tan^{-1}(B/A)$$

In addition to the three common forms considered above, we can have two other cases. First, the term may involve e^{-sT}. For example, $X(s)e^{-sT}$ corresponds to $x(t-T)u(t-T)$ using the time-shifting property. Second, for $X(s) = N(s)/D(s)$, when the degree of the numerator $N(s)$ is higher than the degree of the denominator $D(s)$, we use long division to express $X(s)$. These two cases are best illustrated with Examples 3.13 and 3.14.

Example 3.8

Find the inverse Laplace transform of

$$X(s) = 1 + \frac{2}{s} + \frac{4}{s-1} - \frac{3s}{s^2+9}$$

Solution
The inverse transform is obtained by consulting Table 3.2 for each term.

$$\begin{aligned} x(t) &= \mathcal{L}^{-1}(1) + \mathcal{L}^{-1}\left(\frac{2}{s}\right) + \mathcal{L}^{-1}\left(\frac{4}{s-1}\right) - 3\mathcal{L}^{-1}\left(\frac{s}{s^2+9}\right) \\ &= \delta(t) + (2 + 4e^t - 3\cos 3t)u(t) \end{aligned}$$

Practice Problem 3.8 Determine the inverse Laplace transform of

$$X(s) = \frac{1}{s^2} - \frac{2}{s+3} + \frac{4}{s^2+4}$$

Answer: $(t-2e^{-3t} + 2\sin 2t)u(t)$

Example 3.9

Obtain $h(t)$ given that

$$H(s) = \frac{4}{(s+1)(s+3)}$$

Solution
In the previous example, the partial fraction was provided. Here we determine the partial fraction. We let

$$H(s) = \frac{4}{(s+1)(s+3)} = \frac{A}{s+1} + \frac{B}{s+3} \tag{3.9.1}$$

where A and B are constants. The constants can be determined in two ways:

Method 1: Residue method

$$A = (s+1)H(s)\Big|_{s=-1} = \left.\frac{4}{(s+3)}\right|_{s=-1} = \frac{4}{(2)} = 2$$

$$B = (s+3)H(s)\Big|_{s=-3} = \left.\frac{4}{(s+1)}\right|_{s=-3} = \frac{4}{(-2)} = -2$$

Method 2: Algebraic method
Multiplying both sides of Equation 3.9.1 by $(s + 1)(s + 3)$ and equating the numerator, we get

$$4 = A(s+3) + B(s+1)$$

Equating the coefficients of like powers of s,

constant: $4 = 3A + B$
$s: 0 = A + B \quad B = -A$

Solving these gives $A = 2$, $B = -2$

Thus,

$$H(s) = \frac{4}{(s+1)(s+3)} = \frac{2}{s+1} - \frac{2}{s+3}$$

By taking the inverse transform of each term, we obtain

$$h(t) = \left(2e^{-t} - 2e^{-3t}\right)u(t)$$

Practice Problem 3.9 Find $x(t)$ if

$$X(s) = \frac{10s}{(s+1)(s+2)(s+3)}$$

Answer: $x(t) = (-5e^{-t} + 20e^{-2t} - 15e^{-3t})u(t)$

Example 3.10

Determine $x(t)$ given that

$$X(s) = \frac{2s^2 + 4s + 1}{(s+1)(s+2)^3}$$

Solution

The previous example is on simple roots, while this example is on repeated roots. Let

$$X(s) = \frac{2s^2 + 4s + 1}{(s+1)(s+2)^3} = \frac{A}{s+1} + \frac{B}{s+2} + \frac{C}{(s+2)^2} + \frac{D}{(s+2)^3} \tag{3.10.1}$$

We can determine the constants A, B, C, and D in two ways:

Method 1: Residue method

$$A = F(s)(s+1)\big|_{s=-1} = \frac{2-4+1}{-1+2} = -1$$

$$D = F(s)(s+2)^3\big|_{s=-2} = \frac{8-8+1}{-2+1} = -1$$

$$C = \frac{d}{ds}\left[(s+2)^3 X(s)\right]\Big|_{s=-2} = \frac{d}{ds}\left[\frac{2s^2 + 4s + 1}{(s+1)}\right]\Bigg|_{s=-2}$$

$$= \frac{(s+1)(4s+4) - (2s^2 + 4s + 1)\times 1}{(s+1)^2}\Bigg|_{s=-2} = \frac{2s^2 + 4s + 3}{(s+1)^2}\Bigg|_{s=-2}$$

$$= \frac{8-8+3}{1} = 3$$

$$B = \frac{1}{2}\frac{d^2}{ds^2}\left[(s+2)^3 X(s)\right]\bigg|_{s=-2} = \frac{1}{2}\frac{d}{ds}\left[\frac{2s^2+4s+3}{(s+1)^2}\right]\bigg|_{s=-2}$$

$$= \frac{(s+1)^2(4s+4)-(2s^2+4s+3)\times 2(s+1)}{2(s+1)^4}\bigg|_{s=-2} = \frac{-2}{2(s+1)^3}\bigg|_{s=-2}$$

$$= \frac{-2}{-2} = 1$$

Method 2: Algebraic method

Multiplying Equation 3.10.1 by $(s + 1)(s + 2)^3$ and equating the numerator, we obtain

$$2s^2+4s+1 = A(s+2)^3 + B(s+1)(s+2)^2 + C(s+1)(s+2) + D(s+1)$$

or

$$2s^2+4s+1 = A(s^3+6s^2+12s+8) + B(s^3+5s^2+8s+4) + C(s^2+3s+2) + D(s+1)$$

Equating coefficients,

$$s^3 : 0 = A + B \rightarrow B = -A$$

$$s^2 : 2 = 6A + 5B + C \rightarrow 2 = A + C$$

$$s : 4 = 12A + 8B + 3C + D \rightarrow 4 = 4A + 3C + D$$

$$\text{Constant} : 1 = 8A + 4B + 2C + D \rightarrow 1 = 4A + 2C + D$$

Solving these simultaneously gives $A =- 1$, $B = 1$, $C = 3$, $D = -1$ so that Equation 3.10.1 becomes

$$X(s) = \frac{-1}{s+1} + \frac{1}{s+2} + \frac{3}{(s+2)^2} - \frac{1}{(s+2)^3}$$

By taking the inverse transform of each term, we obtain

$$x(t) = \left(-e^{-t} + e^{-2t} + 3te^{-2t} - \frac{t^2}{2}e^{-2t}\right)u(t)$$

Practice Problem 3.10 Obtain $y(t)$ if

$$Y(s) = \frac{12}{(s+2)^2(s+4)}$$

Answer: $(3e^{-4t} - 3e^{-2t} + 6te^{-2t})u(t)$

Example 3.11

Find the inverse transform of

$$G(s) = \frac{2+1}{(s+2)(s^2+2s+5)}$$

Solution

$G(s)$ has a pair of complex poles at $s^2 + 2s + 5 = 0$ or $s = -2 \pm j2$. We let

$$G(s) = \frac{s+1}{(s+2)(s^2+2s+5)} = \frac{A}{s+2} + \frac{Bs+C}{s^2+2s+5}$$

$$A = G(s)(s+2)\Big|_{s=-2} = \frac{-1}{5}$$

$$s+1 = A(s^2+2s+5) + B(s^2+2s) + C(s+2)$$

Equating coefficients:

$$s^2 : 0 = A + B \rightarrow B = -A = \frac{1}{5}$$

$$s^1 : 1 = 2A + 2B + C = 0 + C \rightarrow C = 1$$

$$s^0 : 1 = 5A + 2C = -1 + 2 = 1$$

$$G(s) = \frac{-1/5}{s+2} + \frac{1/5 \cdot s + 1}{(s+1)^2 + 2^2} = \frac{-1/5}{s+2} + \frac{1/5(s+1)}{(s+1)^2+2^2} + \frac{4/5}{(s+1)^2+2^2}$$

$$g(t) = (-0.2e^{-2t} + 0.2e^{-t}\cos(2t) + 0.4e^{-t}\sin(2t))u(t)$$

Practice Problem 3.11 Find $v(t)$ given that

$$V(s) = \frac{2s+26}{s(s^2+4s+13)}$$

Answer: $v(t) = \left(2 - 2e^{-2t}\cos 3t - \frac{2}{3}e^{-2t}\sin 3t\right)u(t)$.

Example 3.12

Consider the rectangular pulse or gate function $x(t) = u(t) - u(t-2)$. Obtain $y(t) = x(t)*x(t)$, that is, the convolution of the rectangular pulse with itself.

Solution

The Laplace transform of $x(t)$ is

$$X(s) = \frac{1}{s} - \frac{e^{-2s}}{s}$$

Using the convolution property,

$$Y(s) = X(s)X(s) = X^2(s) = \frac{1}{s^2}\left(1 - 2e^{-2s} + e^{-4s}\right)$$

Taking the inverse Laplace transform of each term,

$$\begin{aligned} y(t) = x(t) * x(t) &= tu(t) - 2(t-2) + (t-4)u(t-4) \\ &= r(t) - 2r(t-2) + r(t-4) \end{aligned}$$

which is a triangular pulse, as expected.

Practice Problem 3.12 Given the unit ramp function $r = tu(t)$ and the gate function $g(t) = u(t) - u(t - 1)$, use the convolution property to find $y(t) = r(t)*g(t)$.

Answer: $y(t) = 0.5t^2u(t) - 0.5(t-1)^2u(t-1)$

Example 3.13

Determine the inverse Laplace transform of

$$X(s) = \frac{se^{-2s} + e^{-3s}}{s(s^2 + 5s + 4)}$$

Solution
Let

$$X(s) = X_1(s)e^{-2s} + X_2(s)e^{-3s} \tag{3.13.1}$$

where

$$X_1(s) = \frac{s}{s(s^2 + 5s + 4)} = \frac{1}{(s+1)(s+4)} \tag{3.13.2}$$

$$X_2(s) = \frac{1}{s(s^2 + 5s + 4)} = \frac{1}{s(s+1)(s+4)} \tag{3.13.3}$$

We now use partial fractions to obtain $x_1(t)$ and $x_2(t)$ from $X_1(s)$ and $X_2(s)$, respectively. Let

$$X_1(s) = \frac{1}{(s+1)(s+4)} = \frac{A}{s+1} + \frac{B}{s+4}$$

where

$$A = \left.\frac{1}{s+4}\right|_{s=-1} = \frac{1}{-1+4} = \frac{1}{3}$$

$$B = \left.\frac{1}{s+1}\right|_{s=-4} = \frac{1}{-4+1} = -\frac{1}{3}$$

Hence,

$$X_1(s) = \frac{1}{3}\left[\frac{1}{s+1} - \frac{1}{s+4}\right]$$

Taking the inverse of each term gives

$$x_1(t) = \frac{1}{3}\left(e^{-t} - e^{-4t}\right)u(t) \tag{3.13.4}$$

Similarly,

$$X_2(s) = \frac{1}{s(s+1)(s+4)} = \frac{C}{s} + \frac{D}{s+1} + \frac{E}{s+4}$$

where

$$C = \left.\frac{1}{(s+1)(s+4)}\right|_{s=0} = \frac{1}{(1)(4)} = \frac{1}{4}$$

$$D = \left.\frac{1}{s(s+4)}\right|_{s=-1} = \frac{1}{(-1)(3)} = -\frac{1}{3}$$

$$E = \left.\frac{1}{s(s+1)}\right|_{s=-4} = \frac{1}{(-4)(-3)} = \frac{1}{12}$$

Thus,

$$X_2(s) = \frac{1}{s(s+1)(s+4)} = \frac{1/4}{s} - \frac{1/3}{s+1} + \frac{1/12}{s+4}$$

From this, we obtain

$$x_2(t) = \left[\frac{1}{4} - \frac{1}{3}e^{-t} + \frac{1}{12}e^{-4t}\right]u(t) \tag{3.13.5}$$

From Equation 3.13.1,

$$x(t) = x_1(t-2)u(t-2) + x_2(t-3)u(t-3) \tag{3.13.6}$$

Substituting Equations 3.13.4 and 3.13.5 into Equation 3.13.6 finally gives

$$x(t) = \frac{1}{3}\left(e^{-(t-2)} - e^{-4(t-2)}\right)u(t-2) + \left[\frac{1}{4} - \frac{1}{3}e^{-(t-3)} + \frac{1}{12}e^{-4(t-3)}\right]u(t-3)$$

Practice Problem 3.13 Determine the inverse of

$$Y(s) = \frac{1 - se^{-4s}}{(s+1)(s+2)}$$

Answer: $(e^{-t} - e^{-2t})u(t) + (e^{-(t-4)} - 2e^{-2(t-4)})u(t-4)$.

Example 3.14

Obtain $g(t)$ given that

$$G(s) = \frac{s^3 + 5s^2 + 10}{s^2 + 3s + 2}$$

Solution

The degree of the numerator polynomial is greater than the degree of the denominator. In this case, the fraction is a proper one. The numerator is divided by the denominator by long division. We use long division until the remainder is one degree less than the denominator polynomial.

$$\begin{array}{r|l} & s+2 \\ \hline s^2+3s+2 & s^3+5s^2+0s+10 \\ & s^3+3s^2+2s \\ \hline & \quad 2s^2-2s+10 \\ & \quad 2s^2+6s+4 \\ \hline & \qquad -8s+6 \end{array}$$

Thus,

$$G(s) = s + 2 + \frac{-8s+6}{s^2+3s+2}$$

Let $G(s) = s + 2 + Y(s)$, where $Y(s) = \dfrac{-8s+6}{s^2+3s+2}$

Let $Y(s) = \dfrac{-8s+6}{(s+1)(s+2)} = \dfrac{A}{s+1} + \dfrac{B}{s+2}$

where

$$A = \left.\frac{-8s+6}{s+2}\right|_{s=-1} = \frac{8+6}{-1+2} = 14$$

$$B = \left.\frac{-8s+6}{s+1}\right|_{s=-2} = \frac{16+6}{-2+1} = -22$$

$$Y(s) = \frac{14}{s+1} - \frac{22}{s+2}$$

Thus,

$$G(s) = s + 2 + \frac{14}{s+1} - \frac{22}{s+2}$$

Taking the inverse of each term gives

$$g(t) = \frac{d}{dt}\delta(t) + 2\delta(t) + \left(14e^{-t} - 22e^{-2t}\right)u(t)$$

Practice Problem 3.14 Find $y(t)$ given that

$$Y(s) = \frac{s(s+1)}{(s-1)(s+2)}$$

Answer: $y(t) = \delta(t) + \frac{2}{3}\left(e^{t} - e^{-2t}\right)u(t)$

3.5 TRANSFER FUNCTION

Now that we have mastered finding Laplace transform and its inverse, we begin the application of the Laplace transform to the study of continuous-time systems. We begin with the concept of transfer function.

We know from Chapter 2 that the output $y(t)$ of a continuous-time system is the convolution of the input $x(t)$ and the impulse response $h(t)$, that is,

$$y(t) = x(t) * h(t) \tag{3.83}$$

By applying the convolution property, we obtain

$$Y(s) = X(s)H(s)$$

or

$$\boxed{H(s) = \frac{Y(s)}{X(s)}} \tag{3.84}$$

The Laplace $H(s)$ (which is the Laplace transform of $h(t)$) is known as the *transfer function* or system function of the system. The transfer function of a system describes how the output behaves with respect to the input.

The **transfer function** $H(s)$ is defined as the ratio of the output response $Y(s)$ to the input excitation $X(s)$, assuming all initial conditions are zero.

Figure 3.7 shows the relationship between the impulse response $h(t)$ and transfer function $H(s)$. Given the system transfer function, we can perform extensive system analysis and design without having to apply specific signals to the systems.

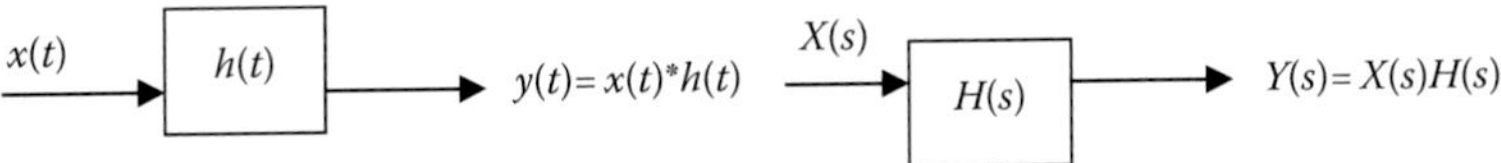

FIGURE 3.7 Impulse response $h(t)$ and transfer function $H(s)$.

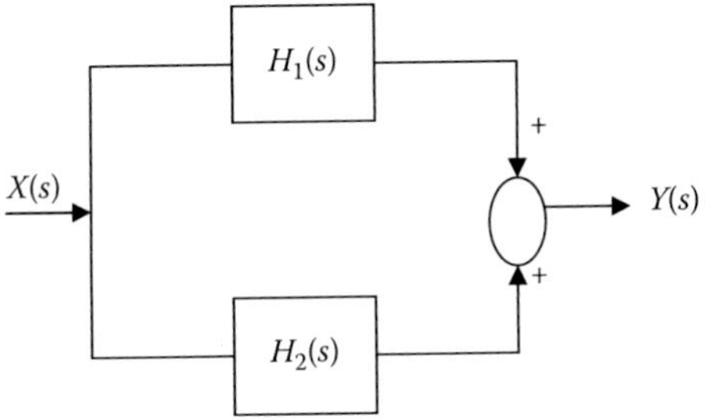

FIGURE 3.8 Parallel interconnection of two systems.

A linear system is often specified in terms of some block diagram. Here we consider three types of interconnections.

Consider a parallel interconnection of two systems with transfer function $H_1(s)$ and $H_1(s)$ with single-input $X(s)$ and single-output $Y(s)$, as shown in Figure 3.8.

$$Y(s) = \left[H_1(s) + H_2(s)\right]X(s) \tag{3.85}$$

or

$$H(s) = \frac{Y(s)}{X(s)} = H_1(s) + H_2(s) \tag{3.86}$$

where $H(s)$ is the transfer function for the overall system.

We next consider a series interconnection (or cascade connection) of two systems with transfer function $H_1(s)$ and $H_1(s)$ with single-input $X(s)$ and single-output $Y(s)$, as shown in Figure 3.9.

$$Y(s) = H_1(s)H_2(s)X(s)$$

or

$$H(s) = \frac{Y(s)}{X(s)} = H_1(s)H_2(s) \tag{3.87}$$

where $H(s)$ is the transfer function for the overall system.

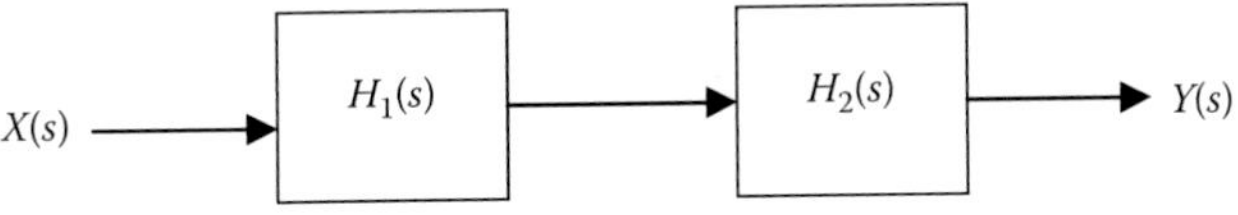

FIGURE 3.9 Series interconnection of two systems.

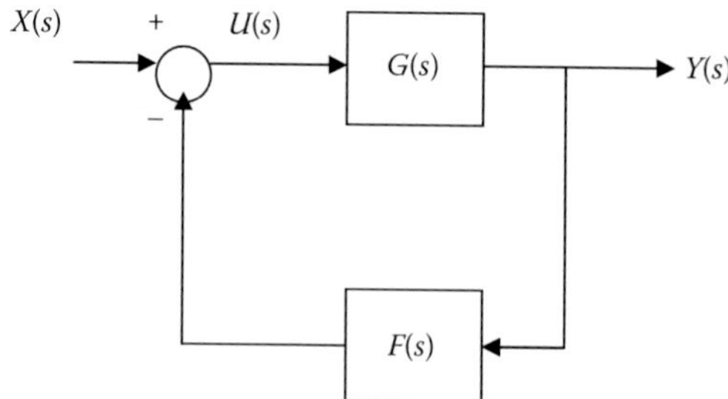

FIGURE 3.10 Feedback interconnection of systems.

Finally, we consider the feedback configuration in Figure 3.10. In this case, the output is feedback to the input. Negative feedback is common in practice because positive feedback can cause system instability. From Figure 3.10,

$$Y(s) = G(s)U(s) \tag{3.88}$$

But

$$U(s) = X(s) - F(s)Y(s) \tag{3.89}$$

Substituting Equation 3.89 into Equation 3.88 gives

$$Y(s) = G(s)(X(s) - F(s)Y(s))$$

or

$$Y(s) + G(s)F(s)Y(s) = G(s)X(s) \tag{3.90}$$

Dividing both sides by $X(s)$ gives the transfer function for the overall system as

$$H(s) = \frac{G(s)}{1 + F(s)G(s)} \tag{3.91}$$

There are two special cases of this. First, if the subtractor in Figure 3.10 is an adder instead, then Equation 3.91 becomes

$$H(s) = \frac{G(s)}{1 - F(s)G(s)} \tag{3.92}$$

Second, for unity feedback $F(s) = 1$, and Equation 3.91 becomes

$$H(s) = \frac{Y(s)}{X(s)} = \frac{G(s)}{1 + G(s)} \tag{3.93}$$

Example 3.15

The output of a linear system is $y(t) = e^{-t}\cos 4t\ u(t)$ when the input is $x(t) = 10e^{-t}u(t)$. Find the transfer function of the system.

Solution

If $x(t) = e^{-t}u(t)$ and $y(t) = 10e^{-t}\cos 4t\ u(t)$, then

$$X(s) = \frac{1}{s+1} \quad \text{and} \quad Y(s) = \frac{10(s+1)}{(s+1)^2 + 4^2}$$

Hence,

$$H(s) = \frac{Y(s)}{X(s)} = \frac{10(s+1)^2}{(s+1)^2 + 16} = \frac{10(s^2 + 2s + 1)}{s^2 + 2s + 17}$$

To find $h(t)$, which is the inverse Laplace transform of $H(s)$, we write $H(s)$ as

$$H(s) = 10 - 40\frac{4}{s^2 + 2s + 17}$$

Taking the Laplace inverse of each and using Table 3.2, we obtain

$$h(t) = 10\delta(t) - 40e^{-t}\sin 4t\ u(t)$$

Practice Problem 3.15 The transfer function of a linear system is

$$H(s) = \frac{2s}{s+6}$$

Find the output $y(t)$ due to the input $e^{-3t}u(t)$ and its impulse response.

Answers: $(-2e^{-3t} + 4e^{-6t})u(t)$, $2\delta(t) - 12e^{-6t}u(t)$

Example 3.16

Find the transfer function for the feedback system shown in Figure 3.11.

Solution

We apply Equation 3.87 to obtain

$$E_2 = G_1G_2E_1$$

$$Y = G_3E_3$$

Applying Equation 3.85,

$$E_1 = X - F_2Y$$

$$E_3 = E_2 - F_1Y$$

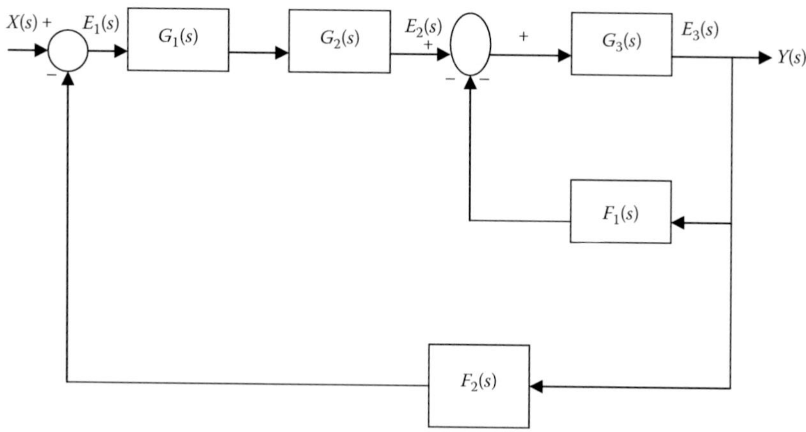

FIGURE 3.11 For Example 3.16.

If we eliminate E_1, E_2, and E_3, we get

$$\frac{Y}{G_3} = G_1G_2\left[X - F_2Y\right] - F_1Y$$

Rearranging things leads to the transfer function:

$$H(s) = \frac{Y(s)}{X(s)} = \frac{G_1(s)G_2(s)G_3(s)}{1 + F_1(s)G_3(s) + F_2(s)G_1(s)G_2(s)G_3(s)}$$

Practice Problem 3.16 Obtain the overall transfer function for the feedback system shown in Figure 3.12.

Answer: $H(s) = \dfrac{2s}{s^3 + 2s^2 + 3s + 3}$

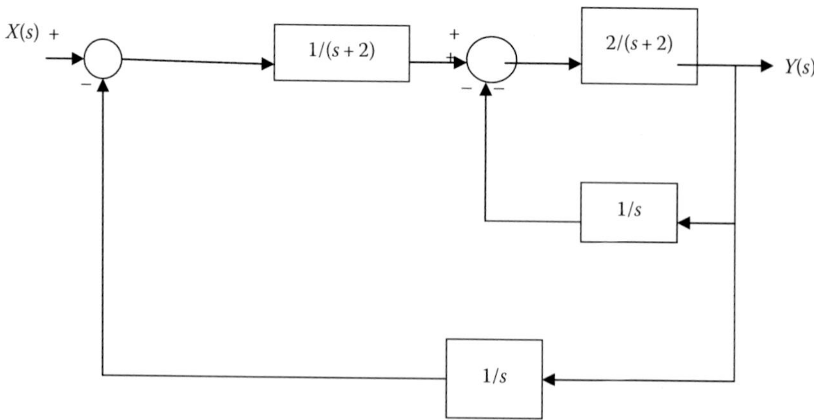

FIGURE 3.12 For Practice Problem 3.16.

3.6 APPLICATIONS

Besides applying Laplace transform to determine the transfer function of a system, Laplace transform finds applications in many areas such as circuit analysis, control systems, communication systems, speech control, and radar systems. In this section, we will apply Laplace transform to three areas: integro-differential equations, circuit analysis, and control systems.

3.6.1 Integro-Differential Equations

We use the Laplace transform in solving linear **integro-differential equations** with constant coefficients. This basically involves employing the differentiation and integration properties of Laplace transform to the terms in the integro-differential equation. Initial conditions are automatically incorporated. We solve the resulting algebraic equation in the s-domain. Using the inverse Laplace transfer, we convert the solution back to the time domain. We illustrate the process with the following examples:

Example 3.17

Consider the system characterized by

$$\frac{d^2y}{dt^2}+3\frac{dy}{dt}-4y=e^{-t}u(t)$$

Use the Laplace transform to solve the differential equation subject to

$$y(0)=1,\quad \frac{dy(0)}{dt}=0.$$

Solution

Take the Laplace transform of each term:

$$\left[s^2Y(s)-sy(0)-y'(0)\right]+3\left[sY(s)-y(0)\right]-4Y(s)=\frac{1}{s+1}$$

We substitute the initial conditions

$$s^2Y(s)-s-0+3sY(s)-3-4Y(s)=\frac{1}{s+1}$$

$$\left(s^2+3s-4\right)Y(s)=s+3+\frac{1}{s+1}=\frac{s^2+4s+4}{s+1}$$

$$Y(s)=\frac{s^2+4s+4}{(s^2+3s-4)(s+1)}=\frac{s^2+4s+4}{(s-1)(s+1)(s+4)}$$

Using partial fractions, we let

$$Y(s) = \frac{A}{(s-1)} + \frac{B}{(s+1)} + \frac{C}{(s+4)}$$

where

$$A = (s-1)Y(s)\Big|_{s=1} = \frac{1+4+4}{2(5)} = \frac{9}{10}$$

$$B = (s+1)Y(s)\Big|_{s=-1} = \frac{1-4+4}{(-2)(3)} = \frac{1}{-6}$$

$$C = (s+4)Y(s)\Big|_{s=-4} = \frac{16-16+4}{(-5)(-3)} = \frac{4}{15}$$

Thus,

$$Y(s) = \frac{9/10}{(s-1)} - \frac{1/6}{(s+1)} + \frac{4/15}{(s+4)}$$

Taking the inverse Laplace transform of each term,

$$y(t) = \left(\frac{9}{10}e^{t} - \frac{1}{6}e^{-t} + \frac{4}{15}e^{-4t}\right)u(t)$$

Practice Problem 3.17 A casual LTI system is described by

$$y''(t) + y'(t) + 2y(t) = x(t)$$

For system input $x(t) = 2u(t)$, find the output $y(t)$. Assume zero initial conditions.

Answer: $y(t) = (1 - 2e^{-t} + e^{-2t})u(t)$

Example 3.18

Solve the integro-differential equation

$$\frac{dv}{dt} + 2v + 5\int_0^t v(\lambda)d\lambda = 4u(t)$$

with $v(0) = -1$ and determine $v(t)$ for $t > 0$.

Solution

Take the Laplace transform of each term of the integro-differential equation. We take the Laplace transform of each term:

$$\left[sV(s) - v(0)\right] + 2V(s) + \frac{5}{s}V(s) = \frac{4}{s}$$

$$\left[sV(s) + 1\right] + 2V(s) + \frac{5}{s}V(s) = \frac{4}{s} \rightarrow V(s) = \frac{4-s}{s^2 + 2s + 5}$$

$$V(s) = \frac{-(s+1)+5}{(s+1)^2+2^2} = \frac{-(s+1)}{(s+1)^2+2^2} + \frac{5}{2}\frac{2}{(s+1)^2+2^2}$$

Taking the inverse of each term leads to

$$v(t) = (-e^{-t}\cos 2t + 2.5e^{-t}\sin 2t)u(t)$$

Practice Problem 3.18 Use Laplace transform to solve for $x(t)$ in

$$x(t) = \cos t + \int_0^t e^{\lambda - t} x(\lambda) d\lambda$$

Answer: $x(t) = \frac{1}{2}e^t + \frac{1}{2}\cos t + \frac{1}{2}\sin t, \quad t > 0.$

3.6.2 Circuit Analysis

The Laplace transform is a powerful tool in analyzing electric circuits. The use of Laplace transform in circuit analysis facilitates the use of various signal sources such as impulse, step, ramp, exponential, and sinusoidal. Applying Laplace transform to analyze circuits usually involves three steps:

1. Laplace transform the circuit from the time-domain to the frequency domain (or s-domain).
2. The circuit in s-domain is solved using circuit analysis techniques (such as voltage division, current division, nodal analysis, mesh analysis, source transformation, and superposition) and we obtain the desired quantity $X(s)$.
3. Obtain the inverse Laplace transform of $X(s)$ to get the desired solution $x(t)$ in the time domain.

The three steps are illustrated in Figure 3.13. Steps 2 and 3 are familiar. Only the first step is new and will be discussed here. We transform a circuit in the time-domain to frequency or s-domain by Laplace transforming each element of the circuit.

For a resistor, the voltage–current relationship in the time-domain is

$$v(t) = Ri(t) \tag{3.94}$$

Laplace transforming this gives

$$V(s) = RI(s) \tag{3.95}$$

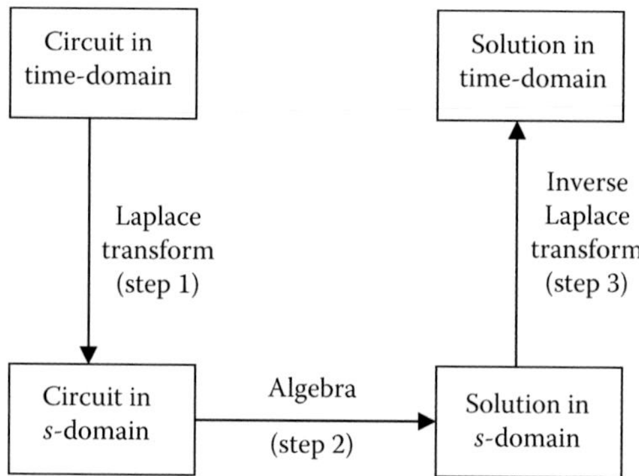

FIGURE 3.13 The three steps involved in analyzing a circuit with Laplace transformation.

For the inductor,

$$v(t) = L\frac{di}{dt} \tag{3.96}$$

Laplace transforming both sides yields

$$V(s) = L\left[sI(s) - i(0^-)\right] = sLI(s) - Li(0^-)$$

or

$$I(s) = \frac{1}{sL}V(s) + \frac{i(0^-)}{s} \tag{3.97}$$

For the capacitor,

$$i(t) = C\frac{dv(t)}{dt} \tag{3.98}$$

This transforms into the s-domain as

$$I(s) = C\left[sV(0s) - v(0^-)\right] = sCV(s) - Cv(0^-)$$

or

$$V(s) = \frac{1}{sC}I(s) + \frac{v(0^-)}{s} \tag{3.99}$$

Notice in Equations 3.97 and 3.99 that the initial conditions are incorporated in the transformation process. This is one of the key advantages of using Laplace transform in circuit analysis. Another advantage is that a complete response (natural and steady state) of a network is obtained at once.

Assuming zero initial conditions for the inductor and the capacitor, the previous equations reduce to:

$$\text{Resistor}: V(s) = I(s)R \tag{3.100}$$

$$\text{Inductor}: V(s) = sLI(s) \tag{3.101}$$

$$\text{Capacitor}: V(s) = \frac{1}{sC} I(s) \tag{3.102}$$

Example 3.19

Find $v_o(t)$ in the circuit in Figure 3.14.

Solution

We first transform the circuit from the time-domain to the s-domain.

$$2u(t) \Rightarrow \frac{2}{s}$$

$$1\,\text{H} \Rightarrow sL = s$$

$$0.2\,\text{F} \Rightarrow \frac{1}{sC} = \frac{1}{0.2s} = \frac{5}{s}$$

Hence the frequency-domain equivalent circuit is in Figure 3.15. Using current division,

$$I_o = \frac{5/s}{5/s + s + 4} \times \frac{2}{s} = \frac{5}{s^2 + 4s + 5} \times \frac{2}{s}$$

$$V_o = 4I_o = \frac{40}{s(s^2 + 4s + 5)} = \frac{A}{s} + \frac{Bs + C}{s^2 + 4s + 5}$$

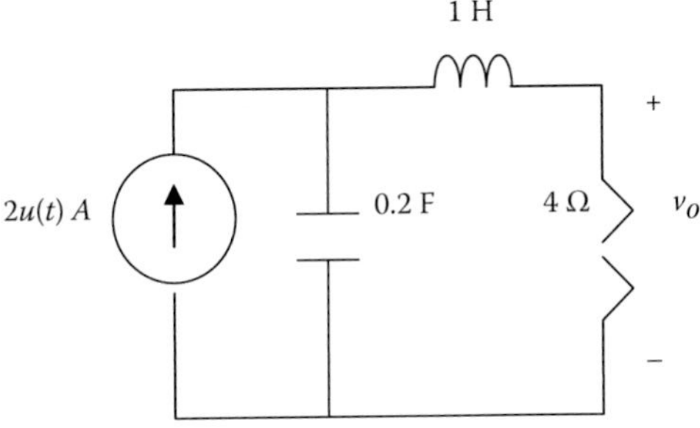

FIGURE 3.14 For Example 3.19.

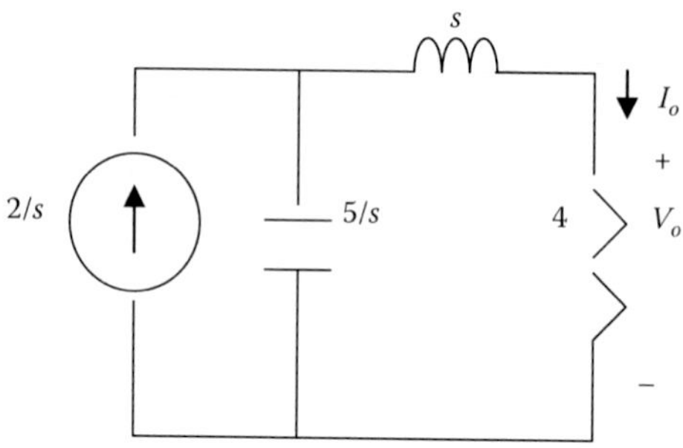

FIGURE 3.15 s-Domain circuit for Example 3.19.

Equating the denominators,

$$40 = A(s^2 + 4s + 5) + Bs^2 + Cs$$

Equating coefficients:

$$\text{Constant: } 40 = 5A \rightarrow A = 8$$

$$s^1: 0 = 4A + C \rightarrow C = -4A = -32$$

$$s^2: 0 = A + B \rightarrow B = -A = -8$$

$$V_o = \frac{8}{s} - \frac{8s + 32}{(s+2)^2 + 1}$$

$$= \frac{8}{s} - \frac{8(s+2)}{(s+2)^2 + 1} - \frac{16}{(s+2)^2 + 1}$$

Taking the inverse Laplace transform of each term, we obtain

$$v_o(t) = \left(8 - 8e^{-2t}\cos t - 8e^{-2t}\sin t\right)u(t)$$

Practice Problem 3.19 Using Laplace transformation, find $v_o(t)$ in the circuit in Figure 3.16.

Answer: $v_o(t) = 3.2(e^{-t} - e^{-6t})u(t)$

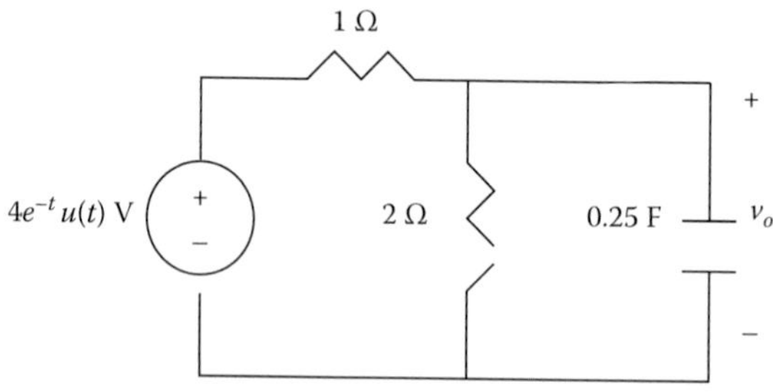

FIGURE 3.16 For Practice Problem 3.19.

Example 3.20

For the circuit in Figure 3.17, find $H(s) = V_o(s)/V_s(s)$. Assume zero initial conditions.

Solution

Transform the circuit to s-domain as shown in Figure 3.18. Let

$$Z = \frac{1}{s} \,||\, (s+2) = \frac{1/s(s+2)}{1/s + s + 2} = \frac{(s+2)}{(s^2+2s+1)}$$

Using voltage division,

$$V_1 = \frac{Z}{Z+1} V_s$$

$$V_o = \frac{2}{s+2} V_1 = \frac{2}{s+2} \frac{Z}{Z+1} V_s$$

$$H(s) = \frac{V_o}{V_s} = \frac{2}{s+2} \times \frac{\dfrac{s+2}{s^2+2s+1}}{\dfrac{s+2}{s^2+2s+1}+1}$$

$$H(s) = \frac{2(s+2)}{s^2+3s+3}$$

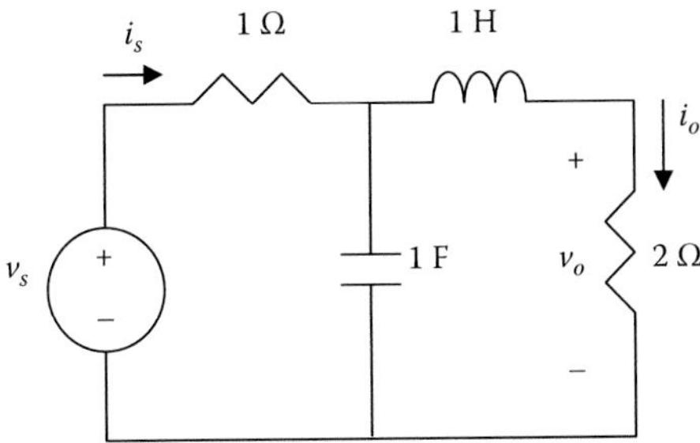

FIGURE 3.17 For Example 3.20 and Practice Problem 3.20.

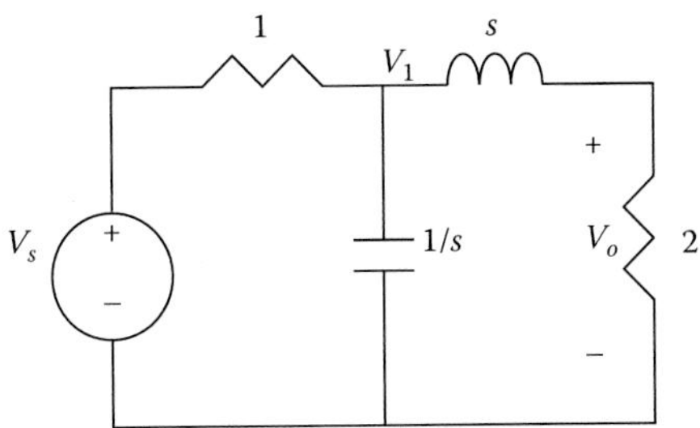

FIGURE 3.18 For Example 3.20.

Practice Problem 3.20 For the circuit in Figure 3.17, obtain $H(s) = I_o(s)/I_s(s)$. Assume zero initial conditions.

Answer: $H(s) = \dfrac{1}{s^2 + 2s + 1}$

3.6.3 Control Systems

A major application of Laplace transform is found in the study of control systems. An important consideration of control systems is their stability. In order for a control system to be useful, it must be stable. We examine the bounded-input bounded-output **(BIBO) stability** of systems.

A system is **BIBO stable** when the output remains bounded for all time for every bounded input.

We can study the stability of a system either through its impulse response $h(t)$ or through its transfer function $H(s)$. An LTI system is BIBO stable if and only if its impulse response $h(t)$ is absolutely integrable, that is,

$$\int_{-\infty}^{\infty} |h(t)|\, dt < \infty \tag{3.103}$$

This condition is satisfied if the impulse response decays exponentially with increase in time.

We can determine whether a system is BIBO stable if its transfer function meets some conditions. The transfer function can be written as

$$H(s) = \frac{N(s)}{D(s)} = \frac{N(s)}{(s+p_1)(s+p_2)\cdots(s+p_n)} \tag{3.104}$$

$H(s)$ must meet two requirements for the system to be stable. First, the numerator degree of $H(s)$ must not be larger than the denominator degree; otherwise, long division would produce

$$H(s) = k_n s^n + k_{n-1} s^{n-1} + \cdots + k_1 s + k_0 + \frac{R(s)}{D(s)} \tag{3.105}$$

where the degree of $R(s)$, the remainder of the long division, is less than the degree of $D(s)$. The inverse of $H(s)$ in Equation 3.105 does not meet the condition in Equation 3.103. Second, all the poles of the transfer function must be in the left half of the

s plane. This means that $H(s)$ must not have poles in the right half of the s plane. Each pole must satisfy $p = -\alpha \pm j\omega$, where $\alpha > 0$, that is, the real part of the pole must be negative.

A system is **BIBO stable** if and only if- all the poles of its transfer function $H(s)$ lie in the left half of the s plane.

The system is considered unstable if one or more of its poles appear in the right half s plane. If the transfer function has poles on the imaginary axis (i.e., with zero real parts), the system is said to be marginally stable.

Example 3.21

Check the stability of a system with transfer function

$$H(s) = \frac{s^2 + s - 10}{s^3 + 3s^2 + 4s + 2}$$

Solution

$H(s)$ can be expressed in terms of its numerator $N(s)$ and denominator $D(s)$.

$$H(s) = \frac{N(s)}{D(s)} = \frac{s^2 + s - 10}{s^3 + 3s^2 + 4s + 2}$$

To determine the poles, we are only interested in the roots of $D(s)$.

$$D(s) = s^3 + 3s^2 + 4s + 2 = (s+1)(s+1-j)(s+1+j)$$

The roots of $D(s) = 0$ or the poles of $H(s)$ are

$$p_1 = -1, \quad p_2 = -1 + j, \quad p_3 = -1 - j$$

All the three poles lie in the left half of the s plane and we conclude that the system is stable.

Practice Problem 3.21 A system has its transfer function as

$$H(s) = \frac{s+6}{s^3 + s^2 + 2s + 8}$$

Check whether it is stable.

Answer: Unstable; it has two poles in the right half of the s plane.

3.7 COMPUTING WITH MATLAB®

MATLAB is a powerful tool that can be used to do most of what we cover in this chapter. The impulse, step, ramp, sinusoidal, and exponential responses of continuous-time systems can be examined using MATLAB. In this section, we will examine how MATLAB can be used to do the following:

- *Find Laplace transform*: MATLAB has a Laplace transform command X = **laplace** (x). If the expression produced by the command is complicated, we can use a MATLAB command **simplify** (x) to reduce the complexity.
- *Find inverse Laplace transform*: This can be done in two ways. We may use the command **ilaplace** to invert $X(s)$. We may also use the command **residue**, which will produce a vector consisting of the residues of $X(s)$. We then apply Table 3.2 to find $x(t)$.
- *Find the roots of a polynomial*: By using the command **roots**, MATLAB can be used to find the roots a polynomial or the zeros, and poles of a transfer function.
- *Find the time response*: MATLAB can be used to find the impulse response of a system by using the command **impulse**. We can also find the step response of the system using the command **step**. For finding ramp response or any other time response, we can use the command **lsim**.
- *Find the frequency response*: The command **bode** produces Bode plots (both magnitude and phase) of a given transfer function $H(s)$. The frequency range and the number of points are automatically selected.

We will illustrate these with examples. There are many other commands (such as **conv, deconv, factor, pretty, lsim, freqs**, etc.), which are related to what we cover in this chapter; they can be examined by typing *help control.*

Example 3.22

Use MATLAB to find the Laplace transform of

$$x(t) = 2\delta(t) + e^{-3t}$$

Solution

We recall that the commands **dirac**(t) and **heaviside**(t) are used to represent the unit impulse $\delta(t)$ and unit step $u(t)$, respectively. The MATLAB commands are:

```
syms x t
x = 2*dirac(t) + exp(-3*t);
X = laplace(x)
```

This produces the following result:

```
X =
1/(s+3) + 2
```

From this, we obtain

$$X(s) = 2 + \frac{1}{s+3}$$

Practice Problem 3.22 Use MATLAB to obtain the Laplace transform of

$$x(t) = u(t-1) - 2e^{-t}.$$

Answer: $X(s) = \frac{1}{s}e^{-s} - \frac{2}{s+1}$

Example 3.23

Use MATLAB to find the inverse Laplace transform of

$$V(s) = \frac{10s^2 + 4}{s(s+1)(s+2)^2}$$

Solution

The MATLAB code is as follows:

```
syms s v V
V =(10*s^2 +4)/( s*(s+1)*(s+2)^2 );
v = ilaplace(V)
```

This produces the following result:

```
v =
1-14*exp(-t)+(13+22*t)*exp(-2*t)
```

so that

$$v(t) = 1 - 14e^{-t} + 13e^{-2t} + 22te^{-2t}, \quad t > 0$$

Practice Problem 3.23 Use MATLAB to obtain $g(t)$ if

$$G(s) = \frac{s^3 + 2s + 6}{s(s+1)^2(s+3)}$$

Answer: $(2 - 3.25e^{-t} - 1.5te^{-t} + 2.25e^{-t})u(t)$

Example 3.24

Use the **residue** command to find the Laplace inverse of

$$X(s) = \frac{4s^5 + 20s^4 + 16s^3 + 10s^2 - 12}{s^4 + 5s^3 + 8s^2 + 4s}$$

Solution

This is an indirect way of finding the inverse Laplace transform. We specify the numerator (num) and the denominator (den) of the transfer function $X(s)$. We find the residues of $X(s)$ using the following code.

```
num = [4 20 16 10 0 -12]; % numerator coefficients in descending
powers of s
den = [1 5 8 4 0]; % denominator coefficients in descending
powers of s
[r,p,k] = residue(num,den); % call residue
r =
-15.0000
46.0000
2.0000
-3.0000
p =
-2.0000
-2.0000
-1.0000
0
k =
4 0
```

This produces a vector r that has the residues and a vector p that has the corresponding poles.

Notice that pole −2 is repeated. Also, since the order of the numerator of $X(s)$ is one greater than the order of the denominator of $X(s)$, k contains two values. From r, p, and k, we can write $X(s)$ as

$$X(s) = 4s + 0s^0 + \frac{-15}{s-(-2)} + \frac{46}{[s-(-2)]^2} + \frac{2}{s-(-1)} + \frac{-3}{s-0}$$

$$= 4s - \frac{15}{s+2} + \frac{46}{(s+2)^2} + \frac{2}{s+1} - \frac{3}{s}$$

Using Table 3.2, we obtain the inverse $v(t)$ as

$$x(t) = 4\delta'(t) - 15e^{-2t} + 46te^{-2t} + 2e^{-t} - 3u(t), \quad t \geq 0$$

Practice Problem 3.24 Given that $X(s) = \frac{5s^2}{(s+1)(s+3)}$, use the **residue** command to find $x(t)$.

Answer: $5\delta(t) - 22.5e^{-3t} + 2.5e^{-t}, \quad t \geq 0$

Example 3.25

Use MATLAB to find the zeros and poles of

$$H(s) = \frac{s^2 + 3s + 1}{s^3 + 4s^2 + 3s}$$

Solution

We use the command **roots** to find the roots of the numerator to get the zeros, and denominator to get the poles.

```
num = [1 3 1];
den = [1 4 3 0];
z=roots(num);
p=roots(den);
This result is:
z =
-2.6180
-0.3820
p =
0
-3
-1
```

Practice Problem 3.25 Repeat Example 3.25 for $H(s) = \frac{s^2 + 4s - 2}{s^3 + 4s^2 + 13s}$

Answer: Zeros: −4.4495, 0.4495, poles: 0, −2 + 3*j*, −2 − 3*j*.

Example 3.26

Use MATLAB to plot the step response of a system whose transfer function is $H(s) = \frac{s+1}{(s+2)(s+3)}$

Solution

We first need to expand the denominator:

$$H(s) = \frac{s+1}{s^2 + 5s + 6}$$

By definition, the step response is the response when the input to the system is the unit step *u*(*t*). Using the MATLAB script below, we obtain the response as plotted in Figure 3.19.

```
num = [1 1];
den = [1 5 6];
t = 0: 0.1: 4;
y=step(num,den,t);
plot(t,y)
```

Practice Problem 3.26 Use **MATLAB** to plot the impulse response of a system whose transfer function is given in Example 3.26. *Hint*: Use impulse (num, den, *t*).

Answer: See Figure 3.20.

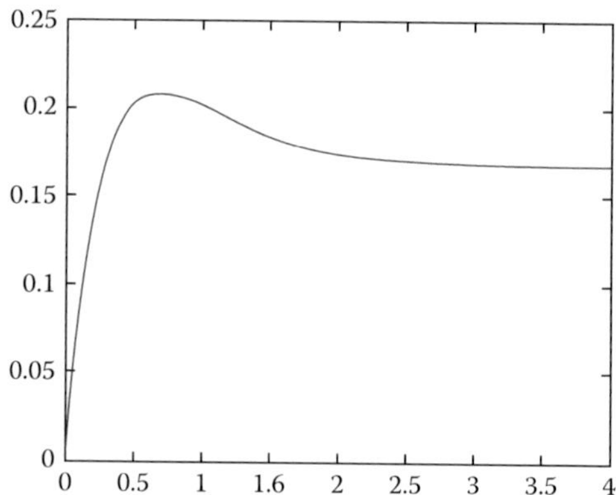

FIGURE 3.19 For Example 3.26.

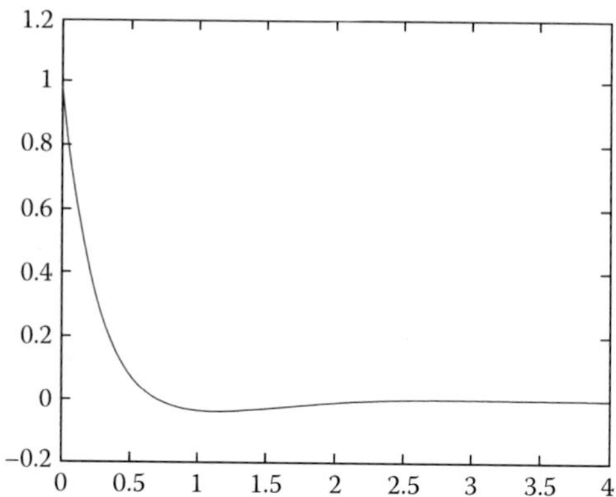

FIGURE 3.20 For Practice Problem 3.26.

Example 3.27

Use MATLAB to obtain the Bode plots for the transfer function

$$H(s) = \frac{100s}{s^2 + 12s + 20}$$

Solution

The MATLAB script for the transfer is shown below, while the Bode plots are in Figure 3.21.

```
num = [100 0];
den = [1 12 20];
bode(num,den); %determines and draws Bode plots
```

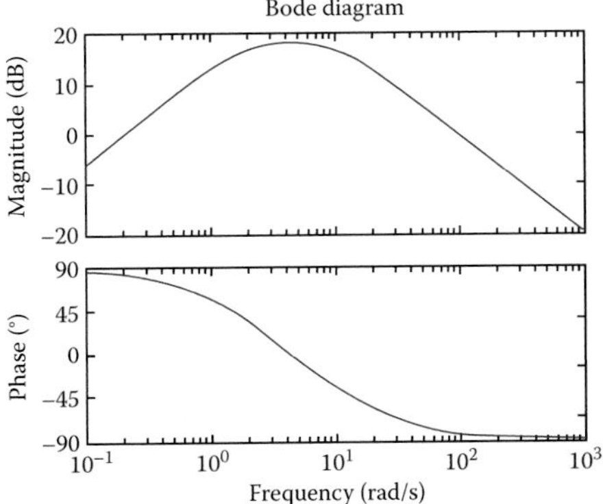

FIGURE 3.21 Magnitude and phase Bode plots for Example 3.27.

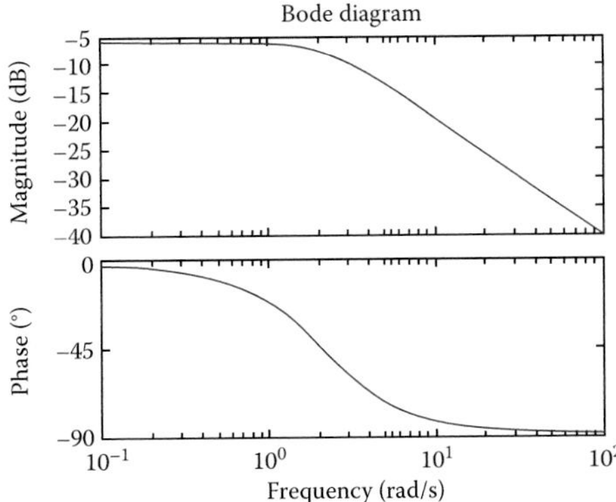

FIGURE 3.22 For Practice Problem 3.27.

Practice Problem 3.27 Repeat Example 3.27 for the transfer function

$$H(s) = \frac{s+4}{s^2+5s+8}$$

Answer: See Figure 3.22.

3.8 SUMMARY

1. The Laplace transform converts a signal in the time-domain to the frequency domain. Its unilateral form (used throughout this chapter) is defined as

$$\mathcal{L}[x(t)] = X(s) = \int_{0^-}^{\infty} x(t)e^{-st}dt$$

The values of s for which $X(s)$ exists form the region of convergence.

2. Laplace transform properties are provided in Table 3.1, while the Laplace transform of basic common functions are listed in Table 3.2.
3. The inverse Laplace transform, $\mathcal{L}^{-1}[X(s)]$, can be found by using partial fraction expansions along with the Laplace transform pairs in Table 3.2. The partial fraction expansion of $X(s)$ can be obtained when the poles of $X(s)$ are simple, repeated, complex conjugates, or a combination of these.
4. The transfer function $H(s)$ of a system is the Laplace transform of the impulse response $h(t)$. $H(s)$ relates the output response $Y(s)$ and an input excitation $X(s)$, that is, $H(s) = Y(s)/X(s)$, assuming all initial conditions are zero.
5. There are several applications of the Laplace transform. Some include the solution of integro-differential equations, electric circuit analysis, and analysis of control systems.
6. Using the Laplace transform to analyze a circuit involves converting each element from time-domain to s-domain, solving the problem using circuit techniques, and converting the result to time-domain using inverse Laplace transform.
7. A system is considered stable if all the poles of the transfer function $H(s)$ lie in the left half of the s plane. Values of s for which $H(s) = 0$ and $H(s) = \infty$ are, respectively, zeros and poles of $H(s)$.
8. MATLAB is a convenient tool for finding the Laplace transform of a signal or its inverse. It is also used in obtaining the time response and frequency response of systems.

REVIEW QUESTIONS

3.1 A Laplace transform exists only in its region of convergence.
(a) True, (b) false

3.2 If $X(s)$ is the Laplace transform of $x(t)$, the variable s is known as:
(a) Laplacian, (b) complex frequency, (c) transfer function, (d) zero, (e) pole

3.3 The one-sided Laplace transform depends only on the value of the signal for $t \geq 0$.
(a) True, (b) false

3.4 The Laplace transform of $t^2u(t)$ is:

(a) $\frac{2}{s}$ (b) $\frac{2}{s^2}$ (c) $\frac{2}{s^3}$ (d) $\frac{e^{-2s}}{s}$

3.5 Let $X(s) = \frac{s+4}{(s+1)(s+2)(s+3)}$

The zero of $X(s)$ is at
(a) −4, (b) −3, (c) −2, (d) −1

3.6 For the function given the previous question, the poles of $X(s)$ are at
(a) −4, (b) −3, (c) −2, (d) −1

3.7 Given

$$X(s) = \frac{s+3}{(s+1)(s+2)}$$

the initial value of $x(t)$ with transform is
(a) Nonexistent, (b) ∞, (c) 0, (d) 1, (e) 1/6

3.8 Let $X(s) = \frac{e^{-2s}}{s+4}$. The inverse Laplace transform $x(t)$ is
(a) $e^{-2(t-1)}u(t-1)$, (b) $e^{-4(t-2)}u(t-2)$, (c) $e^{-4t}u(t-2)$, (d) $e^{-(t-2)}u(t-4)$

3.9 A network has the following transfer function:

$$H(s) = \frac{s+1}{(s+2)(s-3)}$$

Is the network stable?
(a) Yes, (b) no

3.10 The MATLAB command for finding zeros and poles of a transfer function is
(a) Laplace, (b) ilaplace, (c) roots, (d) step, (e) impulse

Answers: 3.1a, 3.2b, 3.3a, 3.4c, 3.5a, 3.6b,c,d, 3.7d, 3.8b, 3.9b, 3.10c.

PROBLEMS

Sections 3.2 and 3.3—Definition of the Laplace Transform and Properties of the Laplace Transform

3.1 Find the Laplace transform of

(a) $x(t) = \begin{cases} 10, & 0 < t < 1 \\ -5, & 1 < t < 2 \\ 0, & \text{otherwise} \end{cases}$

(b) $r(t) = t\,u(t)$.

3.2 Find $F(s)$ given that

$$f(t) = \begin{cases} 2t, & 0 < t < 1 \\ 2, & 1 < t < 2 \\ 0, & \text{otherwise} \end{cases}$$

3.3 Find the Laplace $X(s)$ given that $x(t)$ is
(a) $2tu(t-4)$
(b) $5\cos t\ \delta(t-2)$
(c) $e^{-t}u(t-\tau)$
(d) $\sin 2t\ u(t-\tau)$

3.4 Find the Laplace transform of these signals:

(a) $x_1(t) = te^{2t}u(t)$

(b) $x_2(t) = t\cos 4t$

(c) $x_3(t) = \delta(t + 2)$

(d) $x_4(t) = 5\cos(2t + \pi/4)$

(e) $x_5(t) = t^2u(t-1)$

3.5 Determine the Laplace transform of

(a) $x(t) = u(t) - u(t - 2)$

(b) $y(t) = te^{-3t}u(t)$

3.6 Show that

(a) $\mathcal{L}\left[cos^2(\omega t)u(t)\right] = \dfrac{s^2 + 2\omega^2}{s(s^2 + 4\omega^2)}$

(b) $\mathcal{L}\left[sin^2(\omega t)u(t)\right] = \dfrac{2\omega^2}{s(s^2 + 4\omega^2)}$

3.7 Find the Laplace transform of $x(t)$ shown in Figure 3.23.

3.8 Using the initial value theorem, find $g(0)$ if

$$G(s) = \frac{s+1}{s^3 + 6s^2 + 11s + 5}$$

3.9 Given that

$$F(s) = \frac{s^2 + 10s + 6}{s(s+1)^2(s+2)}$$

evaluate $f(0)$ and $f(\infty)$ if they exist.

3.10 Using Laplace transform, find the convolution $x(t)*y(t)$ for the following cases:

(a) $x(t) = e^{-2t}u(t)$, $y(t) = tu(t)$

(b) $x(t) = e^{-t}u(t)$, $y(t) = e^{-t}u(t)$

(c) $x(t) = (\sin t)u(t)$, $y(t) = (\cos t)u(t)$

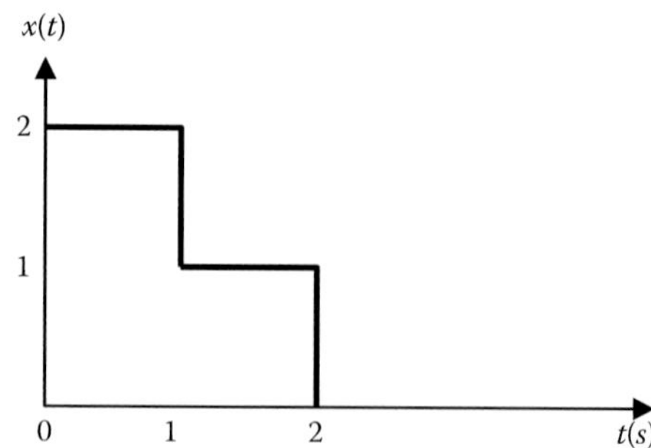

FIGURE 3.23 For Problem 3.7.

Section 3.4—The Inverse Laplace Transform

3.11 Obtain the inverse Laplace transform of the following functions:

(a) $X(s) = \frac{1}{s} + \frac{2}{s}e^{-s}$

(b) $Y(s) = \frac{10}{s^2 - 5s + 4}$

(c) $\frac{s-2}{s^2 + 2s + 10}$

3.12 Find the inverse Laplace transform of

(a) $G(s) = \frac{s+2}{s^3 + 4s^2 + 3s}$

(b) $H(s) = \frac{10e^{-2s}}{(3s+5)^3}$

3.13 Find the inverse Laplace transform for

(a) $X(s) = \frac{3s+1}{s^2 + 2s + 5}$

(b) $Y(s) = \frac{3s+7}{s^2 + 3s + 2}$

(c) $Z(s) = \frac{s^2 - 8}{s^2 - 4}$

(d) $H(s) = \frac{12}{(s+2)^4}$

3.14 Find the inverse Laplace transform of the following functions:

(a) $F(s) = \frac{20(s+2)}{s(s^2 + 6s + 25)}$

(b) $P(s) = \frac{6s^2 + 36s + 20}{(s+1)(s+2)(s+3)}$

3.15 Find $y(t)$ that corresponds to

$$Y(s) = \frac{8e^{-s} - 4e^{-2s} - 2se^{-2s} - 6se^{-3s}}{s^2}$$

3.16 Find the inverse Laplace transform of these functions:

(a) $X(s) = \frac{s^2 + 2s + 2}{s+1}$

(b) $Y(s) = \frac{s^3 + 2s^2 + 3s + 1}{s^2(s+1)}$

3.17 Find the inverse Laplace transform of

(a) $H(s)=\dfrac{s+4}{s(s+2)}$

(b) $G(s)=\dfrac{s^2+4s+5}{(s+3)(s^2+2s+2)}$

(c) $F(s)=\dfrac{e^{-4s}}{s+2}$

(d) $D(s)=\dfrac{10s}{(s^2+1)(s^2+4)}$

3.18 Let $F(s)=\dfrac{5(s+1)}{(s+2)(s+3)}$

(a) Use the initial and final value theorems to find $f(0)$ and $f(\infty)$.
(b) Verify your answer in part (a) by finding $f(t)$ using partial fractions.

Section 3.5— Transfer Function

3.19 Consider a system for which

$$\frac{dy(t)}{dt}+5y(t)=\frac{dx(t)}{dt}+x(t)$$

Assuming that the system is causal, find the impulse response $h(t)$.

3.20 Obtain the impulse response of the system represented by the following equation:

$$\frac{d^2y(t)}{dt^2}+2\frac{dy(t)}{dt}+y(t)=x(t)$$

Assume zero initial conditions.

3.21 A causal LTI system has the transfer function

$$H(s)=\frac{s+2}{s^2+2s+5}$$

Find the response $y(t)$ due to the input $x(t) = 2e^{-3t}u(t)$.

3.22 Find the transfer function of the system shown in Figure 3.24.

3.23 Find the transfer function of the feedback system shown in Figure 3.25.

3.24 Determine the transfer function of the system shown in Figure 3.26.

3.25 For the system shown in Figure 3.27, obtain the transfer function $H(s) = Y(s)/X(s)$.

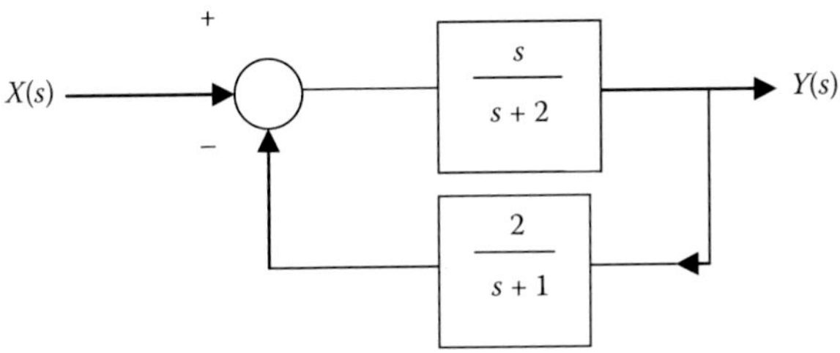

FIGURE 3.24 For Problem 3.22.

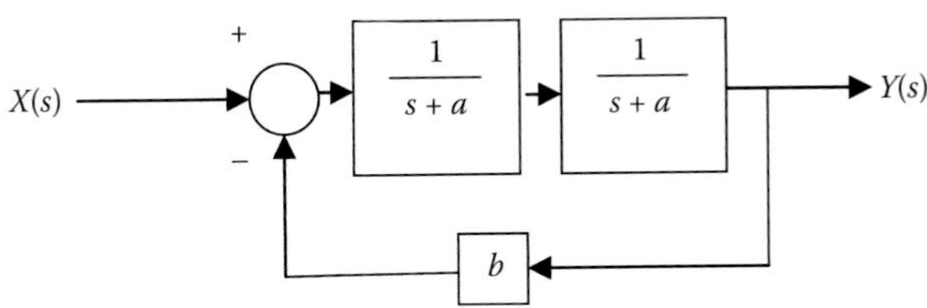

FIGURE 3.25 For Problem 3.23.

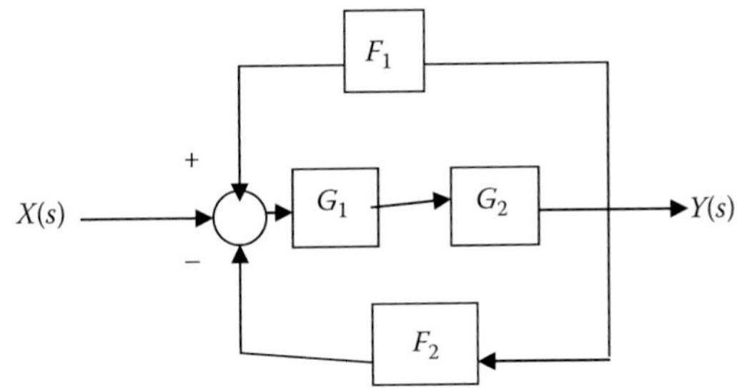

FIGURE 3.26 For Problem 3.24.

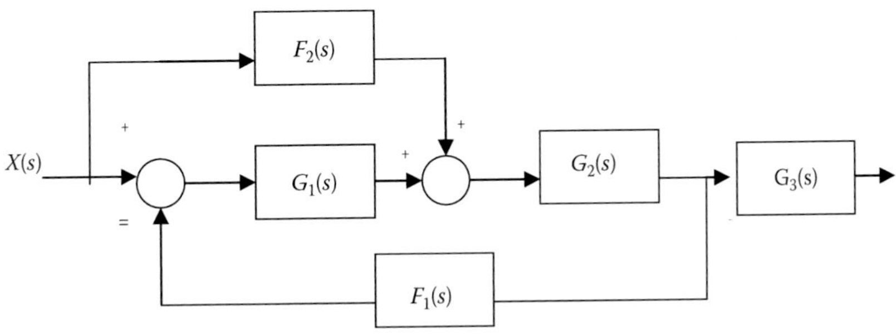

FIGURE 3.27 For Problem 3.25.

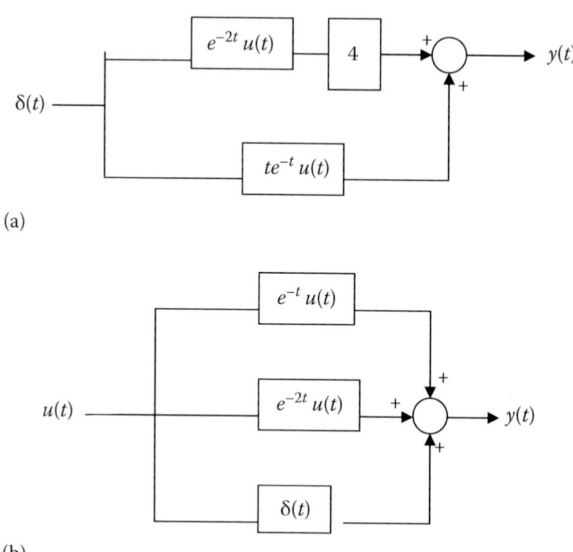

FIGURE 3.28 For Problem 3.26.

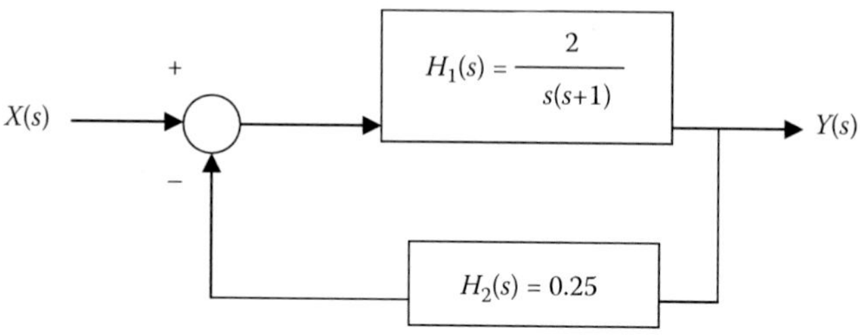

FIGURE 3.29 For Problem 3.27.

3.26 For each of the systems shown in Figure 3.28, use Laplace transform to find $y(t)$.

3.27 A manufacturing process is represented by the feedback control system shown in Figure 3.29.
(a) Obtain the system transfer function
(b) Determine the system response $y(t)$ when $x(t) = u(t)$.

3.28 Determine the transfer function for the system shown in Figure 3.30.

Section 3.6— Applications

3.29 Solve the differential equation

$$y''(t) + 7y'(t) + 12y(t) = e^{-t}u(t)$$

subject to $y(0) = -1$, $y'(0) = 2$

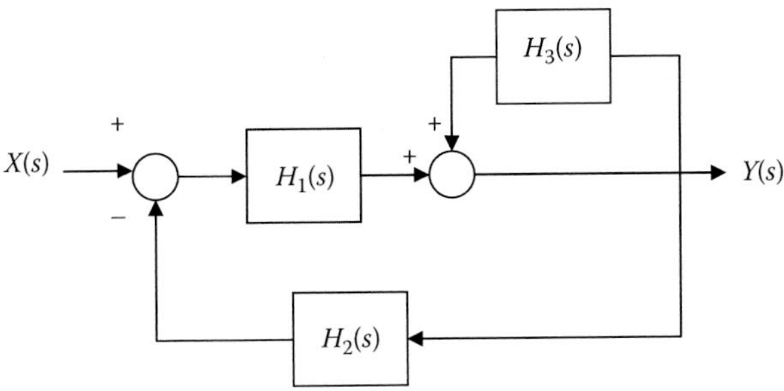

FIGURE 3.30 For Problem 3.28.

3.30 Determine the complete response of the systems described by the following differential equation:

$$y^{(2)}(t)+2y^{(1)}(t)+y(t)=\cos 2tu(t),\quad y(0^-)=0,\quad y^{(1)}(0^-)=1$$

3.31 Obtain the solution to this differential equation:

$$y^{(2)}(t)+2y^{(1)}(t)+y(t)=u(t),\quad y(0^-)=1=y^{(1)}(0^-)$$

3.32 A Butterworth high-pass filter with input $x(t)$ and output $y(t)$ is described by

$$\frac{d^2y(t)}{dt^2}+1.41\frac{dy(t)}{dt}+y(t)=\frac{d^2x(t)}{dt^2}$$

Determine the impulse response.

3.33 Using Laplace transform, solve the following differential equation for $t>0$

$$\frac{d^2i}{dt^2}+4\frac{di}{dt}+5i=2e^{-2t}$$

subject to $i(0) = 0$, $i'(0) = 2$.

3.34 Obtain the transfer function of the circuit in Figure 3.31.

3.35 Consider the circuit shown in Figure 3.32. Find the transfer function $H(s)$.

3.36 Obtain the current $i_o(t)$ in the circuit of Figure 3.33.

3.37 Using nodal analysis, find $v_o(t)$ in the circuit of Figure 3.34.

3.38 Obtain the impedance $Z(s)$ of the network in Figure 3.35.

3.39 Use Laplace transformation to determine the mesh currents $I_1(s)$ and $I_2(s)$ in Figure 3.36. Let $v_s(t) = 2e^{-t}$ V.

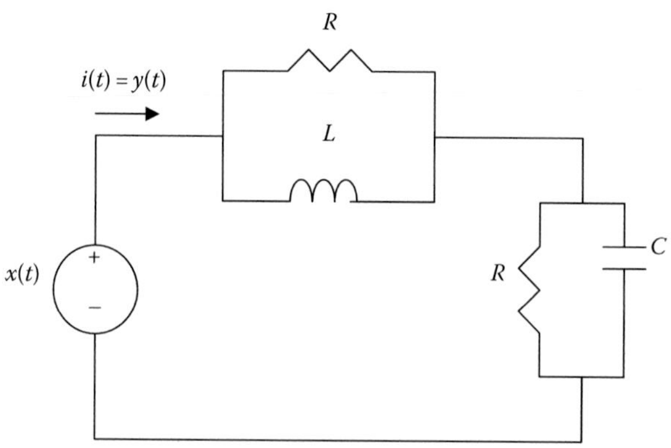

FIGURE 3.31 For Problem 3.34.

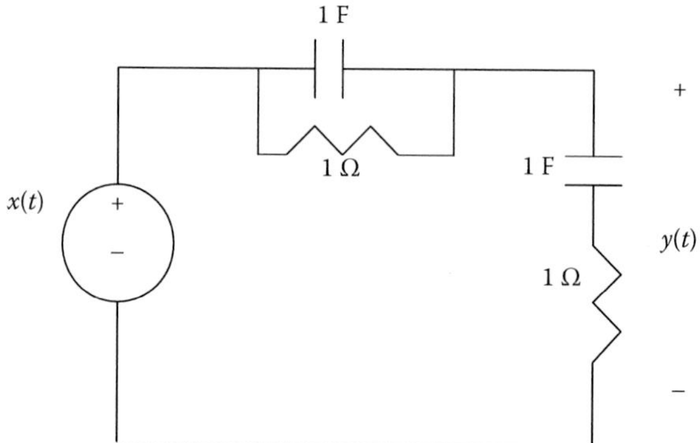

FIGURE 3.32 For Problem 3.35.

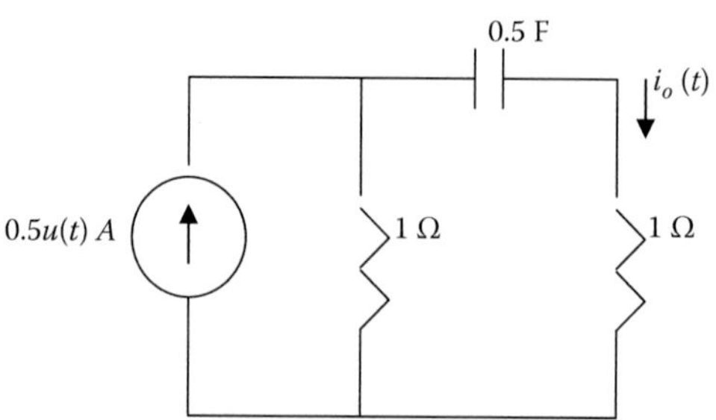

FIGURE 3.33 For Problem 3.36.

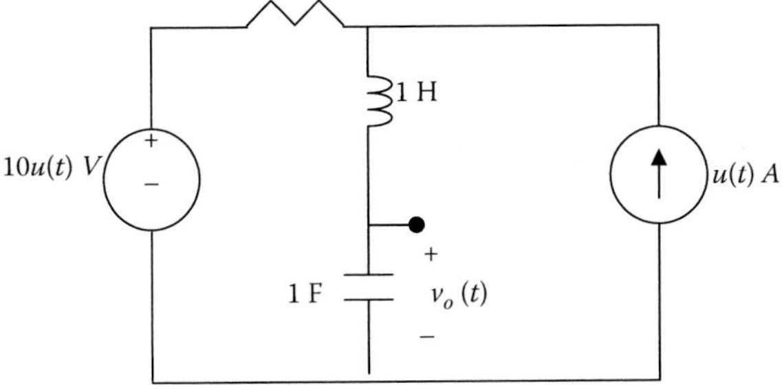

FIGURE 3.34 For Problem 3.37.

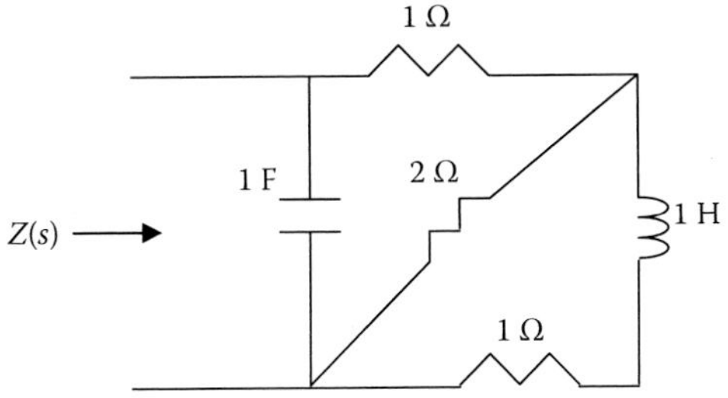

FIGURE 3.35 For Problem 3.38.

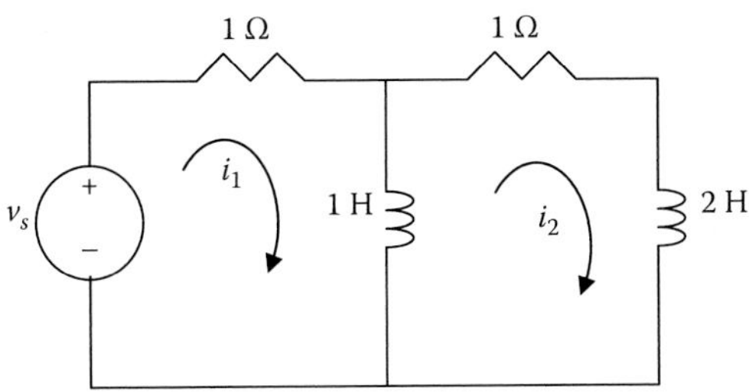

FIGURE 3.36 For Problem 3.39.

3.40 Investigate the stability of these systems:

(a) $H(s) = \dfrac{s+2}{s^2 - 5s - 6}$

(b) $H(s) = \dfrac{s+2}{s^2 + 9}$

(c) $H(s) = \dfrac{s^2 + 6}{s^2 + 5s + 4}$

(d) $H(s) = \dfrac{s^3 + 10}{s^2 + 5s + 6}$

3.41 Determine the stability of the systems represented by the following transfer functions:

(a) $H(s) = \dfrac{4}{s(s+1)}$

(b) $H(s) = -\dfrac{10s}{s^2 + 5s + 6}$

(c) $H(s) = \dfrac{s^2 + 1}{s^2 - 6s + 25}$

3.42 Given the impulse response $h(t)$ of an LTI system, determine if the system is BIBO stable.

(a) $h(t) = (t^3 + 2t^2 - 3t + 1)u(t)$

(b) $h(t) = e^{-t}\sin 4t, \quad t \geq 0$

(c) $h(t) = \delta(t) + 2u(t)$

3.43 An LTI system is described by

$$\frac{d^2 y(t)}{dt^2} + 2\frac{dy(t)}{dt} + 2y(t) = \frac{dx(t)}{dt} - 3x(t)$$

(a) Determine the transfer function of the system.

(b) Obtain the impulse response of the system.

Section 3.7— Computing with MATLAB®

3.44 Use MATLAB to find the Laplace transform of these functions for $t > 0$.

(a) $t^3 + t^2 + 2t - 4$

(b) $t \cos 3t$

(c) $e^{-t}\sin 2t$

3.45 Use MATLAB to find the inverse Laplace transform of the following functions:

(a) $\dfrac{1}{s} + \dfrac{1}{s}e^{-2s}$

(b) $\dfrac{2}{(s+1)^3}$

(c) $\dfrac{s^2 + 3s + 1}{s^2 + 4s + 8}$

3.46 Consider the function

$$H(s) = \frac{s^2 + 6s + 10}{s^3 + 7s^2 + 11s + 5}$$

Use the MATLAB residue function to obtain the inverse Laplace transform of $H(s)$.

3.47 Use MATLAB to find the inverse Laplace transform

$$X(s)=\frac{e^{-3s}+s^2e^{-2s}}{s^3+3s^2+2s}$$

3.48 Use MATLAB command residue to find the inverse Laplace transform of each of these functions:

(a) $X(s)=\dfrac{3s^2+6s-10}{s^2+7s+12}$

(b) $Y(s)=\dfrac{s^2+2s+1}{s^3+4s^2+6s+5}$

(c) $Z(s)=\dfrac{2s^3+12s^2+10s+1}{s^5+16s^4+8s^3+2s}$

3.49 A linear system is represented by its transfer function

$$H(s)=\frac{2s+5}{(s+2)(s+3)}$$

Use MATLAB to determine and plot:
(a) The impulse response
(b) The step response.

3.50 Find the impulse response and the step response for each of the following systems:

(a) $H(s)=\dfrac{s+1}{s^2+5s+6}$

(b) $H(s)=\dfrac{5s}{s^3+10s^2+10s+4}$

3.51 Use MATLAB to find the zeros and poles of these functions:

(a) $\dfrac{s-2}{(s+1)^2+9}$

(b) $\dfrac{s^2+2s+5}{s(s^2+4s+13)}$

(c) $\dfrac{s^2+10s+5}{s^3+4s^2+10s+6}$

3.52 Use MATLAB to determine if the systems represented by the following transfer functions are stable:

(a) $H(s)=\dfrac{s^3+12s^2+3s+10}{2s^4+20s^3+25s^2+15s+12}$

(b) $H(s)=\dfrac{s^4+2s^3-3s^2+s+1}{s^5+s^4+3s^3+3s^2+s+1}$

(c) $H(s)=\dfrac{4s^5+15s^4+25s^3+19s^2-14s-20}{s^6+8s^5+32s^4+64s^3+50s^2+24s+30}$

3.53 Obtain the Bode plots for the following transfer functions using MATLAB:

(a) $H(s) = \dfrac{s(s+10)}{(s+20)(s+50)}$

(b) $H(s) = \dfrac{s+1}{(s+2)(s^2+22.5s+16)}$

4 Fourier Series

Imagination is more important than knowledge.

—Albert Einstein

HISTORICAL PROFILE

Jean Baptiste Joseph Fourier

Jean Baptiste Joseph Fourier (1768–1830), French mathematician, known also as an Egyptologist and administrator, was the first to present the series and transform that bear his name. He was the first to propose that the periodic waveforms could be represented by a sum of sinusoids. His major work, *The Analytic Theory of Heat in 1822*, stated the equations governing heat transfer in solids. Although Fourier's results were not enthusiastically received by the scientific world, his work provided the impetus for later work on trigonometric series and the theory of functions of a real variable.

Fourier was born in France in 1768, the son of a tailor. Joseph's mother died when he was nine years old and his father died the following year. Fourier first attended the local military school conducted by Benedictine monks. He showed proficiency in mathematics and later became a teacher in mathematics at the same school. In 1793, he became involved in politics and joined the local Revolutionary Committee. Due to his political involvement, he narrowly escaped death twice. He died in his bed in Paris on May 16, 1830.

4.1 INTRODUCTION

This chapter is concerned with a means of analyzing systems with periodic excitations. The notion of periodic functions was introduced in Chapter 1, where the importance of sinusoids (or cosinusoids) was emphasized. In this chapter, we introduce Fourier series, a premier tool for analyzing periodic signals

Although the Fourier series has a long history involving many individuals, the Fourier series is named after Jean Baptiste Joseph Fourier (1768–1830), a French mathematician and physicist. In 1822, Fourier was the first to suggest that any periodic function can be represented as a sum of sinusoids. Such a representation provides a powerful tool for analyzing signals and systems. System responses to periodic signals are of practical interest because these signals are common.

Fourier analysis (Fourier series in this chapter and Fourier transform in the next chapter) plays a major role in system analysis for a number of reasons. First, Fourier analysis leads to the frequency spectrum of a continuous-time signal. The frequency spectrum displays the various sinusoidal components that make up the signal. Engineers think of signals in terms of their frequency spectra and of systems in terms of their frequency response. Second, Fourier analysis converts time-domain signals into frequency-domain representation that lends new insight into the nature and the properties of the signals and systems. For many purposes, the frequency-domain representations are more convenient to analyze, synthesize, and process. Third, in the frequency domain linear systems are described by linear algebraic equations that can be easily solved, in contrast to the time-domain representation, where they are described by linear differential equations. It is for these reasons that Fourier analysis is used extensively today in science and engineering.

This chapter begins with the trigonometric Fourier series and how to determine the coefficients of the series. Then, we consider the complex exponential Fourier series. We will discuss some properties of Fourier series. We will cover circuit analysis, filtering, and spectrum analyzers as engineering applications of Fourier series. We will finally see how we can use MATLAB® to plot line spectra.

4.2 TRIGONOMETRIC FOURIER SERIES

The Fourier series can be represented in three ways, the sine–cosine, amplitude–phase, and complex exponential.

A periodic signal is one that repeats itself every T s. In other words, a continuous time signal $x(t)$ satisfies

$$x(t) = x(t + nT) \tag{4.1}$$

where

n is an integer

T is the fundamental period of $x(t)$

Any periodic function can be expressed as an infinite series consisting of sine or cosine functions. Thus, $x(t)$ can be expressed as

$$x(t) = a_0 + a_1 \cos \omega_0 t + b_1 \sin \omega_0 t + a_2 \cos 2\omega_0 t + b_2 \sin 2\omega_0 t$$
$$+ a_3 \cos 3\omega_0 t + b_3 \sin 3\omega_0 t + \cdots \tag{4.2}$$

This can be written as

$$\boxed{x(t) = \underbrace{a_0}_{\text{dc}} + \underbrace{\sum_{n=1}^{\infty} (a_n \cos n\omega_0 t + b_n \sin n\omega_0 t)}_{\text{ac}}} \tag{4.3}$$

where

$$\omega_0 = \frac{2\pi}{T} \tag{4.4}$$

ω_0 is known as the fundamental frequency (in rad/s). The coefficients as are called the Fourier cosine coefficients including a_0, the constant term, which is in reality the 0th cosine term, and bs are called the Fourier sine coefficients. The Fourier coefficient a_0 in the earlier equation is the constant or dc component of $x(t)$. The Fourier coefficients a_n and b_n (for $n \neq 0$) are the amplitudes of the sinusoids in the ac component of $x(t)$.

The **Fourier series** of a periodic signal $x(t)$ is a decomposition of $x(t)$ into a dc component and an ac component that consists of an infinite series of harmonic sinusoids.

A periodic function $x(t)$ can be expanded as a Fourier series only if it fulfills the **Dirichlet conditions** given as

1. $x(t)$ should be integrable over any period; that is,

$$\int_{t_0}^{t_0+T} |x(t)|\,dt < \infty$$

2. $x(t)$ has only a finite number of maxima and minima over any period
3. $x(t)$ has only a finite number of discontinuities over any period

The process of determining the Fourier coefficients a_0, a_n, and b_n is called **Fourier analysis**. The coefficients can be determined as follows:

$$\boxed{a_0 = \frac{1}{T}\int_0^T x(t)dt} \tag{4.5}$$

$$\boxed{a_n = \frac{2}{T}\int_0^T x(t)\cos n\omega_0 t\, dt} \tag{4.6}$$

$$\boxed{b_n = \frac{2}{T}\int_0^T x(t)\sin n\omega_0 t\, dt} \tag{4.7}$$

The fact that sine and cosine functions are orthogonal over a period T leads to the following trigonometric integrals:

$$\int_0^T \sin n\omega_0 t\, dt = 0 = \int_0^T \cos n\omega_0 t\, dt \tag{4.8}$$

$$\int_0^T \sin n\omega_0 t \cos n\omega_0 t\, dt = 0 \tag{4.9}$$

$$\int_0^T \sin n\omega_0 t \sin m\omega_0 t\, dt = 0 = \int_0^T \cos n\omega_0 t \cos m\omega_0 t\, dt, \quad m \neq n \tag{4.10}$$

$$\int_0^T \sin^2 n\omega_0 t\, dt = \frac{T}{2} = \int_0^T \cos^2 n\omega_0 t\, dt \tag{4.11}$$

These integrals will be used in finding the Fourier coefficients.

To find a_0, integrate both sides of Equation 4.3 over one period.

$$\int_0^T x(t)dt = \int_0^T a_0 dt + \sum_{n=1}^{\infty}\left[\int_0^T a_n \cos n\omega_0 t\, dt\right] + \sum_{n=1}^{\infty}\left[\int_0^T b_n \sin n\omega_0 t\, dt\right]$$
$$= a_0 T + 0 + 0 \tag{4.12}$$

Using the identities in Equation 4.8, the two integrals involving the ac terms become zero. Therefore,

$$\int_0^T x(t)dt = a_0 T$$

or

$$a_0 = \frac{1}{T}\int_0^T x(t)dt \tag{4.13}$$

To find a_n, multiply both sides of Equation 4.3 by cos $m\omega_0 t$ and integrate over one period.

$$\int_0^T x(t)\cos m\omega_0 t\ dt = \int_0^T a_0 \cos m\omega_0 t\ dt + \sum_{n=1}^{\infty}\left[\int_0^T \int \ a_n \cos n\omega_0 t \cos m\omega_0 t\ dt\right]$$

$$+\sum_{n=1}^{\infty}\left[\int_0^T b_n \sin n\omega_0 t \cos m\omega_0 t\ dt\right]$$

$$= 0 + \frac{T}{2}a_n + 0 \tag{4.14}$$

The integral containing a_0 and b_n becomes zero according to Equations 4.8 and 4.9, respectively. The integral containing a_n will be zero except when $m = n$, in which case it is $T/2$ according to Equation 4.11. Thus,

$$\int_0^T x(t)\cos m\omega_0 t\ dt = a_n \frac{T}{2}, \quad m = n$$

or

$$a_n = \frac{2}{T}\int_0^T x(t)\cos n\omega_0 t\ dt \tag{4.15}$$

To find b_n, multiply both sides of Equation 4.3 by sin $m\omega_0 t$ and integrate over one period.

$$\int_0^T x(t)\sin m\omega_0 t\ dt = \int_0^T a_0 \sin m\omega_0 t\ dt + \sum_{n=1}^{\infty}\left[\int_0^T a_n \cos n\omega_0 t \sin m\omega_0 t\ dt\right]$$

$$+\sum_{n=1}^{\infty}\left[\int_0^T b_n \sin n\omega_0 t \sin m\omega_0 t\ dt\right]$$

$$= 0 + 0 + \frac{T}{2}b_n \tag{4.16}$$

The integral containing a_0 and a_n becomes zero according to Equations 4.8 and 4.10, respectively. The integral containing b_n will be zero except when $m = n$, in which case it is $T/2$ according to Equation 4.11. Thus,

$$\int_0^T x(t)\sin n\omega_0 t\, dt = b_n \frac{T}{2}, \quad m = n$$

or

$$b_n = \frac{2}{T}\int_0^T x(t)\sin n\omega_0 t\, dt \tag{4.17}$$

Since $x(t)$ is periodic the integration of Equation 4.5 over one full period from t_o to $t_o + T$ or from $-T/2$ to $T/2$ instead of 0 to T gives the same results.

Equation 4.3 is the sine–cosine form of Fourier series. An alternative representation is the amplitude-phase (or polar) form:

$$\boxed{x(t) = a_0 + \sum_{n=1}^{\infty} A_n \cos(n\omega_0 t + \phi_n)} \tag{4.18}$$

This combines the sine–cosine pair at frequency $n\omega_0$ into a single sinusoid. We can apply the trigonometric identity

$$\cos(A + B) = \cos A \cos B - \sin A \sin B \tag{4.19}$$

to the ac terms in Equation 4.18 to get

$$\begin{aligned} a_0 + \sum_{n=1}^{\infty} A_n \cos(n\omega_0 t + \phi_n) &= a_0 + \sum_{n=1}^{\infty} (A_n \cos\phi_n)\cos n\omega_0 t \\ &\quad + \sum_{n=1}^{\infty} (-A_n \sin\phi_n)\sin n\omega_0 t \end{aligned} \tag{4.20}$$

When we equate the coefficients of the series expansions in Equations 4.3 and 4.20, we obtain

$$a_n = A_n \cos\phi_n, \quad b_n = -A_n \sin\phi_n \tag{4.21}$$

$$\boxed{A_n = \sqrt{a_n^2 + b_n^2}, \quad \phi_n = -\tan^{-1}\frac{b_n}{a_n}} \tag{4.22}$$

These may also be related in a compact complex form as

$$A_n < \phi_n = a_n - jb_n \tag{4.23}$$

The frequency components of the signal $x(t)$ can be displayed in terms of the amplitude and phase spectra. The frequency spectrum of $x(t)$ is the combination of both the amplitude and phase spectra.

Example 4.1

Given the periodic train of pulses shown in Figure 4.1, obtain Fourier coefficients a_o, a_n, and b_n. Plot the amplitude and phase spectra.

Solution

The signal can be expressed as

$$x(t) = \begin{cases} 10, & 0 < t < 1 \\ 0, & 1 < t < 5 \end{cases}$$

With $T = 5$, $\omega_0 = 2\pi/5$ Using Equation 4.5,

$$a_0 = \frac{1}{T}\int_0^T x(t)dt = \frac{1}{5}\left[\int_0^1 10dt + \int_1^5 0dt\right] = \frac{1}{5}10 = 2$$

Next, we use Equation 4.6,

$$a_n = \frac{2}{T}\int_0^T x(t)\cos n\omega_o t\, dt = \frac{2}{5}\int_0^1 10\cos n\omega_o t\, dt$$

$$= 4\left.\frac{\sin n\omega_o t}{n\omega_o}\right|_0^1 = 4\frac{\sin n\omega_o}{n\omega_o}$$

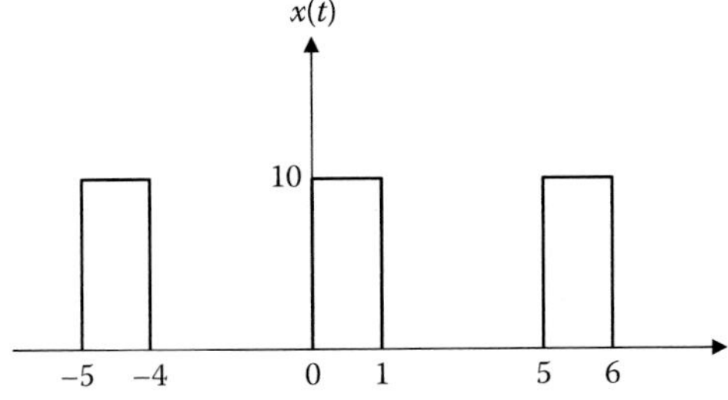

FIGURE 4.1 For Example 4.1.

Similarly, using Equation 4.7 gives

$$b_n = \frac{2}{T}\int_0^T x(t)\sin n\omega_o t\,dt = \frac{2}{5}\int_0^1 10\sin n\omega_o t\,dt$$

$$= -4\frac{\cos n\omega_o t}{n\omega_o}\Bigg|_0^1 = 4\left[\frac{1-\cos n\omega_o}{n\omega_o}\right]$$

The Fourier series expansion of the signal is

$$x(t) = 2 + 4\sum_{n=1}^{\infty}\left(\frac{\sin n\omega_o}{n\omega_o}\cos n\omega_o t + \left[\frac{1-\cos n\omega_o}{n\omega_o}\right]\sin n\omega_o t\right)$$

Finally, we obtain the amplitude and phase spectra for the signal in Figure 4.1 by using

$$A_n = \sqrt{a_n^2 + b_n^2}, \quad \phi_n = -\tan^{-1}\frac{b_n}{a_n}$$

Applying these formulas for values of $n = 1, 2, \ldots, 8$ produces the result in Table 4.1 From the table, the amplitude and phase spectra are plotted as shown in Figure 4.2.

Practice Problem 4.1 Obtain the Fourier series expansion for the waveform shown in Figure 4.3.

Answer: $y(t) = -0.5 + \sum_{\substack{n=1 \\ n=\text{odd}}}^{\infty} \frac{6}{n\pi}\sin n\pi t$

TABLE 4.1
For Example 4.1

n	a_n	b_n	A_n	ϕ_n (Degrees)
1	3.0273	3.0164	4.2735	−44.89
2	0.9355	5.2876	5.3697	−79.97
3	−0.6237	4.8584	4.8983	82.68
4	−0.7568	3.7541	3.8296	78.60
5	0	3.3634	3.3634	90
6	0.5046	3.8361	3.8691	−82.51
7	0.2673	4.3679	4.3861	−86.50
8	−0.2339	4.3219	4.3282	86.90

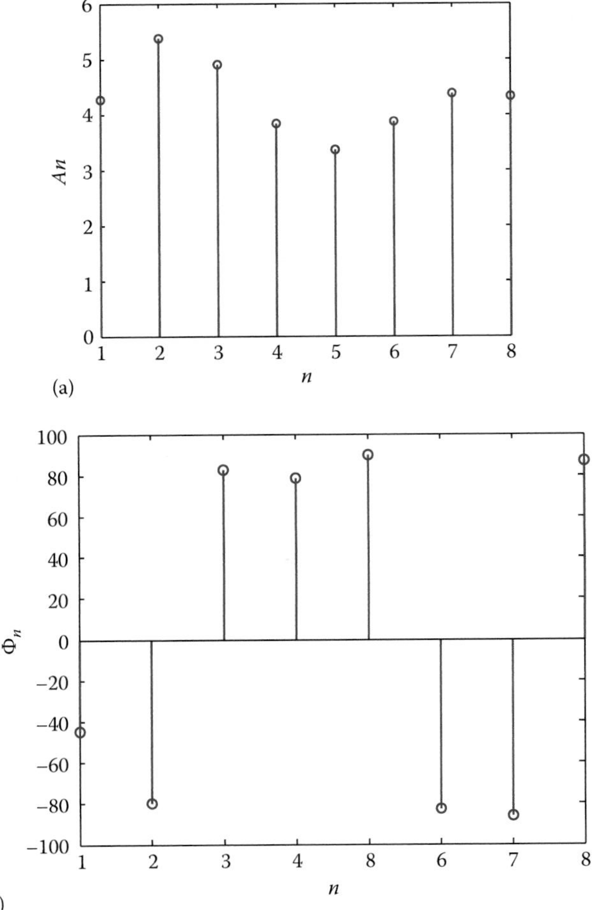

FIGURE 4.2 For Example 4.1: (a) amplitude spectrum and (b) phase spectrum.

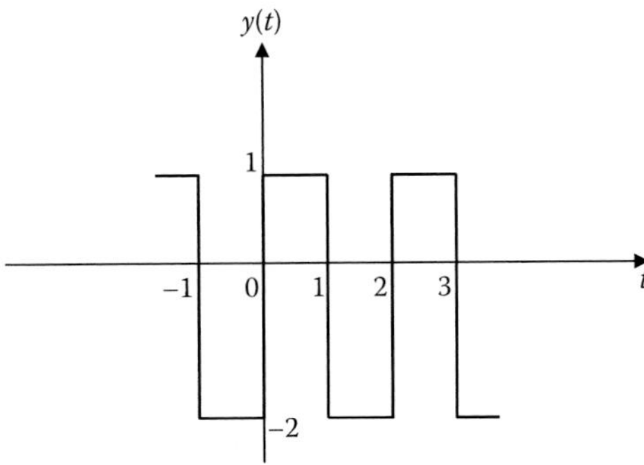

FIGURE 4.3 For Practice Problem 4.1.

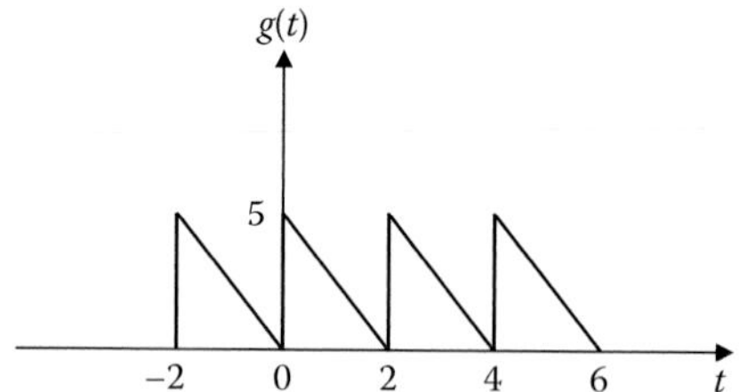

FIGURE 4.4 For Example 4.2.

Example 4.2

Find the Fourier series expansion of the backward sawtooth waveform of Figure 4.4.

Solution

$$g(t) = 5 - 2.5t,\ 0 < t < 2,$$

$$T = 2,\ \omega_0 = 2\pi/T = \pi$$

$$a_0 = \frac{1}{T}\int_0^T g(t)dt = \frac{1}{2}\int_0^2 (5 - 2.5t)dt = 0.5[5t - 2.5t^2/2]\Big|_0^2 = 2.5$$

$$a_n = \frac{2}{T}\int_0^T g(t)\cos n\omega_o t\ dt = \frac{2}{2}\int_0^2 (5 - 2.5t)\cos n\pi t\ dt$$

Using the formulas from Appendix A.5,

$$\int \cos at = \frac{1}{a}\sin at$$

$$\int t\cos at = \frac{1}{a^2}\cos at + \frac{t}{a}\sin at$$

$$a_n = \frac{5}{n\pi}\sin n\pi t\Big|_0^2 + \left[\frac{2.5}{n^2\pi^2}\cos n\pi t + \frac{-2.5t}{n\pi}\sin n\pi t\right]_0^2$$

$$= 0 + \frac{2.5}{n^2\pi^2}(\cos 2n\pi - 1) - 0 = 0$$

Similarly,

$$b_n = \frac{2}{T}\int_0^T g(t)\sin n\omega_o t\ dt = \frac{2}{2}\int_0^2 (5 - 2.5t)\sin n\pi t\ dt$$

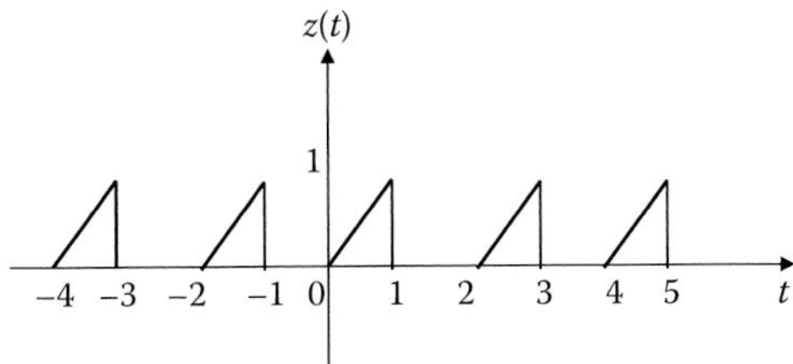

FIGURE 4.5 For Practice Problem 4.2.

We use the formulas

$$\int \sin at = -\frac{1}{a}\cos at$$

$$\int t\sin at = \frac{1}{a^2}\sin at - \frac{t}{a}\cos at$$

$$b_n = -\frac{5}{n\pi}\cos n\pi t\Big|_0^2 + \left[\frac{-2.5}{n^2\pi^2}\sin n\pi t + \frac{2.5t}{n\pi}\cos n\pi t\right]\Bigg|_0^2$$

$$= 0 + 0 + \frac{5}{n\pi}(\cos 2n\pi - 0) = \frac{5}{n\pi}$$

Hence,

$$g(t) = 2.5 + \frac{5}{\pi}\sum_{n=1}^{\infty}\frac{1}{n}\sin(n\pi t)$$

Practice Problem 4.2 Obtain the trigonometric Fourier series for the signal in Figure 4.5.

Answer: $z(t) = 0.5 - \sum_{n=1}^{\infty}\left[\frac{\{1-(-1)^n\}}{n\pi}\cos n\pi t + \frac{(-1)^n}{n\pi}\sin n\pi t\right]$

4.3 EXPONENTIAL FOURIER SERIES

The sine and cosine functions can be represented in terms of complex exponentials. It turns out that we can use complex exponentials to represent Fourier series. In many respects, this makes for a simpler representation.

The Fourier series representation of $x(t)$ can be expressed in complex exponential form as

$$\boxed{x(t) = \sum_{n=-\infty}^{\infty} c_n e^{jn\omega_0 t}} \tag{4.24}$$

where

$\omega_0 = 2\pi/T$ is the fundamental frequency
the coefficients c_n are given by

$$\boxed{c_n = \frac{1}{T}\int_0^T x(t)e^{-jn\omega_0 t}dt} \tag{4.25}$$

The **exponential Fourier series** of a periodic signal $x(t)$ is a representation that is the sum of the complex exponentials at positive and negative harmonic frequencies.

By substituting the following Euler's identities in Equation 4.3

$$\cos n\omega_0 t = \frac{1}{2}\left[e^{jn\omega_0 t} + e^{-jn\omega_0 t}\right] \tag{4.26}$$

$$\sin n\omega_0 t = \frac{1}{2j}\left[e^{jn\omega_0 t} - e^{-jn\omega_0 t}\right] = -\frac{j}{2}\left[e^{jn\omega_0 t} - e^{-jn\omega_0 t}\right] \tag{4.27}$$

we get

$$x(t) = a_0 + \frac{1}{2}\sum_{n=1}^{\infty} a_n\left(e^{jn\omega_0 t} + e^{-jn\omega_0 t}\right) - jb_n\left(e^{jn\omega_0 t} - e^{-jn\omega_0 t}\right)$$

$$= a_0 + \frac{1}{2}\sum_{n=1}^{\infty}\left[(a_n - jb_n)e^{jn\omega_0 t} + (a_n + jb_n)e^{-jn\omega_0 t}\right] \tag{4.28}$$

So the new coefficient c_n will be

$$c_0 = a_0,$$

$$c_n = \frac{(a_n - jb_n)}{2} = |c_n| < \theta_n, \tag{4.29}$$

$$c_{-n} = c_n^* = \frac{(a_n + jb_n)}{2}$$

so that $x(t)$ becomes

$$x(t) = c_0 + \sum_{n=1}^{\infty}\left[c_n e^{jn\omega_0 t} + c_{-n}e^{-jn\omega_0 t}\right] \tag{4.30}$$

The concise form is

$$x(t) = \sum_{n=-\infty}^{\infty} c_n e^{jn\omega_0 t} = \sum_{n=-\infty}^{\infty} |c_n| e^{j(n\omega_0 t + \theta_n)} \tag{4.31}$$

The complex Fourier coefficients c_n can be readily obtained as follows using Equations 4.6 and 4.7 for a_n, b_n.

$$c_0 = a_0 = \frac{1}{T}\int_0^T x(t)dt \tag{4.32}$$

For $n = 1, 2, 3, \ldots$, we have

$$c_n = \frac{(a_n - jb_n)}{2} = \frac{1}{T}\int_0^T x(t)(\cos n\omega_0 t - j\sin n\omega_0 t)dt = \frac{1}{T}\int_0^T x(t)e^{-jn\omega_0 t}dt \tag{4.33}$$

Similarly,

$$c_{-n} = c_n^* = \frac{(a_n + jb_n)}{2} = \frac{1}{T}\int_0^T x(t)e^{jn\omega_0 t}dt$$

The last expression is equivalent to stating that for $n = -1, -2, -3, \ldots$

$$c_n = \frac{1}{T}\int_0^T x(t)e^{-jn\omega_0 t}dt \tag{4.34}$$

The three Equations 4.32, 4.33, 4.34 can be combined into one expression

$$c_n = \frac{1}{T}\int_0^T x(t)e^{-jn\omega_0 t}dt \quad \text{for } n = 0, \pm 1, \pm 2, \pm 3, \ldots \tag{4.35}$$

Therefore, the complex Fourier series is

$$x(t) = \sum_{n=-\infty}^{\infty} c_n e^{jn\omega_0 t} \tag{4.36}$$

Equation 4.36 is also called complex frequency spectrum of $x(t)$, which is formed by both the complex amplitude spectrum (even symmetry) and the phase spectrum (odd symmetry).

The three forms of Fourier series are now summarized as follows:

Sine–Cosine Form

$$x(t) = a_0 + \sum_{n=1}^{\infty} (a_n \cos n\omega_0 t + b_n \sin n\omega_0 t) \tag{4.37}$$

Amplitude–Phase Form

$$x(t) = a_0 + \sum_{n=1}^{\infty} A_n \cos(n\omega_0 t + \phi_n) \tag{4.38}$$

Complex Exponential Form

$$x(t) = \sum_{n=-\infty}^{\infty} c_n e^{jn\omega_0 t} \tag{4.39}$$

The coefficients of the three forms of Fourier series (sine–cosine form, amplitude–phase form, and exponential form) are related by

$$A_n \angle \phi_n = a_n - jb_n = 2c_n \tag{4.40}$$

or

$$c_n = |c_n| \angle \theta_n = \frac{\sqrt{a_n^2 + b_n^2}}{2} \angle - \tan^{-1} \frac{b_n}{a_n} = A_n \angle \phi_n \tag{4.41}$$

if only $a_n > 0$. Note that the phase θ_n of c_n is equal to ϕ_n.

Example 4.3

Find the exponential Fourier series for $x(t) = t$, $-1 < t < 1$, with $f(t + 2n) = f(t)$ and n is an integer. Plot the amplitude and phase spectra.

Solution

Since $T = 2$, $\omega_0 = 2\pi/T = \pi$. Hence,

$$c_n = \frac{1}{T} \int_{-T/2}^{T/2} f(t) e^{-j\omega_0 nt} dt = \frac{1}{2} \int_{-1}^{1} t e^{-jn\pi t} dt$$

Using integration by parts,

$$u = t \to du = dt$$

$$dv = \frac{1}{2}e^{-jn\pi t}dt \to v = -\frac{1}{2jn\pi}e^{-jn\pi t}$$

$$c_n = -\frac{t}{2jn\pi}e^{-jn\pi t}\Big|_{-1}^{1} + \frac{1}{2jn\pi}\int_{-1}^{1} e^{-jn\pi t}dt$$

$$= \frac{j}{2n\pi}\left[e^{-jn\pi} + e^{jn\pi}\right] + \frac{1}{2n^2\pi^2(-j)^2}e^{-jn\pi t}\Big|_{-1}^{1}$$

$$= \frac{j}{n\pi}\cos n\pi - \frac{1}{2n^2\pi^2}\left(e^{-jn\pi} - e^{jn\pi}\right)$$

$$c_n = \frac{j(-1)^n}{n\pi} + \frac{2j}{2n^2\pi^2}\sin(n\pi) = \frac{j(-1)^n}{n\pi}, \quad n \neq 0$$

We treat $n = 0$ as a special case.

$$c_0 = \frac{1}{2}\int_{-1}^{1} te^{-0}dt = \frac{1}{2}\frac{t^2}{2}\Big|_{-1}^{1} = \frac{1}{4}(1-1) = 0$$

Thus,

$$x(t) = \sum_{n=-\infty}^{\infty} c_n e^{jn\omega_0 t} = \sum_{\substack{n=-\infty \\ n\neq 0}}^{\infty} (-1)^n \frac{j}{n\pi} e^{jn\pi t}$$

For different values of n, we calculate $|c_n|$ and θ_n as shown in Table 4.2. By plotting $|c_n|$ and θ_n for different n, we obtain the amplitude spectrum and the phase spectrum shown in Figure 4.6. Notice that the amplitude spectrum is an even function, while the phase spectrum is odd, as expected.

TABLE 4.2
For Example 4.3

N	$\lvert c_n \rvert$	ϕ_n (Degrees)
–3	0.1061	90
–2	0.1592	–90
–1	0.3183	90
0	0	0
1	0.3183	–90
2	0.1592	90
3	0.1061	–90

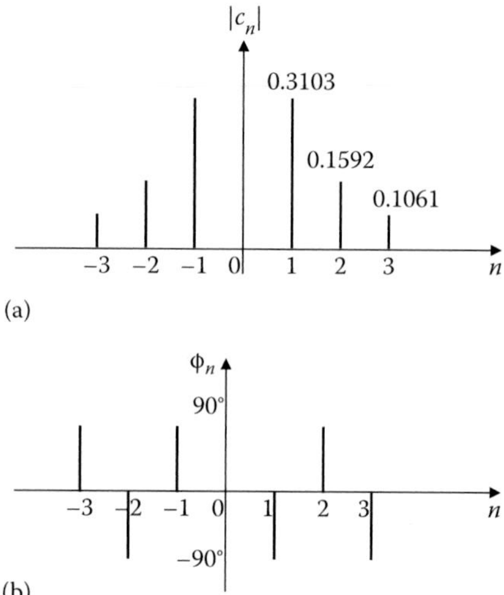

FIGURE 4.6 For Example 4.3.

Practice Problem 4.3 For the periodic function $y(t) = t^2$, $0 < t < T$, obtain the complex Fourier series for the case $T = 2$. Determine the amplitude and phase spectra.

Answer: $y(t) = 1.333 + \sum_{\substack{n=-\infty \\ n\neq 0}}^{\infty} \frac{2}{n^2\pi^2}(1 + jn\pi)e^{jn\pi t}$

See Figure 4.7 for the spectra.

Example 4.4

Find the complex Fourier series of the signal in Figure 4.8.

Solution

$$T = 2, \quad \omega_o = 2\pi/T = \pi$$

$$f(t) = \begin{cases} 5(1-t), & 0 < t < 1 \\ 0, & 1 < t < 2 \end{cases}$$

$$c_n = \frac{1}{T}\int_0^T f(t)e^{-jn\omega_o t}dt = \frac{1}{2}\int_0^1 5(1-t)e^{-jn\pi t}dt$$

$$= \frac{5}{2}\int_0^1 e^{-jn\pi t}dt - \frac{5}{2}\int_0^1 te^{-jn\pi t}dt = \frac{5}{2}\frac{e^{-jn\pi t}}{-jn\pi}\bigg|_0^1 - \frac{5}{2}\frac{e^{-jn\pi t}}{(-jn\pi)^2}(-jn\pi t - 1)\bigg|_0^1$$

$$= \frac{5}{2}\frac{\left[e^{-jn\pi} - 1\right]}{-jn\pi} - \frac{5}{2}\frac{e^{-jn\pi}}{-n^2\pi^2}(-jn\pi - 1) + \frac{5}{2}\frac{(-1)}{-n^2\pi^2}$$

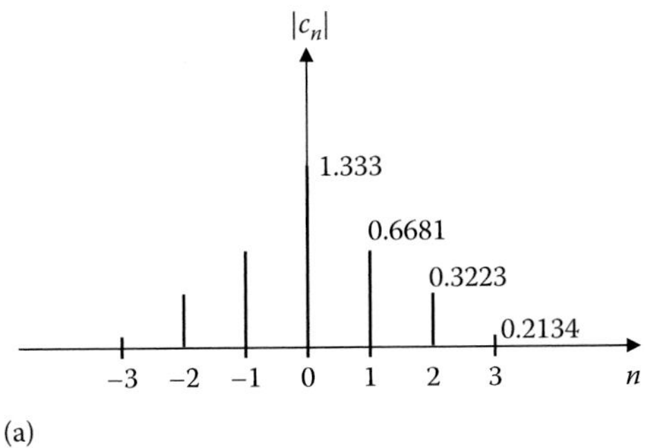

(a)

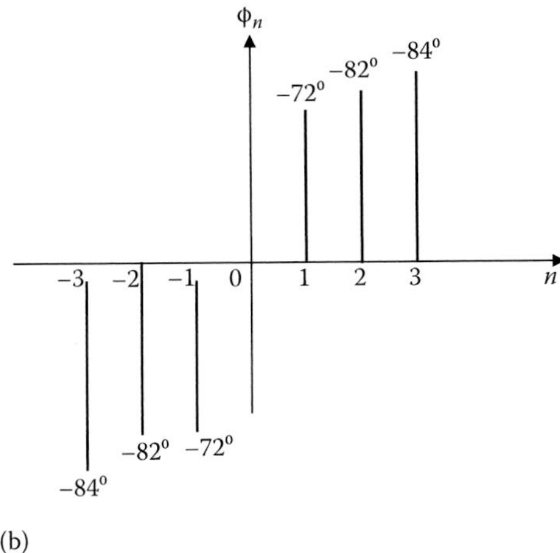

(b)

FIGURE 4.7 For Practice Problem 4.3.

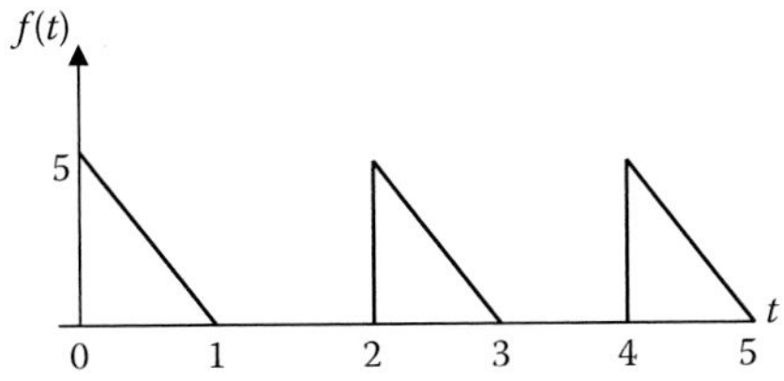

FIGURE 4.8 For Example 4.4.

But $e^{-jn\pi} = \cos\pi n - j\sin n\pi = \cos n\pi + 0 = (-1)^n$

$$c_n = \frac{2.5\left[1-(-1)^n\right]}{jn\pi} - \frac{2.5(-1)^n[1+jn\pi]}{n^2\pi^2} + \frac{2.5}{n^2\pi^2}$$

$$f(t) = \sum_{n=-\infty}^{\infty} c_n e^{jn\pi t}$$

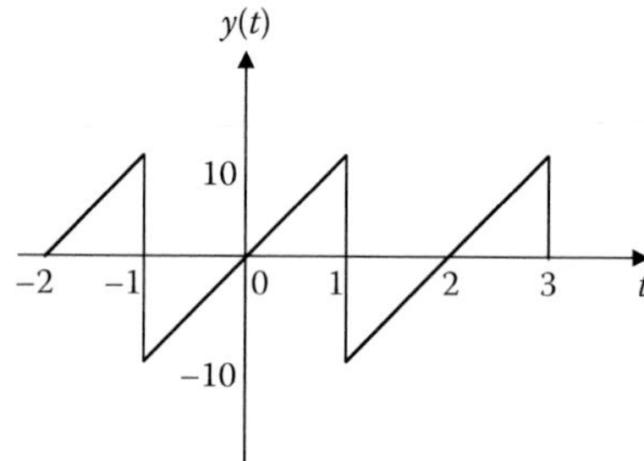

FIGURE 4.9 For Practice Problem 4.4.

Practice Problem 4.4 Obtain the exponential Fourier series expansion of $y(t)$ shown in Figure 4.9.

Answer: $y(t) = \sum_{\substack{n=-\infty \\ n\neq 0}}^{\infty} (-1)^n \frac{j10}{n\pi} e^{jn\pi t}$

4.4 PROPERTIES OF FOURIER SERIES

Some properties of Fourier series are discussed in this section. These properties provide us a better understanding of the Fourier series.

4.4.1 Linearity

The Fourier series expansion for periodic signals $x(t)$ and $y(t)$ with the same period is given by

$$x(t) = \sum_{n=-\infty}^{\infty} \alpha_n \exp[jn\omega_0 t] \tag{4.42}$$

$$y(t) = \sum_{n=-\infty}^{\infty} \beta_n \exp[jn\omega_0 t] \tag{4.43}$$

If $z(t)$ is a linear combination of $x(t)$ and $y(t)$, then

$$z(t) = k_1 x(t) + k_2 y(t) \tag{4.44}$$

where k_1 and k_2 are arbitrary constants. Then we can write

$$z(t) = \sum_{n=-\infty}^{\infty} (k_1\alpha_n + k_2\beta_n)\exp[jn\omega_0 t] = \sum_{n=-\infty}^{\infty} \gamma_n \exp[jn\omega_0 t] \tag{4.45}$$

This implies that the Fourier coefficients are

$$\gamma_n = k_1\alpha_n + k_2\beta_n \tag{4.46}$$

The linearity property is easily extended to a linear combination of an arbitrary number of signals with period T.

4.4.2 Time Shifting

When a time shift is applied to a periodic signal $x(t)$, the period T of the signal is preserved. The Fourier series coefficient γ_n of the resulting signal $x(t - \tau)$ may be expressed as

$$\gamma_n = \frac{1}{T}\int_0^T x(t-\tau)\exp[-jn\omega_0 t]dt = \exp[-j\omega_0\tau]\frac{1}{T}\int_0^T x(\lambda)\exp[-jn\omega_0\lambda]d\lambda$$

$$= \alpha_n \exp[-jn\omega_0\tau] \tag{4.47}$$

where α_n is the nth Fourier series coefficient of $x(t)$. Here, we have changed variables by letting $\lambda = t-\tau$. That is, if

$$x(t) \xrightarrow{FS} \alpha_n$$

then

$$x(t-\tau) \xrightarrow{FS} \alpha_n e^{-jn\omega_0\tau} \tag{4.48}$$

One consequence of this property is that, when a periodic signal is shifted in time, the magnitude of the Fourier coefficient remains unaltered, that is, $|\gamma_n| = |\alpha_n|$.

4.4.3 Time Reversal

Time reversal property, when applied to a continuous-time signal, results in a time reversal of the corresponding sequence of Fourier series coefficients. To determine the Fourier series coefficients of $y(t) = x(-t)$, consider

$$x(-t) = \sum_{n=-\infty}^{\infty} \alpha_n \exp\left[-\frac{jn\pi t}{T}\right] \tag{4.49}$$

By substituting $n = -m$, we get

$$y(t) = x(-t) = \sum_{n=-\infty}^{\infty} \alpha_{-m} \exp\left[\frac{jm\pi t}{T}\right] \tag{4.50}$$

Thus, we conclude that

$$\begin{aligned} x(t) &\xrightarrow{FS} \alpha_n \\ x(-t) &\xrightarrow{FS} \alpha_{-n} \end{aligned} \tag{4.51}$$

An interesting consequence of time reversal is that when $x(t)$ is even, that is, if $x(-t) = x(t)$, then its Fourier series coefficients are also even, that is, $\alpha_{-n} = \alpha_n$. Similarly, when $x(t)$ is odd, that is, if $x(-t) = -x(t)$, then its Fourier series coefficients are also odd, that is, $\alpha_{-n} = -\alpha_n$.

4.4.4 Time Scaling

Time scaling operation applies directly to each of the harmonic component of $x(t)$. If $x(t)$ has the Fourier series representation $x(t) = \sum_{n=-\infty}^{\infty} \alpha_n \exp[jn\omega_0 t]$, then the Fourier series representation of the time-scaled signal $x(at)$ is

$$x(at) = \sum_{n=-\infty}^{\infty} \alpha_n \exp[jn\omega_0 at] \tag{4.52}$$

That is,

$$x(at) \xrightarrow{FS} \alpha_n \tag{4.53}$$

The Fourier series coefficients have not changed, the Fourier series representation has changed because of the change in the fundamental frequency $a\omega_0$.

4.4.5 Even and Odd Symmetries

Three types of symmetries have been discussed—even, odd, and half-wave odd symmetry.

A function $x(t)$ is even if $x(t) = x(-t)$, and odd if $x(t) = -x(-t)$. Any function $x(t)$ is a sum of an even function and an odd function, and this can be done in only one way.

We know that any periodic function $x(t)$, with period 2π, has a Fourier expansion of the form

$$x(t) = a_0 + \sum_{n=1}^{\infty} a_n \cos(n\omega_0 t) + b_n \sin(n\omega_0 t) \tag{4.54}$$

If $x(t)$ is even, then all the b_ns vanish and the Fourier series is simply

$$x(t) = a_0 + \sum_{n=1}^{\infty} a_n \cos(n\omega_0 t) \tag{4.55}$$

If $x(t)$ is odd, then all the a_0, a_ns vanish and the Fourier series is simply

$$x(t) = \sum_{n=1}^{\infty} b_n \sin(n\omega_0 t) \tag{4.56}$$

A periodic function possesses half-wave symmetry if it satisfies the constraints

$$x(t) = -x\left(t - \frac{T}{2}\right) \tag{4.57}$$

If a function is shifted one half period and inverted, it looks identical to the original, then it is called a half-wave symmetry. A half-wave odd symmetry can be even, odd or neither.

$$\boxed{\begin{aligned} &a_0 = 0, \quad a_n = 0 \quad \text{and} \quad b_n = 0 \quad \text{for } n \text{ even} \\ &a_n = \frac{4}{T}\int_0^{\frac{T}{2}} x(t)\cos n\omega_0 t \, dt, \quad \text{for } n \text{ odd} \\ &b_n = \frac{4}{T}\int_0^{\frac{T}{2}} x(t)\sin n\omega_0 t \, dt, \quad \text{for } n \text{ odd} \end{aligned}} \tag{4.58}$$

This shows that the Fourier series of a half-wave symmetric function contains only odd harmonics (Figure 4.10).

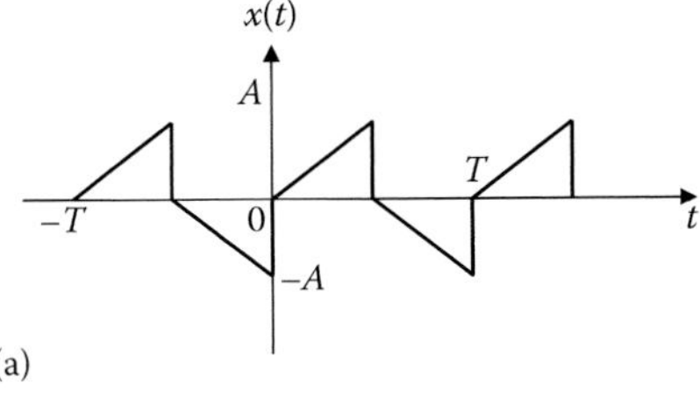

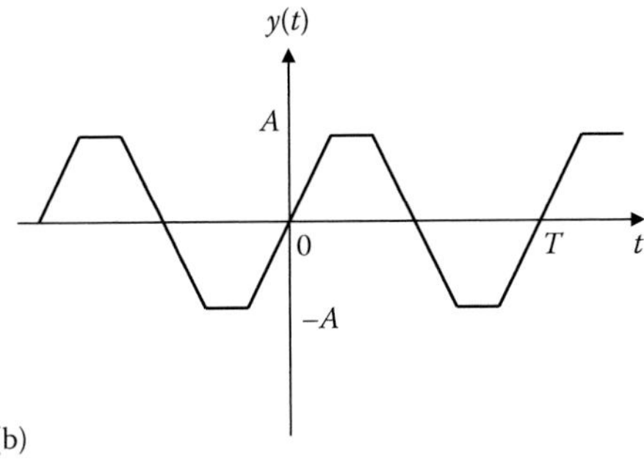

FIGURE 4.10 Typical examples of half-wave odd symmetric functions.

4.4.6 Parseval's Theorem

Parseval's theorem states that if $x(t)$ is a periodic function with period T, then the average power P of the signal is defined by

$$P = \frac{1}{T}\int_0^T x^2(t)dt \tag{4.59}$$

Again let $x(t)$ be an arbitrary periodic signal with period T and consider the Fourier series of $x(t)$ given by Equation 4.24. By Parseval's theorem, the average power P of the signal $x(t)$ is given by

$$P = \sum_{n=-\infty}^{\infty} |c_n|^2 \tag{4.60}$$

Parseval's theorem states that the average power of the signal $x(t)$ over one period equals the sum of the squared magnitudes of all the complex Fourier coefficients.

To prove Parseval's theorem, assume $x(t)$ has a complex Fourier series of the usual form:

$$x(t) = \sum_{n=-\infty}^{\infty} c_n e^{jn\omega_0 t}, \quad \omega_0 = \left(\frac{2\pi}{T}\right)$$

$$c_n = \frac{1}{T}\int_0^T x(t)e^{-jn\omega_0 t}dt$$

$$x^2(t) = x(t)x(t) = x(t)\sum c_n e^{jn\omega_0 t} = \sum c_n x(t)e^{jn\omega_0 t}dt$$

$$\begin{aligned} P &= \frac{1}{T}\int_0^T x^2(t)dt = \frac{1}{T}\int_0^T \sum c_n x(t)e^{jn\omega_0 t}dt \\ &= \frac{1}{T}\sum c_n \int_0^T x(t)e^{jn\omega_0 t}dt \\ &= \sum c_n \cdot c_n^* \\ P &= \sum_{n=-\infty}^{\infty} |c_n|^2 \end{aligned} \tag{4.61}$$

Parseval's theorem can also be written in terms of the Fourier coefficients a_n, b_n of the trigonometric Fourier series. The engineering interpretation of this theorem is

TABLE 4.3
Power Associated with Fourier Series Coefficients

Term	Sine–Cosine Form	Amplitude–Phase Form	Exponential Form
DC term	a_o^2	a_o^2	$\lvert c_0\rvert^2$
Others	$\frac{1}{2}\left(a_n^2+b_n^2\right)$	$\frac{1}{2}A_n$	$2\lvert c_n\rvert^2 = \lvert c_n\rvert^2 + \lvert c_{-n}\rvert^2$

as follows. Suppose $x(t)$ denotes an electrical signal (current or voltage), then from elementary circuit theory $x^2(t)$ is the instantaneous power (in a 1 ohm resistor) so that

$$P = \frac{1}{T}\int_0^T x^2(t)dt \tag{4.62}$$

is the energy dissipated in the resistor during one period. In terms of the trigonometric Fourier series, Equation 4.61 becomes

$$P = a_o^2 + \frac{1}{2}\sum_{n=1}^{\infty}\left(a_n^2 + b_n^2\right) \tag{4.63}$$

In terms of the amplitude–phase Fourier series, we have

$$P = a_o^2 + \frac{1}{2}\sum_{n=1}^{\infty} A_n^2 \tag{4.64}$$

Table 4.3 summarizes the power associated with Fourier series coefficients.

Example 4.5

The waveform in Figure 4.11a has the following Fourier series:

$$x(t) = \frac{A}{2} - \frac{4A}{\pi^2}\sum_{n=1}^{\infty}\frac{1}{(2n-1)^2}\cos(2n-1)\omega_0 t$$

Obtain the Fourier series of $y(t)$ in Figure 4.11b.

Solution

If we take the derivative of $x(t)$ in Figure 4.11a, we obtain $y(t)$ in Figure 4.11b.

$$\frac{dx}{dt} = \frac{4A}{\pi^2}\sum_{n=1}^{\infty}\frac{1}{(2n-1)}\omega_0\sin(2n-1)\omega_0 t,\quad \omega_0 = \frac{2\pi}{T}$$

$$= \frac{8A}{\pi T}\sum_{n=1}^{\infty}\frac{1}{(2n-1)}\sin(2n-1)\omega_0 t \tag{4.5.1}$$

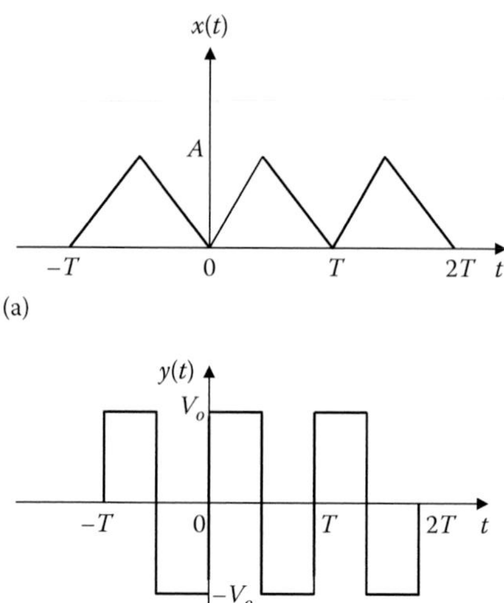

FIGURE 4.11 For Example 4.5.

But the slope of $x(t)$ is the amplitude of $y(t)$, that is,

$$\text{Slope } = \frac{A}{T/2} = \frac{2A}{T} = V_o \longrightarrow A = \frac{V_o T}{2} \tag{4.5.2}$$

Substituting for A into Equation 4.5.1 gives

$$y(t) = \frac{4V_0}{\pi} \sum_{n=1}^{\infty} \frac{1}{(2n-1)} \sin(2n-1)\omega_0 t$$

Practice Problem 4.5 Obtain the Fourier series of the function $x(t)$ in Figure 4.12a from the Fourier series of $y(t)$ in Figure 4.12b, which is given as

$$y(t) = \frac{4A}{\pi} \sum_{n=1}^{\infty} \frac{1}{2n-1} \sin(2n-1)\omega_0 t$$

Answer: $x(t) = \dfrac{V_o}{2} + \dfrac{2V_o}{\pi} \displaystyle\sum_{n=1}^{\infty} \frac{1}{2n-1} \sin(2n-1)t$

Example 4.6

For the waveform shown in Figure 4.13,

(a) Specify the type of symmetry it has.
(b) Calculate a_3 and b_3.

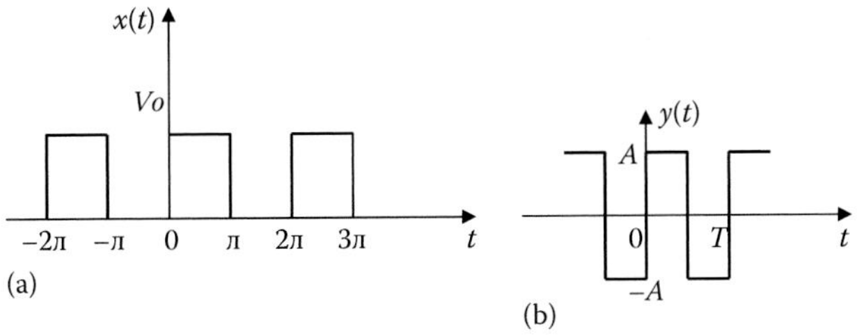

FIGURE 4.12 For Practice Problem 4.5.

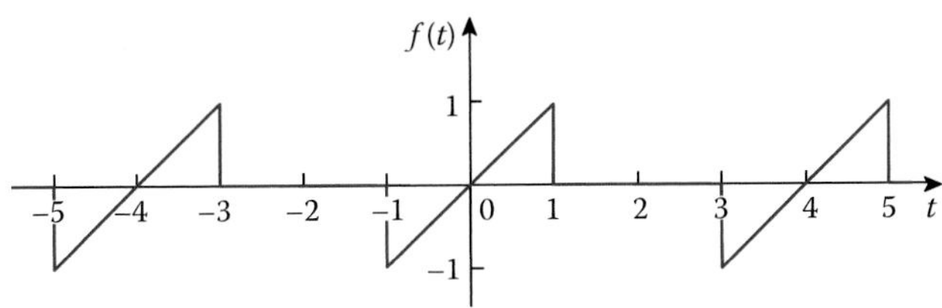

FIGURE 4.13 For Example 4.6.

Solution

(a) This is an odd function so that $a_0 = a_n = 0$.

(b) $T = 4$, $\omega_0 = 2\pi/T = \pi/2$

$$f(t) = \begin{cases} t, & 0 < t < 1 \\ 0, & 1 < t < 2 \end{cases}$$

$$b_n = \frac{4}{4}\int_0^1 t\sin\frac{n\pi t}{2}\,dt = \left[\frac{4}{n^2\pi^2}\sin\frac{n\pi t}{2} - \frac{2t}{n\pi}\cos\frac{n\pi t}{2}\right]_0^1$$

$$= \frac{4}{n^2\pi^2}\sin\frac{n\pi}{2} - \frac{2}{n\pi}\cos\frac{n\pi}{2} - 0$$

$$b_n = \begin{cases} \dfrac{4(-1)^{(n-1)/2}}{n^2\pi^2}, & n = \text{odd} \\ \dfrac{-2(-1)^{n/2}}{n\pi}, & n = \text{even} \end{cases}$$

$$a_3 = 0,\ b_3 = 4(-1)/(9\pi^2) = -0.04503$$

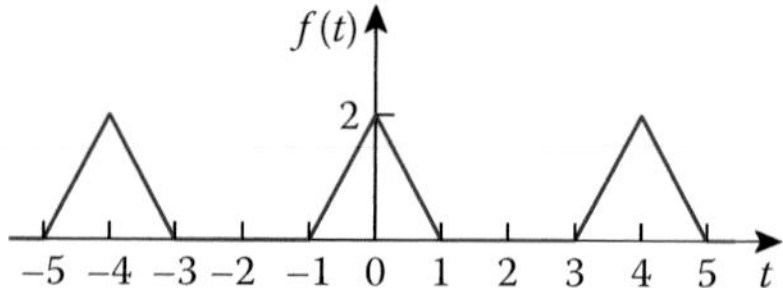

FIGURE 4.14 For Practice Problem 4.6.

Practice Problem 4.6 Determine the trigonometric Fourier series expansion of the signal in Figure 4.14. What type of symmetry does it have?

Answer: $f(t)=\frac{1}{2}+\frac{8}{\pi^2}\sum_{k=1}^{\infty}\frac{1}{n^2}\left[1-\cos\left(\frac{n\pi}{2}\right)\right]\cos\left(\frac{n\pi t}{2}\right),$ even.

Example 4.7

Calculate the average power in the signal

$$x(t)=4+2\cos(2t-30°)+\cos(4t-15°).$$

Solution

Using Equation 4.64, we get

$$P=4^2+\frac{1}{2}(2^2+1^2)=18.5$$

Practice Example 4.7 Determine the average power in the signal

$$x(t)=1+4\cos(60\pi t+10°)+2\cos(120\pi t-30°)$$

Answer: 11.

4.5 TRUNCATED COMPLEX FOURIER SERIES

The Fourier series representation of $x(t)$ can be expressed in complex exponential form as

$$x(t)=\sum_{n=-\infty}^{\infty}c_n e^{jn\omega_0 t} \tag{4.65}$$

It is an infinite summation and to truncate $x(t)$ to finite partial sum of $x_N(t)$ at $n = N$, we can write Equation 4.65 as follows:

$$x_N(t)=\sum_{n=-N}^{N}c_n e^{jn\omega_0 t} \tag{4.66}$$

We expect the signal approximation $x_N(t)$ to converge to $x(t)$ as $N \to \infty$. The behavior of the partial sum in the vicinity of discontinuity exhibits ripples and that the peak amplitude of these ripples does not seem to decrease with increasing N. Gibbs showed that for the discontinuity of unit height, the partial sum exhibits an overshoot of 9% of the height of the discontinuity, no matter how large N becomes. Thus, as N increases, the ripples in the partial sum become compressed toward the discontinuity, but for any finite value of N, the peak amplitude of the ripples remains constant. This is known as **Gibbs phenomenon**. The Gibbs phenomenon (also known ringing artifacts) is named after the American physicist J. Willard Gibbs.

Gibbs discovered that the truncated Fourier series approximation $x_N(t)$ of a signal $x(t)$ with discontinuous will in general exhibit high-frequency ripples and overshoot $x(t)$ near the discontinuities. If such an approximation is used in practice, a large enough value of N should be chosen so as to guarantee that the total energy in this ripple is insignificant. These are shown in the Figure 4.15 for various values of N. We know that the energy in the approximation error vanishes and that the Fourier series representation of a discontinuous signal such as the square wave converges.

4.6 APPLICATIONS

Fourier analysis finds engineering applications in circuit analysis, filtering, amplitude modulation, rectification, harmonic distortion, and signal transmission through a linear system. We will only consider three simple applications: circuit analysis, spectrum analyzers, and filters.

4.6.1 Circuit Analysis

Fourier analysis enables us to find the steady response of a circuit to a periodic excitation. The first step involves finding the Fourier series expansion of the excitation. This Fourier series representation may be regarded as a series-connected sinusoidal sources with each source having its own amplitude and frequency. The second step is finding the response to each term in the Fourier series by the phasor techniques. Finally, following the principle of superposition, we add all the individual responses.

Example 4.8

If the signal $g(t)$ in Figure 4.4 (see Example 4.2) is the voltage source $v_s(t)$ in the circuit of Figure 4.16. Find the response $v_o(t)$ of the circuit.

Solution

From Example 4.2,

$$v_s(t) = 2.5 + \frac{5}{\pi}\sum_{n=1}^{\infty}\frac{1}{n}\sin(n\pi t)$$

where

$T = 2$

$\omega_0 = 2\pi/T = \pi$

$\omega_n = n\omega_0 = n\pi$ rad/s

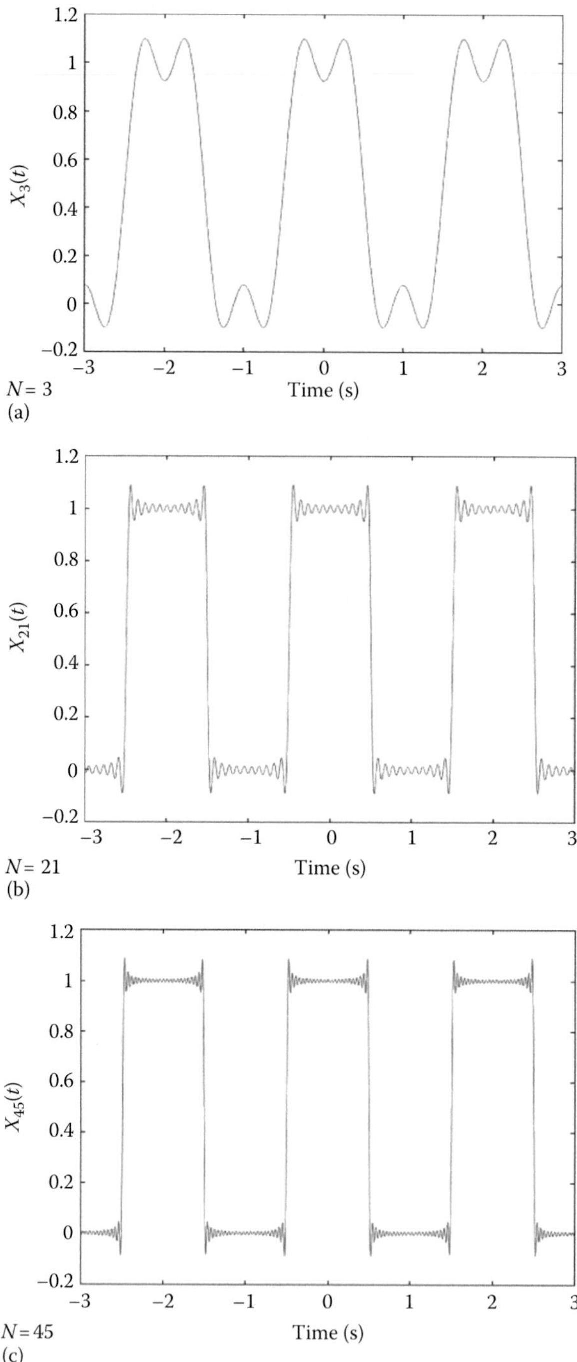

FIGURE 4.15 Convergence of the Fourier series representation of a square wave for several values of N.

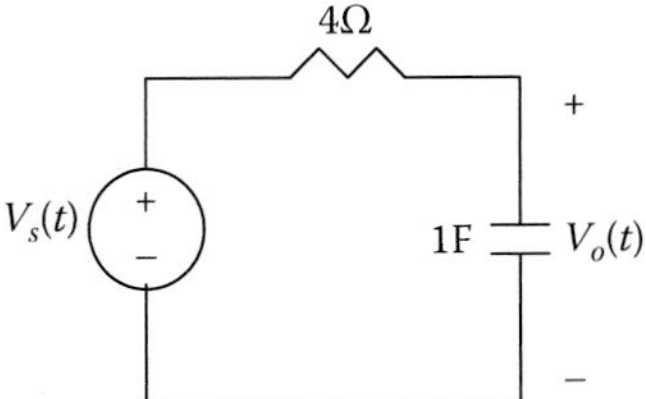

FIGURE 4.16 For Example 4.8.

Using phasors, we obtain the response $\boldsymbol{V}_o$ in the circuit of Figure 4.16 by voltage division, that is,

$$\boldsymbol{V}_o = \frac{1/j\omega_n C}{R + \left(1/j\omega_n C\right)} \boldsymbol{V}_s = \frac{1}{1 + j\omega_n RC} \boldsymbol{V}_s = \frac{\boldsymbol{V}_s}{1 + j4n\pi} \tag{4.8.1}$$

where

$R = 4$

$C = 1$

For the dc component ($\omega_n = 0$ or $n = 0$),

$$\boldsymbol{V}_s = 2.5 \Rightarrow \boldsymbol{V}_o = 2.5$$

This is expected since the capacitor is an open circuit to dc. For the nth harmonic,

$$\boldsymbol{V}_s = \frac{5}{n\pi} \angle 0° \tag{4.8.2}$$

and the corresponding response is obtained from Equation 4.8.1 as

$$\begin{aligned} \boldsymbol{V}_o &= \frac{1}{\sqrt{1 + 16n^2\pi^2} \angle \tan^{-1} 4n\pi} \left(\frac{5}{n\pi} \angle 0° \right) \\ &= \frac{5 \angle (-\tan^{-1} 4n\pi)}{n\pi\sqrt{1 + 16n^2\pi^2}} \end{aligned} \tag{4.8.3}$$

In the time domain,

$$v_o(t) = 2.5 + \sum_{n=1}^{\infty} \frac{5}{n\pi\sqrt{1 + 16n^2\pi^2}} \sin\left(n\pi t - \tan^{-1} 4n\pi\right), \tag{4.8.3}$$

Practice Problem 4.8 Let the signal in Figure 4.3 (see Practice Problem 4.1) be applied to the circuit in Figure 4.17. Find the response $v_o(t)$.

Answer: $v_o(t) = \sum_{n=1}^{\infty} \frac{6}{\sqrt{9 + n^2\pi^2}} \cos(n\pi t - \tan^{-1} n\pi/3)$

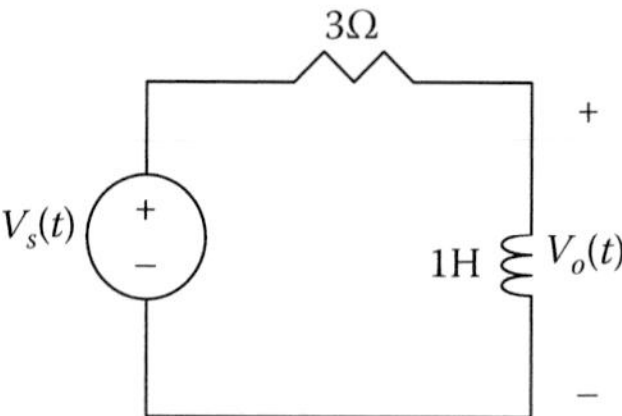

FIGURE 4.17 For Practice Problem 4.8.

4.6.2 Spectrum Analyzers

As we have observed, the Fourier series provides the spectrum of a signal. By providing the spectrum of a signal *x*(*t*), the Fourier series helps us identify which frequencies are playing an important role in the shape of the output and which ones are not. For example, audible sounds have significant components in the frequency range of 20 Hz–15 kHz, while microwaves range from 2.4 to 300 GHz.

We regard a periodic signal as *band-limited* if its amplitude spectrum contains only a finite number of coefficients A_n or c_n. For a band-limited signal, the Fourier series becomes (truncated)

$$x(t) = \sum_{n=-N}^{N} c_n e^{jn\omega_0 t} = a_0 + \sum_{n=1}^{N} A_n \cos(n\omega_0 t + \phi_n) \tag{4.67}$$

From this, we see that we need only $2N + 1$ terms (namely, $a_0, A_1, A_2, \ldots, A_N, \phi_1, \phi_2, \ldots, \phi_N$) to completely specify *x*(*t*) if ω_0 is known. This leads to the sampling theorem.

The **sampling theorem** states that a band-limited periodic signal whose Fourier series contains *N* harmonics is uniquely specified by its values at $2N + 1$ instants in one period.

A spectrum analyzer is an instrument that displays the frequency components (spectra lines) of a signal. Spectrum analyzers are commercially available in various sizes and shapes. Figure 4.18 illustrates a typical one. An oscilloscope shows the signal in the time domain, while the spectrum analyzer shows the signal in the frequency domain. Spectrum analyzer is perhaps the most useful instrument for circuit analysis. It is widely used to measure the frequency response, noise and spurious signal analysis, and more.

4.6.3 Filters

Filters are frequency-selective devices used in electronics and communications systems. Here, we investigate how to design filters to select the fundamental component

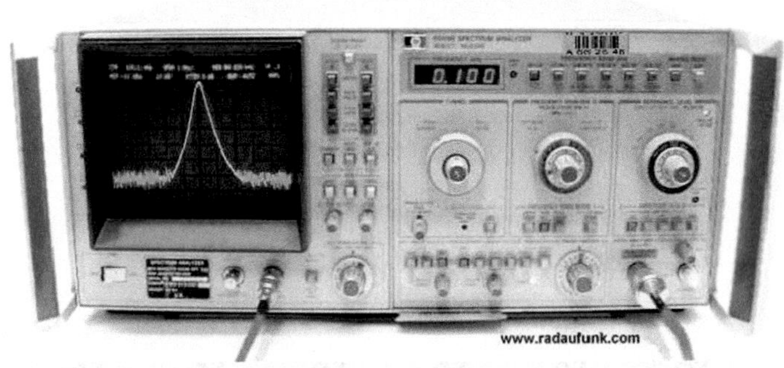

FIGURE 4.18 A spectrum analyzer. (From http://www.radaufunk.com/hp8569b.htm.)

(or any desired harmonic) of the input signal and reject other harmonics. This filtering process cannot be accomplished without the Fourier series expansion of the input signal.

> A **filter** is a device designed to pass signals with desired frequencies and block or attenuate others.

As shown in Figure 4.19, there are four types of filters:

1. A *lowpass filter* (LPF) passes low frequencies and rejects high frequencies.
2. A *highpass filter* (HPF) passes high frequencies and rejects low frequencies.
3. A *bandpass filter* (BPF) passes frequencies within a frequency band and blocks or attenuates frequencies outside the band.
4. A *bandstop filter* (BSF) passes frequencies outside the frequency band and stops or attenuates frequencies within the band.

For the purpose of illustration, we will consider two cases: a LPF and BPF.

A typical LPF is shown in Figure 4.20. The output of a LPF depends on the input signal $v_i(t)$ (with fundamental frequency ω_0), the transfer function $H(\omega)$ of the filter, and the corner or half-power frequency $\omega_c = 1/RC$. The LPF characteristically passes the dc and low-frequency components, while blocking or attenuating the high-frequency components. If we make ω_c sufficiently large ($\omega_c \gg \omega_0$, e.g., making C small), a large number of the harmonics can be passed. On the other hand, if we make ω sufficiently small ($\omega_c \ll \omega_0$), we can block out all the ac components and pass only dc.

A typical BPF is shown in Figure 4.21. The output of a BPF depends on the input signal $v_i(t)$ (with fundamental frequency ω_0), the transfer function $H(\omega)$ of the filter, its bandwidth B, and its center frequency ω_c. The filter passes all the harmonics of the input signal within a band of frequencies ($\omega_1 < \omega < \omega_2$) centered around ω_c and rejects other harmonics.

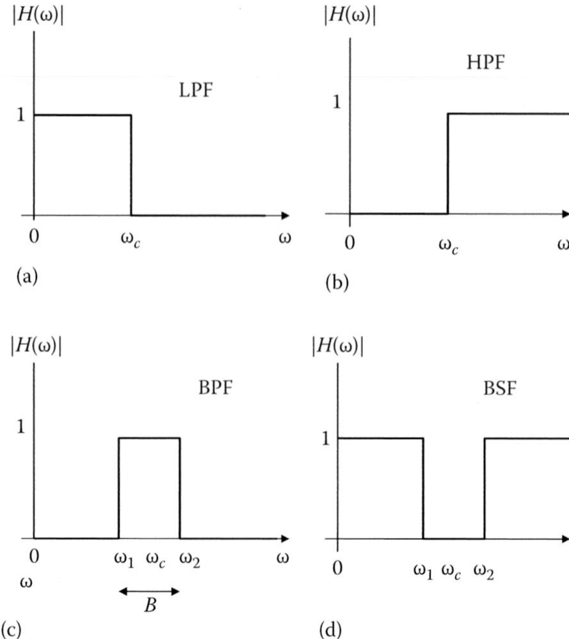

FIGURE 4.19 Ideal frequency response of four types of filters.

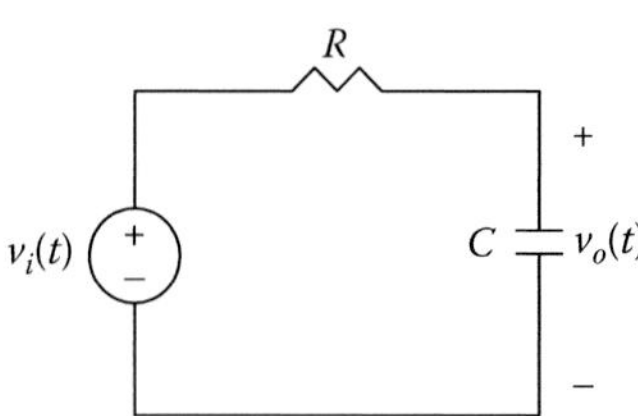

FIGURE 4.20 A typical lowpass filter.

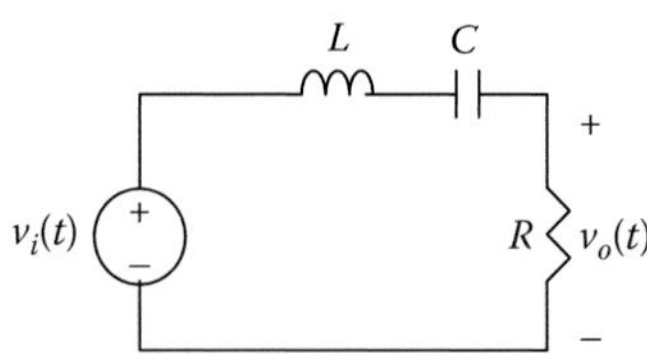

FIGURE 4.21 A typical bandpass filter.

Example 4.9

If the signal in Figure 4.22a is applied to an ideal LPF with the transfer function shown in Figure 4.22b, find the output.

Solution

The input signal in Figure 4.22a is the same as the signal in Figure 4.3 except for the period is 2π instead of 2. Hence, $\omega_0 = 2\pi/T = 1$. From Practice Problem 4.1, we know that the Fourier series expansion is (obtained by time scaling)

$$x(t) = -0.5 + \sum_{\substack{n=1 \\ n=\text{odd}}}^{\infty} \frac{6}{n\pi} \sin nt$$

or

$$x(t) = -\frac{1}{2} + \frac{6}{\pi} \sin t + \frac{6}{3\pi} \sin 3t + \frac{6}{5\pi} \sin 5t + \frac{6}{7\pi} \sin 7t + \cdots$$

Since the corner frequency of the filter is $\omega_c = 6$ rad/s, only the dc component and harmonics with $n\omega_0 < 6$ will be passed. Hence, the output of the filter is

$$y(t) = -\frac{1}{2} + \frac{6}{\pi} \sin t + \frac{6}{3\pi} \sin 3t + \frac{6}{5\pi} \sin 5t$$

Practice Problem 4.9 Rework Example 4.9 if the ideal LPF is replaced by the ideal BPF whose transfer function is shown in Figure 4.23.

Answer: $xy(t) = \frac{6}{5\pi} \sin 5t + \frac{6}{7\pi} \sin 7t + \frac{6}{9\pi} \sin 9t$

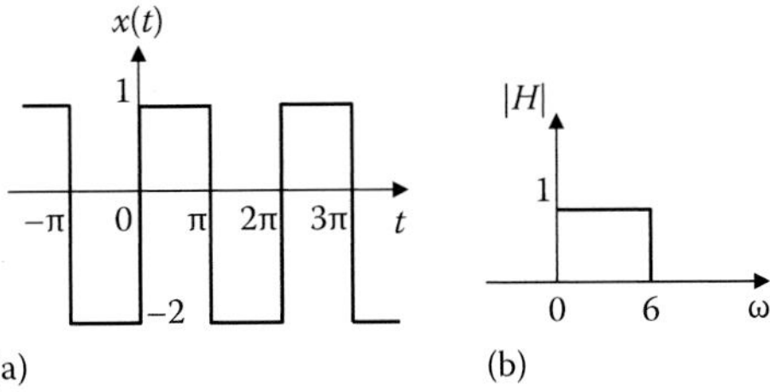

FIGURE 4.22 For Example 4.9.

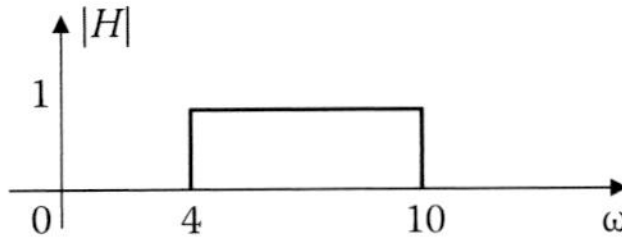

FIGURE 4.23 For Practice Problem 4.9.

4.7 COMPUTING WITH MATLAB®

Although MATLAB does not help in finding the Fourier series of a signal, it can help in plotting line spectra and examining the convergence of a Fourier series. The following examples show us how.

Example 4.10

Refer to the triangular wave in Figure 4.24. Take $A = 10$ and $T = 2$. Truncate the series at $n = N$. Plot $x(t)$ for $N = 3$, 5, and 10.

Our goal is to examine the convergence of this Fourier series using MATLAB.

Solution

We can show that the Fourier series expansion of $x(t)$ is

$$x(t) = \frac{A}{2} - \frac{4A}{\pi^2}\sum_{n=1}^{\infty}\frac{1}{(2n-1)^2}\cos(2n-1)\omega_0 t$$

$A = 10$ and $\omega = 2\pi/T = \pi$. If we truncate the series at $n = N$, then the finite sum becomes

$$x_N(t) = 5 - \frac{40}{\pi^2}\sum_{n=1}^{N}\frac{1}{(2n-1)^2}\cos(2n-1)\pi t$$

The MATLAB script to compute and plot $x_N(t)$ is shown next. Since the period is $T = 2$, we show $x_N(t)$ for only $0 < t < 2$.

```
N = input('Enter the highest harmonic desired'); %3, 5, 10
t=0:0.01:2;
xN = 5*ones(1,length(t)); % dc component
fac = -40/(pi*pi);
for n=1:N
xN = xN + fac*cos( (2*n -1)*pi*t )/( (2*n-1)^2);
end
plot(t,xN)
xlabel('t')
ylabel('partial sum')
```

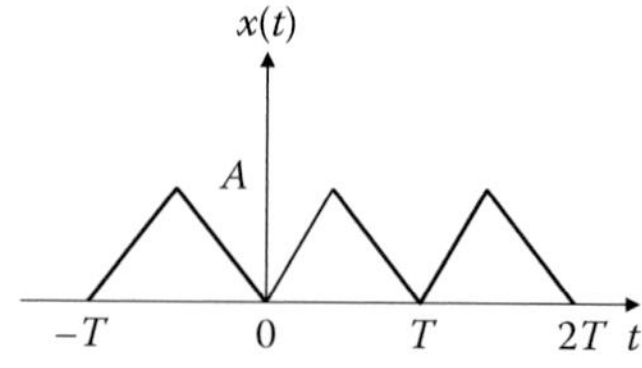

FIGURE 4.24 For Example 4.10.

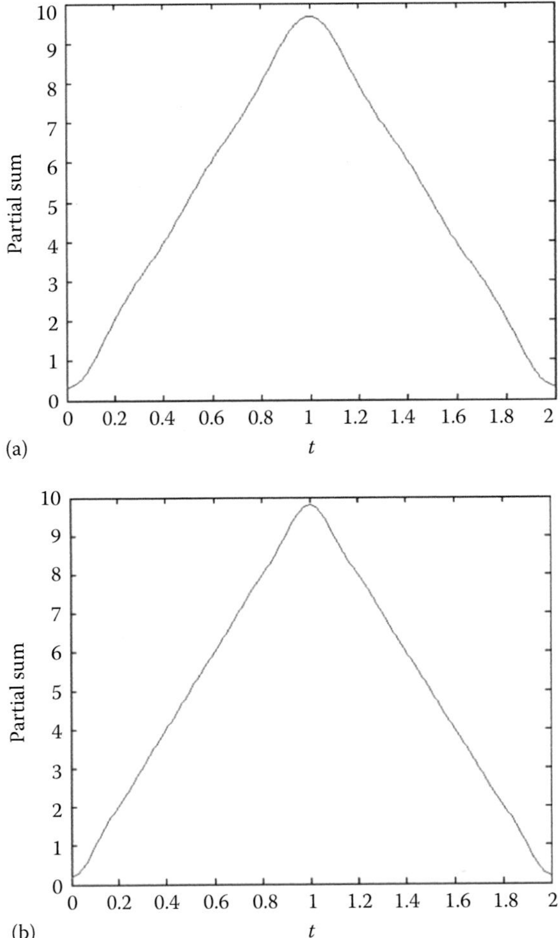

FIGURE 4.25 For Example 4.12: (a) $x_N(t)$ for $N = 3$, (b) $x_N(t)$ for $N = 5$.

The script was run for $N = 3$ and $N = 5$. The results for these two cases are shown in Figure 4.25. There is a noticeable difference for the two cases. However, for $N = 5$ or greater, the result remains the same. We notice that the Fourier series converges exactly to the signal. This confirms Fourier's theorem, which requires that $x_N(t)$ should converge to $x(t)$ as $N \to \infty$.

Practice Problem 4.10 Repeat Example 4.10 for the periodic square wave shown in Figure 4.26. The Fourier series expansion is

$$x(t) = \frac{4A}{\pi} \sum_{n=1}^{\infty} \frac{1}{2n-1} \sin(2n-1)\omega_0 t$$

Take $A = 10$ and $T = 2$. Try $N = 5$ and $N = 10$.

Answer: See Figure 4.27.

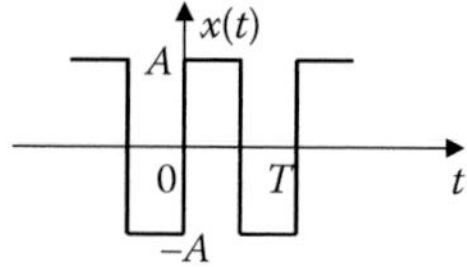

FIGURE 4.26 For Practice Problem 4.10.

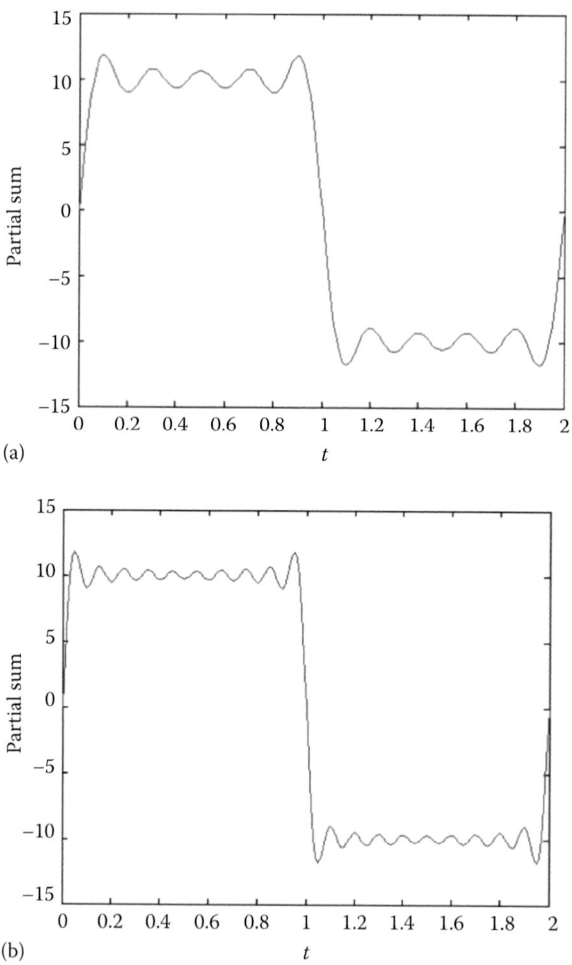

FIGURE 4.27 For Practice Problem 4.12: (a) $x_N(t)$ for $N = 5$, (b) $x_N(t)$ for $N = 10$.

Example 4.11

In Example 4.3, we obtained the exponential Fourier series for $f(t) = t$, $-1 < t < 1$, with $f(t + 2n) = f(t)$ and n is an integer.

$$f(t) = \sum_{\substack{n=-\infty \\ n\neq 0}}^{\infty} (-1)^n \frac{j}{n\pi} e^{jn\pi t}, \quad c_0 = 0. \tag{4.11.1}$$

Use MATLAB to plot the line spectrum for the magnitude and the phase of $x(t)$.

Solution

From Equation 4.11.1, we obtain

$$c_n = (-1)^n \frac{j}{n\pi}, \quad n \neq 0, \quad c_0 = 0. \tag{4.11.2}$$

The following MATLAB script uses Equation 4.11.2. The stem plot is shown in Figure 4.28.

```
clear
clc
n = -5:1:5;
c1 = j*(-1).^n;
c2 = n*pi;
if n==0
   cn = 0;
else
cn=c1./c2;
end
mag = abs(cn);
phase = angle(cn)*180/pi; % converts angle in radians to degrees
subplot(2,1,1); stem(n,mag)
xlabel('n'); ylabel('|cn|')
subplot(2,1,2); stem(n,phase)
xlabel('n'); ylabel('\theta_n')
```

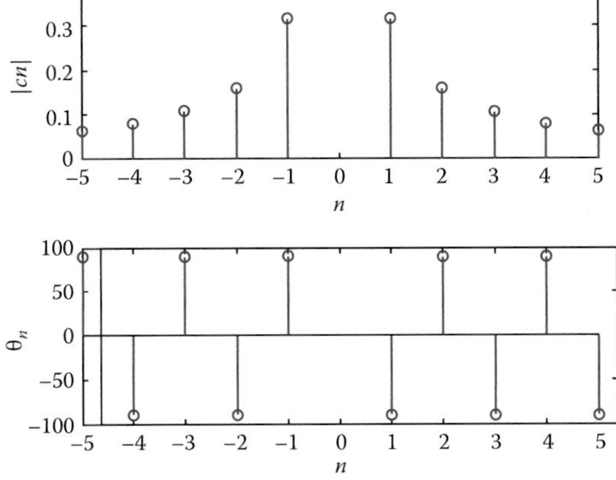

FIGURE 4.28 For Example 4.11.

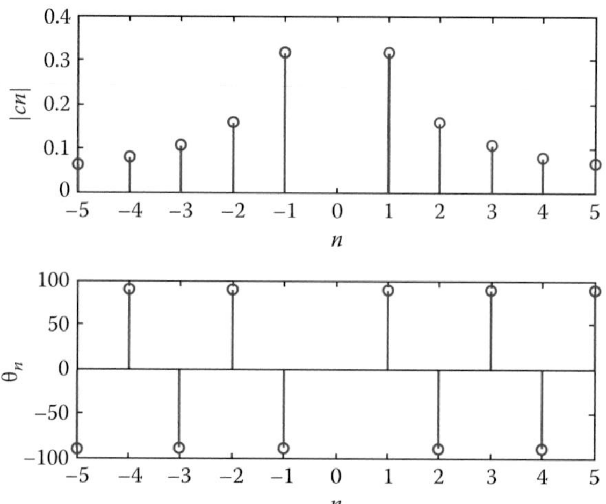

FIGURE 4.29 For Practice Problem 4.11.

Practice Problem 4.11 From Practice Problem 4.3, the periodic function $y(t) = t^2$, $0 < t < T$ has the complex Fourier series

$$y(t) = 1.333 + \sum_{\substack{n=-\infty \\ n\neq 0}}^{\infty} \frac{2}{n^2\pi^2}(1 + jn\pi)e^{jn\pi t}$$

Use MATLAB to obtain the magnitude and phase line spectra for the signal $f(t)$.
Answer: See Figure 4.29.

4.8 SUMMARY

1. A periodic function $x(t)$ repeats itself every period T, that is, $x(t + T) = x(t)$.
2. A periodic function $x(t)$ can expressed in terms of sinusoids using Fourier series:

$$x(t) = \underbrace{a_0}_{\text{dc}} + \underbrace{\sum_{n=1}^{\infty} (a_n \cos n\omega_0 t + b_n \sin n\omega_0 t)}_{\text{ac}}$$

where $\omega_0 = 2\pi/T$ is the fundamental frequency. The Fourier series decomposes the signal into the dc component a_0 and ac components.

There are three basic forms of Fourier series representation: the sine–cosine form, the amplitude–phase form, and the exponential form.

3. The Fourier coefficients of sine–cosine form are determined as

$$a_0 = \frac{1}{T}\int_0^T x(t)dt$$

$$a_n = \frac{2}{T}\int_0^T x(t)\cos n\omega_0 t \, dt$$

$$b_n = \frac{2}{T}\int_0^T x(t)\sin n\omega_0 t \, dt$$

4. The amplitude–phase form of Fourier series is

$$x(t) = a_0 + \sum_{n=1}^{\infty} A_n \cos(n\omega_0 t + \phi_n)$$

where

$$A_n = \sqrt{a_n^2 + b_n^2}, \quad \phi_n = -\tan^{-1}\frac{b_n}{a_n}$$

5. The exponential (or complex) form of Fourier series is

$$x(t) = \sum_{n=-\infty}^{\infty} c_n e^{jn\omega_0 t}$$

where

$$c_n = \frac{1}{T}\int_0^T x(t)e^{-jn\omega_0 t}dt$$

and $\omega_0 = 2\pi/T$.

6. For a Fourier series to converge, it must meet Dirichlet conditions, which require that the signal $x(t)$ must be integrable, have only a finite number of minima and maxima, and have a finite number of discontinuities over a period.
7. The line spectra (amplitude and phase) of $x(t)$ is the plot of A_n and ϕ_n or $|c_n|$ and θ_n versus frequency.

8. Fourier series representation of a periodic function results in an overshoot behavior at points of discontinuity due to partial summation. The overshoot is about 9% of the peak-to-peak value. This is known as Gibbs phenomenon.
9. The average-power of periodic voltage and current is

$$P = \frac{1}{T}\int_0^T |x(t)|^2\, dt = a_0^2 + \frac{1}{2}\sum_{n=1}^{\infty}\left(a_n^2 + b_n^2\right) = a_0^2 + \frac{1}{2}\sum_{n=1}^{\infty} A_n^2 = \sum_{n=-\infty}^{\infty} |c_n|^2$$

This result is known as *Parseval's theorem*. It states that the average power in a periodic signal is the sum of the average power in its dc component and the average of the powers in its harmonics.
10. In this chapter, Fourier series is applied to circuit analysis, spectrum analyzers, and filters.
11. Fourier series expansion allows us to use the phasor method in analyzing circuits with nonsinusoidal periodic source signal. We apply the phasor technique to determine the response due to each harmonic in the series, transform the responses to the time domain, and add them together.
12. Using MATLAB, we can plot the Fourier series and the line spectra for both magnitude and phase.

REVIEW QUESTIONS

4.1 If $x(t) = 2t$, $0 < t < \pi$, $x(t + T) = x(t + \pi) = x(t)$, the value of ω_0 is
(a) 1, (b) 2, (c) π, (d) 2π

4.2 Which of the following signals are even?
(a) $t^2 \cos t$, (b) $t^2 \ln(t)$, (c) $\exp(t^2)$, (d) $t^2 + t^4$, (e) $\sinh t$

4.3 Which of the following signals are odd?
(a) $\sin t$, (b) $t\sin t$, (c) $\exp(t)$, (d) $t^3 \cos t$, (e) $\sinh t$

4.4 If $x(t) = 8\cos 100t + 4\cos 200t + 2\cos 300t + \cos 400t + \cdots$, the signal is:
(a) Odd, (b) even, (c) half-wave symmetric, (d) quarter-wave symmetric

4.5 Since $\cos t$ is an even function, $\cos t + \cos(-t)$ is
(a) $2\cos t$, (b) $0.5\cos t$, (c) $2\sin t$, (d) $0.5\sin 2t$

4.6 If $x(t) = 12 + 8\cos t + 4\cos 3t + 2\cos 5t + \cdots$, the magnitude of the dc component is:
(a) 12, (b) 8, (c) 4, (d) 2, (e) 0.

4.7 If $x(t) = 12 + 8\cos t + 4\cos 3t + 2\cos 5t + \cdots$, the frequency of the sixth harmonic is
(a) 12, (b) 11, (c) 9, (d) 6

4.8 If $x(t) = 10 + 4\cos t + 2\cos 3t$, the power in the signal is
(a) 164, (b) 120, (c) 110, (d) 16

4.9 A Fourier cosine series is one that involves only cosine terms.
(a) True, (b) False

4.10 The plot of θ_n versus $n\omega_0$ is called
(a) Complex frequency spectrum, (b) Complex phase spectrum, (c) Complex amplitude spectrum

Answers: 4.1b; 4.2a,d; 4.3a,d,e; 4.4b; 4.5a; 4.6a; 4.7b; 4.8c; 4.9a; 4.10b.

PROBLEMS

Section 4.2—Trigonometric Fourier Series

4.1 Find the trigonometric Fourier series for

$$x(t) = \begin{cases} 5, & 0 < t < \pi \\ 10, & \pi < t < 2\pi \end{cases}$$
$$x(t + 2\pi) = x(t).$$

4.2 Determine the sine–cosine Fourier series for the periodic signal in Figure 4.30.

4.3 A periodic signal is given by

$$x(t) = \sum_{n=1}^{\infty} \frac{4}{n} \sin^2\left(\frac{n\pi}{2}\right) \cos\,(400n\pi t)$$

(a) Compute the period T and the fundamental frequency ω_0.
(b) Identify the Fourier series coefficients a_n and b_n.

4.4 Obtain the Fourier series expansion for the signal shown in Figure 4.31.

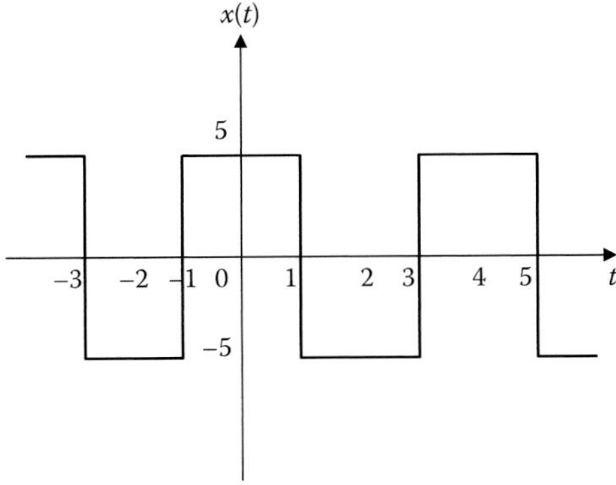

FIGURE 4.30 For Problem 4.2.

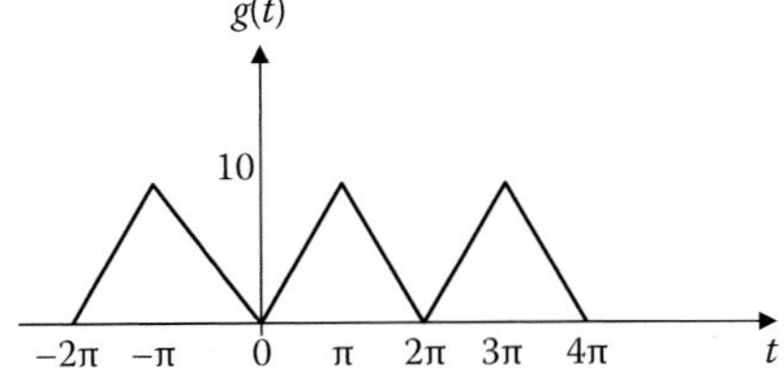

FIGURE 4.31 For Problem 4.4.

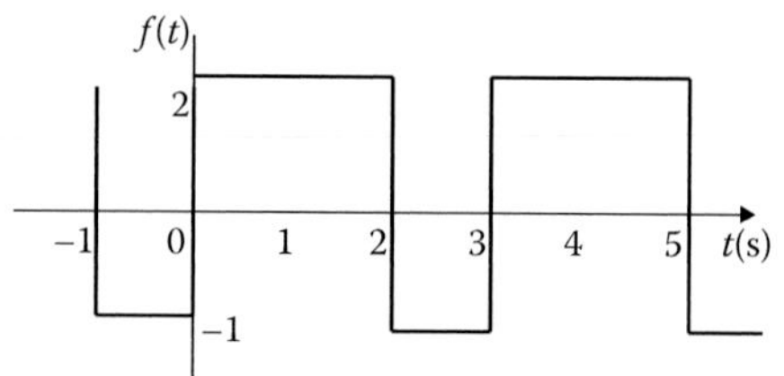

FIGURE 4.32 For Problem 4.7.

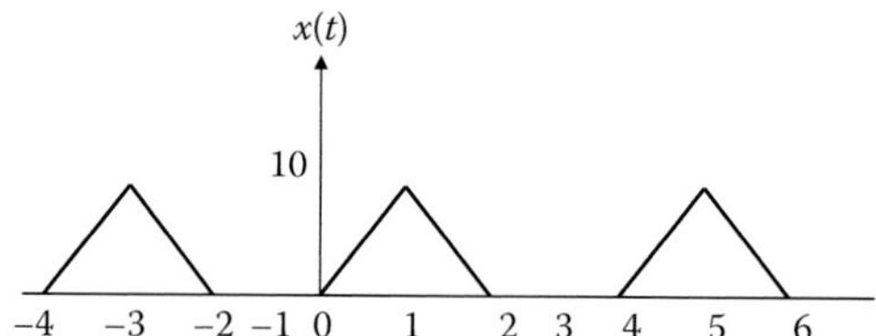

FIGURE 4.33 For Problem 4.8.

4.5 Plot the amplitude and phase spectra for the signal

$$x(t) = 8\cos(100\pi t - 0.2) + 6\cos(200\pi t - 1.2) - 4\cos(300\pi t)$$

4.6 Show that

$$A\cos\omega t + B\sin\omega t = \sqrt{A^2 + B^2}\cos\left(\omega t - \tan^{-1}\frac{B}{A}\right)$$

4.7 Determine the Fourier series of the periodic function in Figure 4.32.

4.8 Expand $x(t)$ in Figure 4.33 in Fourier series.

Section 4.3—Exponential Fourier Series

4.9 Obtain the exponential Fourier series coefficients for

$$x(t) = \sin^3(2\pi t)$$

4.10 Express the sawtooth waveform $x(t)$ in Figure 4.34 as a complex Fourier series.

4.11 Determine the exponential Fourier series coefficients of the signal in Figure 4.35.

4.12 Determine the exponential Fourier series for the impulse train in Figure 4.36.

4.13 The complex exponential Fourier series of a signal $x(t)$ over $0 < t < T$ is

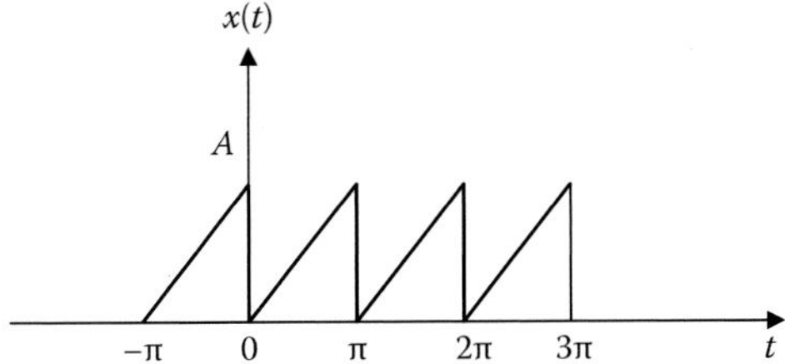

FIGURE 4.34 For Problem 4.10.

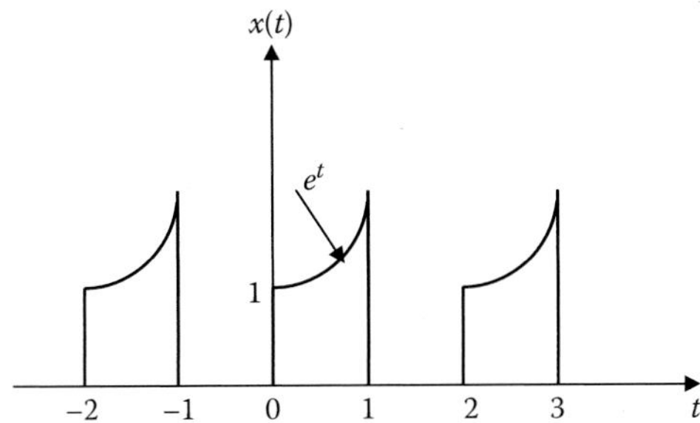

FIGURE 4.35 For Problem 4.11.

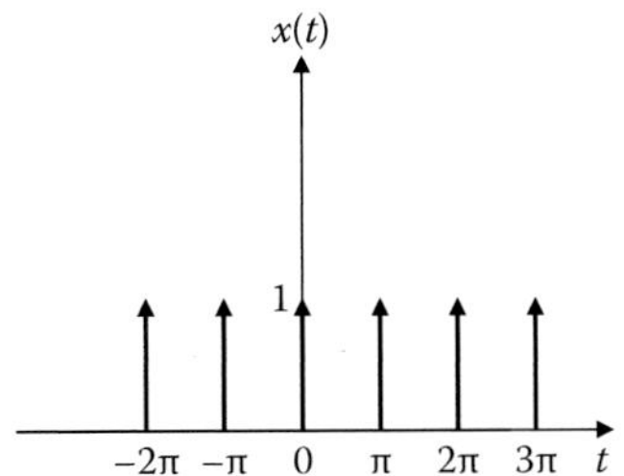

FIGURE 4.36 For Problem 4.12.

$$x(t) = \sum_{n=-\infty}^{\infty} j\sin\left(\frac{n\pi}{2}\right)e^{in\pi t/50}$$

(a) Calculate T.
(b) Determine the average value of $x(t)$.
(c) Find the amplitude of the third harmonic.

4.14 The complex exponential Fourier series of a signal is given as

$$x(t) = (2-j)e^{-j3t} + j4e^{-jt} + 5 - j4e^{jt} + (2+j)e^{j3t}$$

Sketch the amplitude and phase spectra.

4.15 A periodic signal is expressed as

$$x(t) = \sum_{n=-\infty}^{\infty} c_n e^{-jnt}$$

where

$$c_0 = 0.4$$

$$c_1 = c_{-1} = 0.2$$

$$c_2 = c_{-2} = 0.12$$

$$c_3 = c_{-3} = 0.08$$

(a) Determine the period T and the fundamental frequency ω_0.

(b) Write out $x(t)$ in exponential form and use Euler relation to express $x(t)$ in cosine form.

4.16 Suppose $x(t) = 1 + \cos 2t + 0.5\cos 4t$. Find the Fourier series coefficients of the exponential form of $x(t)$.

4.17 Find the exponential Fourier series for the waveform in Figure 4.37.

4.18 Determine the complex Fourier series if $x(t) = 10\cos 3t$.

Section 4.4—Properties of Fourier Series

4.19 A periodic signal is given by

$$x(t) = \sum_{n=-\infty}^{\infty} c_n e^{jn\omega_0 t}$$

where

$\omega_0 = 2\pi/T$

T is the period

Determine the Fourier coefficients in terms of c_n for $y(t)$, where

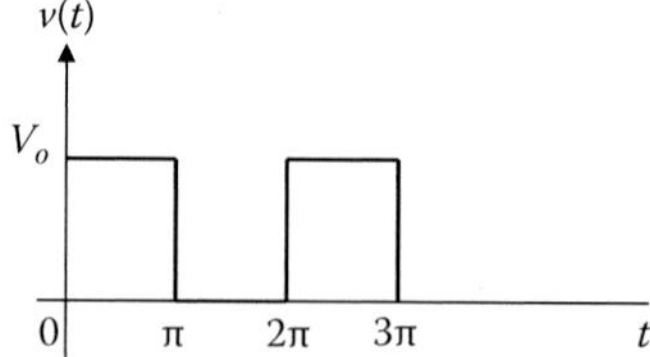

FIGURE 4.37 For Problem 4.17.

(a) $y(t) = x(-t)$
(b) $y(t) = x(2t)$
(c) $y(t) = x(t-3)$
(d) $y(t) = \dfrac{dx(t)}{dt}$

4.20 If c_n is the exponential Fourier series with coefficients of a periodic signal $x(t)$. Determine the coefficients of the following:
(a) $y(t) = x(3t)$
(b) $z(t) = x(-3t)$
(c) $h(t) = 1 + x(2t)$

4.21 Let $w(t) = x(t)y(t)$. Show that if $x(t)$ is even and $y(t)$ is odd, $w(t)$ is odd.

4.22 Let $x(t)$ and $y(t)$ be periodic signals with the same period T such that

$$x(t) = \sum_{n=-\infty}^{\infty} c_n e^{jn\omega_0 t}, \quad y(t) = \sum_{n=-\infty}^{\infty} d_n e^{jn\omega_0 t}$$

where $\omega_0 = 2\pi/T$. Show that the signal $z(t) = x(t)y(t)$ is periodic with the same period T and can be expressed as

$$z(t) = \sum_{n=-\infty}^{\infty} e_n e^{jn\omega_0 t}$$

where $e_n = \sum_{m=-\infty}^{\infty} c_m d_{n-m}$

4.23 Consider the signals shown in Figure 4.38. Which of them have
(a) Even symmetry?
(b) Odd symmetry?
(c) Half-wave odd symmetry?

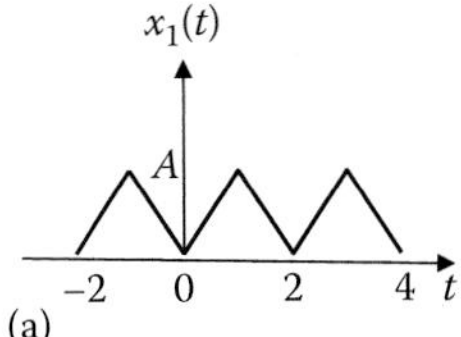

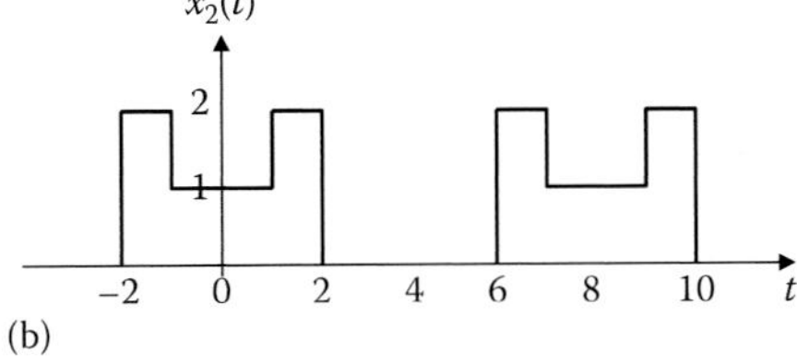

FIGURE 4.38 For Problems 4.23 and 4.24. *(Continued)*

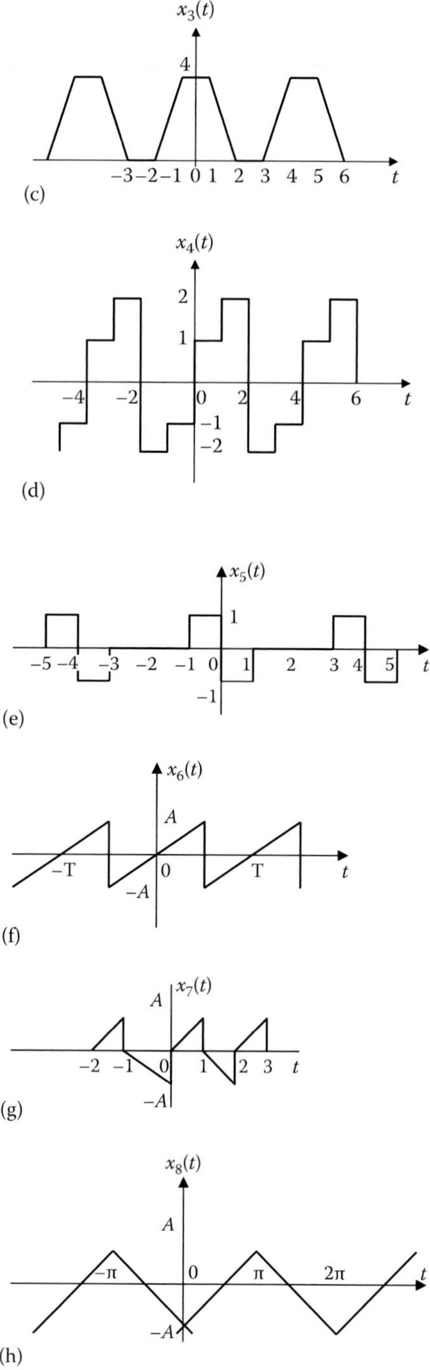

FIGURE 4.38 (*Continued*) For Problems 4.23 and 4.24.

4.24 Without evaluating any Fourier series coefficients, state which of the signals in Figure 4.38 contain

(a) A dc component

(b) Only odd harmonics

(c) Both sine and cosine terms

4.25 Given the exponential Fourier series of a periodic signal as

$$x(t) = \sum_{n=-\infty}^{\infty} c_n e^{jn\omega_0 t}$$

Show that

$$P = \frac{1}{T}\int_0^T |x(t)|^2\, dt = \sum_{n=-\infty}^{\infty} |c_n|^2$$

4.26 The first half cycle of a periodic signal is shown in Figure 4.39. Sketch the signal over $-T/2<t<T/2$ for each of these cases:

(a) $y(t)$ is even

(b) $y(t)$ is odd

4.27 Find the average power dissipated in a 2Ω resistor when the applied current is

$$i(t) = 8\cos\omega t - 6\cos 3\omega t + 3\cos 5\omega t \;\; A$$

4.28 The amplitude and phase spectra of a truncated Fourier series are shown in Figure 4.40.

(a) Find an expression for the periodic voltage using amplitude-phase form.

(b) Is the voltage odd or even function of t?

4.29 The exponential Fourier coefficients of the signal in Figure 4.41a is

$$c_n = \begin{cases} \dfrac{A\sin^2(n\pi/2)}{(n\pi/2)^2}, & n \neq 0 \\ 0 & n = 0 \end{cases}$$

Find the coefficients for the signal in Figure 4.41b.

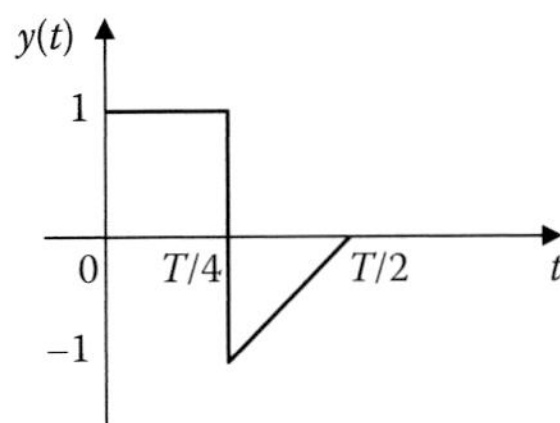

FIGURE 4.39 For Problem 4.26.

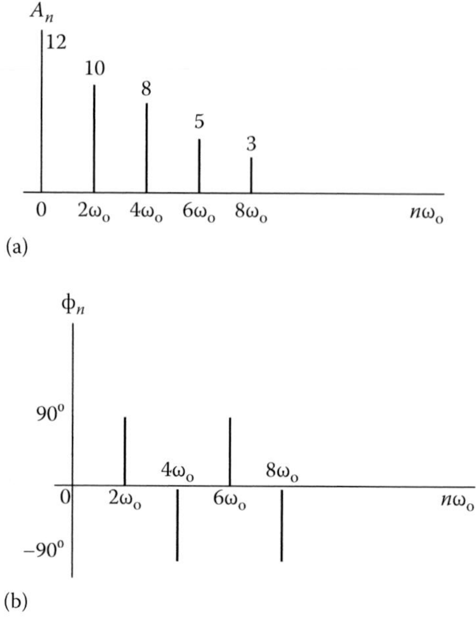

FIGURE 4.40 For Problem 4.28.

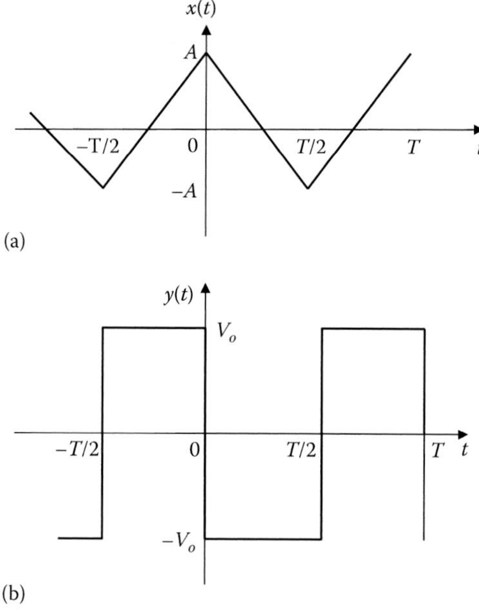

FIGURE 4.41 For Problem 4.29.

Section 4.6—Applications

4.30 Let the input voltage $x(t)$ to the circuit in Figure 4.42 be

$$x(t) = 1 + 4(\cos t + \cos 2t + \cos 3t + \cos 4t).$$

Determine the output voltage $y(t)$.

4.31 The periodic current waveform in Figure 4.43 is applied across a 2 kΩ resistor. Find the percentage of the total average power dissipation caused by the dc component.

4.32 If the periodic current waveform in Figure 4.44a is applied to the circuit in Figure 4.43b, find v_o.

4.33 Find the response $i_o(t)$ for the circuit in Figure 4.45a, where $v_s(t)$ is shown in Figure 4.45b.

Section 4.7—Computing with MATLAB®

4.34 Using MATLAB, write a program to compute and plot the partial sum of the sawtooth waveform in Figure 4.4.

4.35 Use MATLAB to plot

(a) $x(t) = 10 + 6\cos 2\pi t + 4\cos 4\pi t \quad 0 < t < 2$

(b) $y(t) = \sum_{n=-5}^{5} \frac{(-1)^n}{10} \sin\left(\frac{n\pi t}{2}\right), \quad -1/2 < t < 1/2$

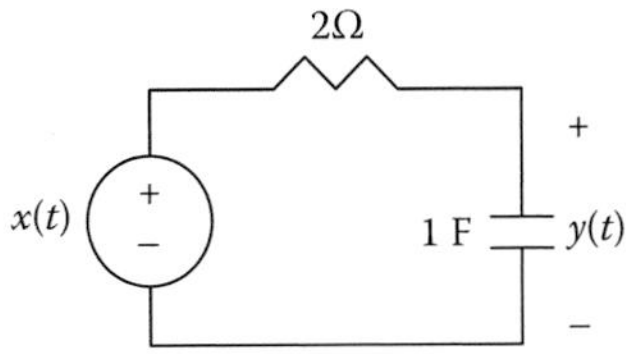

FIGURE 4.42 For Problem 4.30.

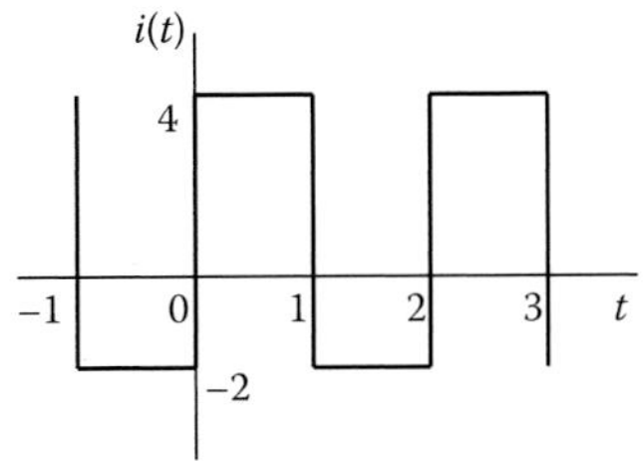

FIGURE 4.43 For Problem 4.31.

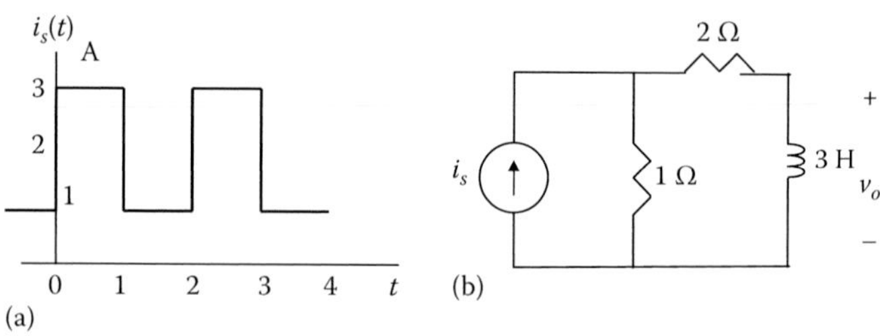

FIGURE 4.44 For Problem 4.32.

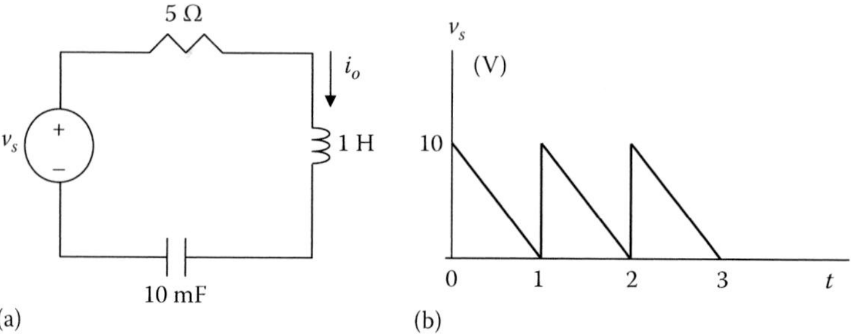

FIGURE 4.45 For Problem 4.33.

4.36 The Fourier series of a signal is

$$x(t) = \frac{1}{2} + \sum_{n=1}^{\infty} \frac{2}{n\pi} \sin(2n\pi t)$$

Use MATLAB to plot the partial sum $x_N(t)$ for $N = 5$ and $N = 9$.

4.37 Use MATLAB to plot the Fourier series expansion

$$x(t) = \frac{10}{\pi^2}\left[\sin t - \frac{1}{9}\sin 3t + \frac{1}{25}\sin 5t - \frac{1}{49}\sin 7t\right]$$

4.38 The exponential Fourier series coefficient of a signal is

$$c_n = -\frac{1}{j2n\pi}(\cos n\pi - 1)$$

Using MATLAB, plot the magnitude and phase spectra. Take $-5 < n < 5$.

4.39 Repeat the previous problem for

$$c_n = \frac{j10}{n\pi}\left((-1)^n - 1\right)$$

5 Fourier Transform

Nothing in the world is difficult for one who sets his mind to do it.

—Chinese saying

CAREER IN CONTROL SYSTEMS

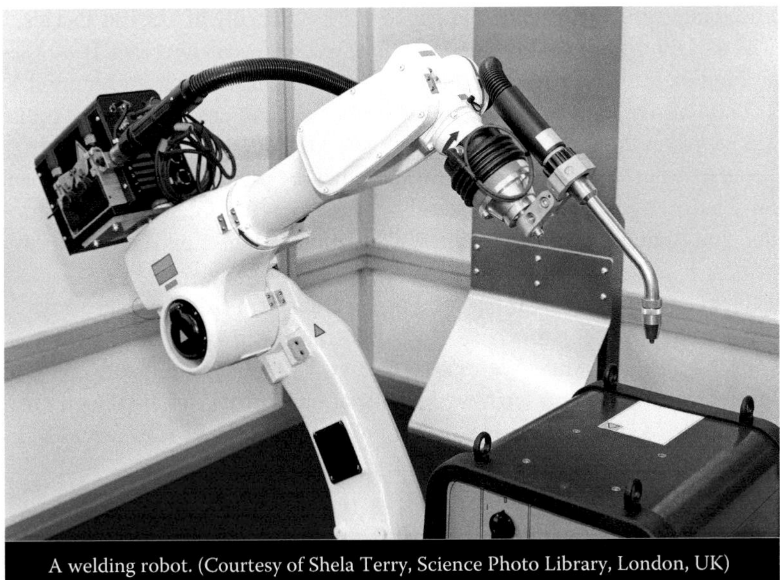

A welding robot. (Courtesy of Shela Terry, Science Photo Library, London, UK)

Control systems are an area of electrical engineering where signals and systems are used. A control system is designed to regulate the behavior of one or more variables in some desired manner. Control systems play major roles in our everyday life. Household appliances such as heating and air-conditioning systems, switch-controlled thermostats, washers and dryers, cruise control in automobiles, elevators, traffic lights, manufacturing plants, navigation systems—all utilize control systems. In the aerospace field, precision guidance of space probes, the wide range of operational modes of the space shuttle, and the ability to maneuver space vehicles remotely from earth all require knowledge of control systems. In the manufacturing sector, repetitive production line operations are increasingly performed by robots, which are programmable control systems designed to operate for many hours without fatigue.

Control engineering applies signals and systems theory. It is not limited to any specific engineering discipline but may involve environmental, chemical, aeronautical, mechanical, civil, and electrical engineering. For example, a typical task for a control engineer might be to design a speed regulator for a disk drive head. A thorough understanding of control systems techniques is essential to the electrical engineer and is of great value for designing control systems to perform the desired task.

5.1 INTRODUCTION

In Chapter 4, we saw that Fourier series enabled us to represent a periodic signal as a linear combination of infinite sinusoids. The Fourier series expansion reveals the frequency content of the periodic signal. It is also possible to analyze the frequency content of nonperiodic signals. The tool that enables us to do this is the Fourier transform (also known as Fourier integral). The transform assumes that a nonperiodic (or aperiodic) signal is a periodic signal with an infinite period. The Fourier transform is important in analyzing and processing signals in science and engineering, especially in medical images, computerized axis tomography (CAT), and magnetic resonance imaging (MRI).

The Fourier transform is similar to Laplace transform. Both are integral transforms. The Fourier transform may be regarded a special case of Laplace transform with $s = j\omega$, when the Laplace transform exists. Just like the bilateral Laplace transform, Fourier transform can deal with systems having inputs for $t < 0$ as well as those for $t > 0$. Fourier transform is very useful in linear systems, filtering, communication systems, and digital signal processing, in situations where Laplace transform does not exist.

The chapter begins by defining the Fourier transform from Fourier series. Then we present and prove the properties of Fourier transform, which are useful in deriving Fourier transform pairs. We discuss Parseval's theorem, compare the Laplace and Fourier transforms, and see how the Fourier transform is applied in circuit analysis, amplitude modulation, and sampling. We finally learn how to use MATLAB® in Fourier analysis.

5.2 DEFINITION OF THE FOURIER TRANSFORM

There is a close connection between the definition of Fourier series and the Fourier transform for functions $x(t)$ that are zero outside of an interval. For such a function, we can calculate its Fourier series on any interval that includes the points where $x(t)$ is not identically zero. The Fourier transform is also defined for such a function. As we increase the length of the interval on which we calculate the Fourier series, the Fourier series coefficients begin to look like the Fourier transform and the sum of the Fourier series of $x(t)$ begins to look like the inverse Fourier transform. To explain this more precisely, suppose T is large enough so that the interval $[-T/2, T/2]$ contains the interval on which x is not identically zero.

Recall that if $x(t)$ is a periodic function with period T, then we can expand it in a complex Fourier Series,

$$x(t) = \sum_{n=-\infty}^{\infty} c_n e^{jn\omega_0 t} \tag{5.1}$$

where

$$c_n = \frac{1}{T}\int_{-\frac{T}{2}}^{\frac{T}{2}} x(t)e^{-jn\omega_0 t}dt \tag{5.2}$$

The spacing between adjacent harmonics is

$$\Delta\omega = (n+1)\omega_0 - n\omega_0 = \omega_0 = \frac{2\pi}{T} \tag{5.3}$$

We substitute Equation 5.2 into Equation 5.1 and putting $\frac{1}{T} = \frac{\omega_0}{2\pi}$ we obtain

$$x(t) = \sum_{n=-\infty}^{\infty}\left[\frac{1}{T}\int_{-\frac{T}{2}}^{\frac{T}{2}} x(t)e^{-jn\omega_0 t}\,dt\right]e^{jn\omega_0 t}$$

$$= \sum_{n=-\infty}^{\infty}\left[\frac{\Delta\omega}{2\pi}\int_{-\frac{T}{2}}^{\frac{T}{2}} x(t)e^{-jn\omega_0 t}\,dt\right]e^{jn\omega_0 t}$$

$$= \frac{1}{2\pi}\sum_{n=-\infty}^{\infty}\left[\int_{-\frac{T}{2}}^{\frac{T}{2}} x(t)e^{-jn\omega_0 t}\,dt\right]\Delta\omega e^{jn\omega_0 t} \tag{5.4}$$

In view of the previous discussion, as $T \to \infty$, we can put ω_0 as $\Delta\omega$ and replace the sum over the discrete frequencies $n\omega_0$ by an integral over all frequencies. We replace $n\omega_0$ by a general frequency variable ω. We then obtain the double integral representation:

$$x(t) = \frac{1}{2\pi}\int_{-\infty}^{\infty}\left[\int_{-\infty}^{\infty} x(t)e^{-j\omega t}\,dt\right]e^{j\omega t}\,d\omega \tag{5.5}$$

The inner integral is known as the **Fourier transform** of $x(t)$ and is represented by $X(\omega)$, that is,

$$\boxed{X(\omega) = \mathcal{F}\left[x(t)\right] = \int_{-\infty}^{\infty} x(t)e^{-j\omega t}\,dt} \tag{5.6}$$

where $\mathcal{F}$ is the Fourier transform operator.

The **Fourier transform** of a signal $x(t)$ is the integration of the product of $x(t)$ and $e^{-j\omega t}$ over the interval from $-\infty$ to $+\infty$.

$X(\omega)$ is an integral transformation of $x(t)$ from the time-domain to the frequency-domain and is generally a complex function. $X(\omega)$ is known as the spectrum of $x(t)$. The plot of its magnitude $|X(\omega)|$ versus ω is called the amplitude spectrum, while that of its phase $\angle X(\omega)$ versus ω is called the phase spectrum. If a signal is real-valued, its magnitude spectrum is even, that is, $|X(\omega)| = |X(-\omega)|$ and its phase spectrum is odd, that is, $\angle X(\omega) = -\angle X(-\omega)$. Both the amplitude and phase spectra provide the physical interpretation of the Fourier transform.

We can write Equation 5.5 in terms of $X(\omega)$ and we obtain the inverse Fourier transform as

$$\boxed{x(t) = \mathcal{F}\left[X(\omega)\right] = \frac{1}{2\pi}\int_{-\infty}^{\infty} X(\omega)\, e^{j\omega t}\, d\omega} \tag{5.7}$$

We say that the signal $x(t)$ and its transform $X(\omega)$ form a Fourier transform pair and denote their relationship by

$$x(t) \Leftrightarrow X(\omega) \tag{5.8}$$

We will denote signals by lowercase letters and their transforms by uppercase letters.

Not all functions have Fourier transforms. The necessary conditions required for a function to be transformable are called the **Dirichlet conditions** similar to the Fourier series. These are listed as:

1. $x(t)$ is bounded.
2. $x(t)$ has a finite number of maxima and minima within any finite interval.
3. $x(t)$ has a finite number of discontinuities within any finite interval. Furthermore, each of these discontinuities must be finite.
4. $x(t)$ is absolutely integrable, that is,

$$\int_{-\infty}^{\infty} |x(t)|\, dt < \infty \tag{5.9}$$

Therefore, absolutely integrable signals that are continuous or that have a finite number of discontinuities have Fourier transforms.

Example 5.1

Find the Fourier transform of the following functions: (a)$\delta(t)$, (b) $e^{j\omega_0 t}$, (c) $\sin\omega_0 t$, (d) $e^{-at}u(t)$.

Solution

(a) For the impulse function,

$$X(\omega) = \mathcal{F}[\delta(t)] = \int_{-\infty}^{\infty} \delta(t) e^{-j\omega t} dt = e^{-j\omega t}\Big|_{t=0} = 1 \tag{5.1.1}$$

where the sifting property of the impulse function in Equation 1.21 has been applied. Thus,

$$\mathcal{F}[\delta(t)] = 1 \tag{5.1.2}$$

This shows that the magnitude of the spectrum of the impulse function is constant; that is, all frequencies are equally represented in the impulse function.

(b) From Equation 5.1.2,

$$\delta(t) = \mathcal{F}^{-1}[1]$$

Using the inverse Fourier transform formula in Equation 5.7,

$$\delta(t) = \mathcal{F}^{-1}[1] = \frac{1}{2\pi}\int_{-\infty}^{\infty} 1e^{j\omega t}\,d\omega$$

or

$$\int_{-\infty}^{\infty} e^{j\omega t}\,d\omega = 2\pi\delta(t) \tag{5.1.3}$$

Interchanging variables t and ω results in

$$\int_{-\infty}^{\infty} e^{j\omega t}\,dt = 2\pi\delta(\omega) \tag{5.1.4}$$

Using this result, the Fourier transform of given function is

$$\mathcal{F}[e^{j\omega_0 t}] = \int_{-\infty}^{\infty} e^{j\omega_0 t}e^{-j\omega t}\,dt = \int_{-\infty}^{\infty} e^{j(\omega_0-\omega)t}\,dt = 2\pi\delta(\omega_0-\omega)$$

Since the impulse function is an even function, $\delta(\omega_0-\omega) = \delta(\omega-\omega_0)$,

$$\mathcal{F}[e^{j\omega_0 t}] = 2\pi\delta(\omega-\omega_0) \tag{5.1.5}$$

By simply changing the sign of ω_0, we readily obtain

$$\mathcal{F}[e^{-j\omega_0 t}] = 2\pi\delta(\omega+\omega_0) \tag{5.1.6}$$

Also, by setting $\omega_0 = 0$,

$$\mathcal{F}[1] = 2\pi\delta(\omega) \tag{5.1.7}$$

(c) By using the result in Equations 5.1.5 and 5.1.6, we get

$$\mathcal{F}[\sin\omega_0 t] = \mathcal{F}\left[\frac{e^{j\omega_0 t} - e^{-j\omega_0 t}}{2j}\right]$$

$$= \frac{1}{2j}\mathcal{F}\left[e^{j\omega_0 t}\right] - \frac{1}{2j}\mathcal{F}\left[e^{-j\omega_0 t}\right]$$

$$= j\pi\left[\delta(\omega+\omega_0) - \delta(\omega-\omega_0)\right] \tag{5.1.8}$$

(d) Let $x(t) = e^{-at}u(t) = \begin{cases} e^{-at}, & t > 0 \\ 0, & t < 0 \end{cases}$

$$X(\omega) = \int_{-\infty}^{\infty} x(t)e^{-j\omega t}dt = \int_{0}^{\infty} e^{-at}e^{-j\omega t}dt = \int_{0}^{\infty} e^{-(a+j\omega)t}dt$$

$$\mathcal{F}\left[e^{-at}u(t)\right] = X(\omega) = \frac{-1}{a+j\omega}e^{-(a+j\omega)t}\Big|_0^{\infty} = \frac{1}{a+j\omega} \tag{5.1.9}$$

Practice Problem 5.1 Determine the Fourier transform of the following functions: (a) the rectangular pulse $\Pi(t/\tau)$, (b) $\delta(t + 3)$, (c) $2\cos\omega_0 t$.

Answer: (a) $\frac{2}{\omega}\sin\frac{\omega\tau}{2} = \tau\,\text{sinc}\frac{\omega\tau}{2}$, (b) $e^{j3\omega}$, (c) $2\pi[\delta(\omega + \omega_0) + \delta(\omega-\omega_0)]$

Example 5.2

Obtain the Fourier transform of the signal shown in Figure 5.1.

Solution

$$X(\omega) = \int_{-\infty}^{\infty} x(t)e^{-j\omega t}dt = \int_{-1}^{0}(-A)e^{-j\omega t}dt + \int_{0}^{1} Ae^{-j\omega t}dt$$

$$= \frac{A}{j\omega}e^{-j\omega t}\Big|_{-1}^{0} - \frac{A}{j\omega}e^{-j\omega t}\Big|_{0}^{1}$$

$$= \frac{-jA}{\omega}\left(1 - e^{j\omega} - e^{-j\omega} + 1\right)$$

$$= \frac{j2A}{\omega}(\cos\omega - 1)$$

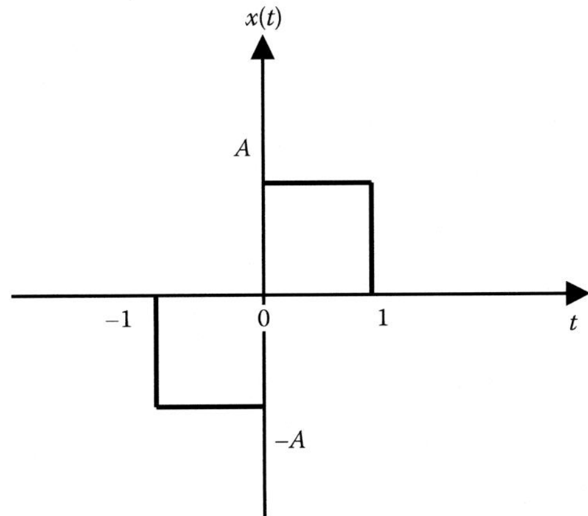

FIGURE 5.1 For Example 5.2.

Practice Problem 5.2 Derive the Fourier transform of the triangular pulse defined as

$$\Lambda(t/\tau) = \begin{cases} 1 - \dfrac{|t|}{\tau}, & |t| \le \tau \\ 0, & |t| > \tau \end{cases}$$

Answer: $\tau \operatorname{sinc}^2\left(\dfrac{\omega\tau}{2}\right)$

Example 5.3

Find the Fourier transform of the two-sided exponential pulse shown in Figure 5.2. Sketch the transform.

Solution

Let $x(t) = e^{-a|t|} = \begin{cases} e^{at}, & t < 0 \\ e^{-at}, & t > 0 \end{cases}$

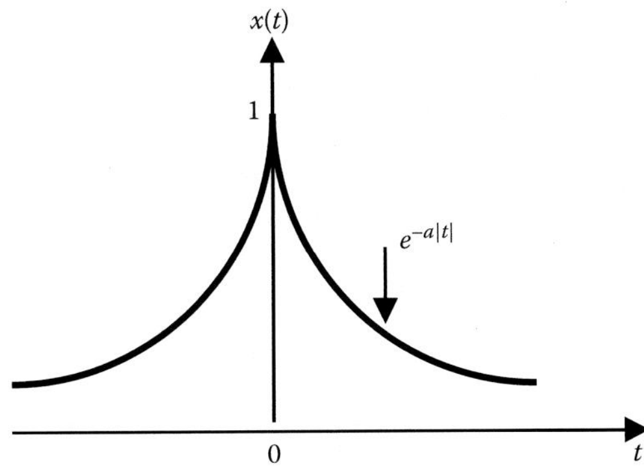

FIGURE 5.2 For Example 5.3.

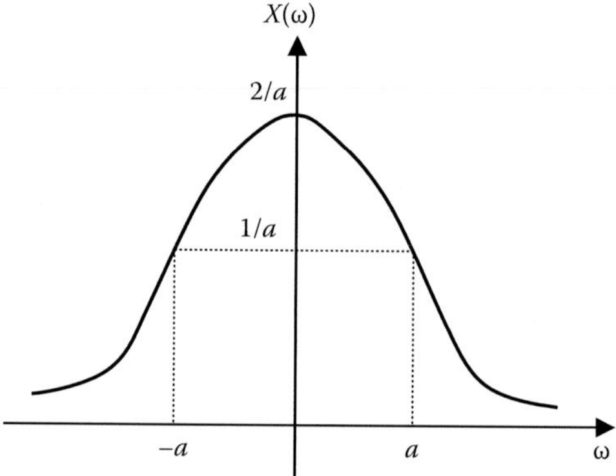

FIGURE 5.3 Fourier transform of $x(t)$ in Figure 5.2.

The Fourier transform is

$$X(\omega) = \int_{-\infty}^{\infty} x(t)e^{-j\omega t}dt = \int_{-\infty}^{0} e^{at}e^{-j\omega t}dt + \int_{0}^{\infty} e^{-at}e^{-j\omega t}dt$$

$$= \frac{1}{a - j\omega} + \frac{1}{a + j\omega}$$

$$= \frac{2a}{a^2 + \omega^2}$$

$X(\omega)$ is real in this case and it is sketched in Figure 5.3.

Practice Problem 5.3 Determine the Fourier transform of the signum function in Figure 5.4, that is, $x(t) = \text{sgn}(t) = \begin{cases} 1, & t > 0 \\ -1, & t < 0 \end{cases}$. Sketch $|X(\omega)|$.

Answer: $\frac{2}{j\omega}$. See Figure 5.5.

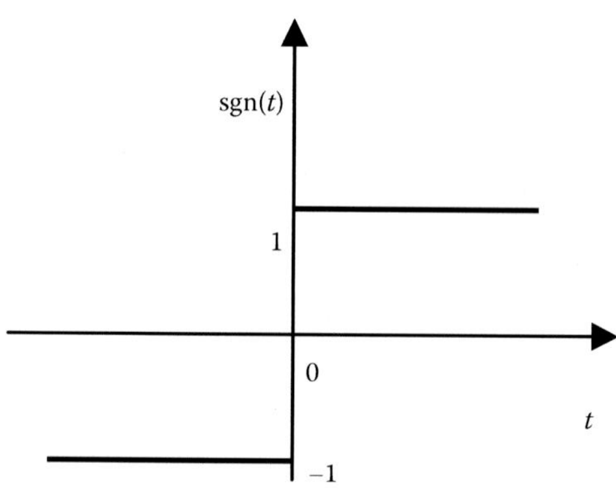

FIGURE 5.4 The signum function of Practice Problem 5.3.

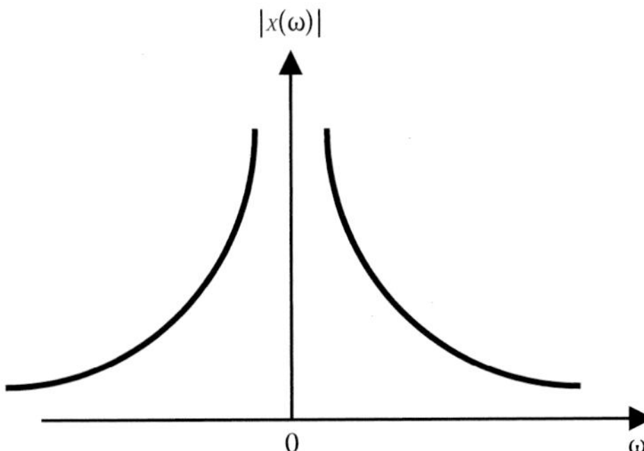

FIGURE 5.5 Fourier transform of the signal in Figure 5.4.

5.3 PROPERTIES OF FOURIER TRANSFORM

The properties of the Fourier transform are summarized as follows. These properties have important interpretations and meanings in the context of signals and signal processing. The properties of the Fourier transform provides valuable insight into how signals operating in the time-domain are described in the frequency-domain. Many properties like linearity, time scaling and shifting, frequency shifting, time differentiation and integration, frequency differentiation, duality, convolution, and Parseval's Theorem are discussed as follows.

5.3.1 LINEARITY

The Fourier transform is a linear transform. That is, suppose we have two functions $x_1(t)$ and $x_2(t)$, with Fourier transform given by $X_1(\omega)$ and $X_2(\omega)$, respectively, then the Fourier transform of $x_1(t)$ and $x_2(t)$ can be easily found as

$$\boxed{\mathcal{F}\left[a_1x_1(t)+a_2x_2(t)\right]=a_1X_1(\omega)+a_2X_2(\omega)} \tag{5.10}$$

where a_1 and a_2 are constants (real or complex numbers). Equation 5.10 can be easily shown to be true via using the definition of the Fourier transform

$$\begin{aligned}\mathcal{F}\left[a_1x_1(t)+a_2x_2(t)\right]&=\int_{-\infty}^{\infty}[a_1x_1(t)+a_2x_2(t)]e^{-j\omega t}\,dt\\&=\int_{-\infty}^{\infty}a_1x_1(t)e^{-j\omega t}\,dt+\int_{-\infty}^{\infty}a_2x_2(t)e^{-j\omega t}\,dt\\&=a_1X_1(\omega)+a_2X_2(\omega)\end{aligned} \tag{5.11}$$

This can be extended to a linear combination of an arbitrary number of signals.

5.3.2 Time Scaling

The scaling property of the Fourier transformation is as follows. If $\mathcal{F}(\omega)=\left[x(t)\right]$ and a is a real constant. Then

$$\boxed{\mathcal{F}\left[x(at)\right]=\frac{1}{|a|}X\left(\frac{\omega}{a}\right)} \tag{5.12}$$

resulting in a new frequency ω/a. Equation 5.12 can be found using the definition

$$\mathcal{F}\left[x(at)\right]=\int_{-\infty}^{\infty}x(at)e^{-j\omega t}\,dt$$

Letting $\lambda = at$, $d\lambda = adt$ so that

$$\mathcal{F}\left[x(at)\right]=\int_{-\infty}^{\infty}x(\lambda)e^{-\frac{j\omega\lambda}{a}}\frac{d\lambda}{a}=\frac{1}{a}X\left(\frac{\omega}{a}\right) \tag{5.13}$$

Now if a is positive, ($a > 0$),

$$\mathcal{F}\left[x(at)\right]=\int_{-\infty}^{\infty}x(\lambda)e^{-\frac{j\omega\lambda}{a}}\frac{d\lambda}{a}=\frac{1}{a}X\left(\frac{\omega}{a}\right)=\frac{1}{|a|}X\left(\frac{\omega}{a}\right)$$

If a is negative ($a > 0$),

$$\mathcal{F}\left[x(at)\right]=\int_{-\infty}^{\infty}x(\lambda)e^{-\frac{j\omega\lambda}{a}}\frac{d\lambda}{a}=-\int_{-\infty}^{\infty}x(\lambda)e^{-\frac{j\omega\lambda}{a}}\frac{d\lambda}{a}=\frac{1}{-a}X\left(\frac{\omega}{a}\right)$$
$$=\frac{1}{|a|}X\left(\frac{\omega}{a}\right)$$

Therefore, we can see that for the general case of scaling with a real number a we get Equation 5.12. The scaling property is sometimes also called the reciprocal spreading property.

5.3.3 Time Shifting

Another simple property of the Fourier transform is the time shifting. If $X(\omega)=\mathcal{F}\left[x(t)\right]$ and t_0 is a constant, then

$$\boxed{\mathcal{F}\left[x(t-t_0)\right]=e^{-j\omega t_0}X(\omega)} \tag{5.14}$$

If the original function $x(t)$ is shifted in time by a constant amount, it should have the same magnitude of the spectrum, $X(\omega)$. That is, a time delay does not cause the frequency content of $X(\omega)$ to change at all. Since the complex exponential always has a magnitude of 1, we see that time delay alters the phase of $X(\omega)$ but not its magnitude.

By definition,

$$\mathcal{F}\left[x(t-t_0)\right]=\int_{-\infty}^{\infty}\left[x(t-t_0)\right]e^{-j\omega t}dt \tag{5.15}$$

Letting $\lambda = t - t_0$, $d\lambda = dt$, and $t = \lambda + t_0$, so that

$$\mathcal{F}\left[x(t-t_0)\right]=\int_{-\infty}^{\infty}x(\lambda)e^{-j\omega(\lambda+t_0)}d\lambda=e^{-j\omega t_0}\int_{-\infty}^{\infty}x(\lambda)e^{-j\omega\lambda}d\lambda=e^{-j\omega t_0}X(\omega)$$

By following similar steps,

$$\mathcal{F}\left[x(t+t_0)\right]=e^{j\omega t_0}X(\omega) \tag{5.16}$$

5.3.4 Frequency Shifting

This property forms a basis for every radio and TV transmitter. This property states that if $X(\omega)=\mathcal{F}\left[x(t)\right]$ and ω_0 is constant, then

$$\boxed{\mathcal{F}\left[x(t)e^{j\omega_0 t}\right]=X(\omega-\omega_0)} \tag{5.17}$$

Note that this is a dual of the time shifting property. The previous equation means that a shifting in the frequency-domain is equivalent to a phase shift in the time-domain. By definition,

$$\mathcal{F}\left[x(t)e^{j\omega_0 t}\right]=\int_{-\infty}^{\infty}x(t)e^{j\omega_0 t}e^{-j\omega t}dt$$

$$=\int_{-\infty}^{\infty}x(t)e^{-j(\omega-\omega_0)t}dt=X(\omega-\omega_0) \tag{5.18}$$

5.3.5 Time Differentiation

If $X(\omega)=\mathcal{F}\left[x(t)\right]$, then the Fourier transform of the derivative of $x(t)$ is given by

$$\boxed{\mathcal{F}\left[x'(t)\right]=j\omega X(\omega)} \tag{5.19}$$

This states that the transform of the derivative of $x(t)$ is obtained by multiplying its transform $X(\omega)$ by $j\omega$. By definition,

$$x(t)=\mathcal{F}^{-1}\left[X(\omega)\right]=\frac{1}{2\pi}\int_{-\infty}^{\infty}X(\omega)e^{j\omega t}\,d\omega \tag{5.20}$$

Taking the derivative of both sides with respect to t gives

$$\begin{aligned}\mathcal{F}\frac{dx(t)}{dt}&=\frac{j\omega}{2\pi}\int_{-\infty}^{\infty}X(\omega)e^{j\omega t}\,d\omega=j\omega^{-1}\left[X(\omega)\right]\\ \mathcal{F}\left[x'(t)\right]&=j\omega X(\omega)\end{aligned} \tag{5.21}$$

The nth derivative of the Equation 5.19 is

$$\mathcal{F}\left[x^{(n)}(t)\right]=(j\omega)^n X(\omega) \tag{5.22}$$

5.3.6 Frequency Differentiation

This property states that if $X(\omega)=\mathcal{F}\left[x(t)\right]$ then

$$\mathcal{F}\left[(-jt)^n x(t)\right]=\frac{d^n}{d\omega^n}X(\omega) \tag{5.23}$$

By definition,

$$\begin{aligned}\frac{d^n}{d\omega^n}X(\omega)&=\frac{d^n}{d\omega^n}\left(\int_{-\infty}^{\infty}x(t)e^{-j\omega t}\,dt\right)=\int_{-\infty}^{\infty}x(t)\frac{d^n}{d\omega^n}e^{-j\omega t}\,dt\\ &=\int_{-\infty}^{\infty}x(t)(-jt)^n e^{-j\omega t}\,dt=\int_{-\infty}^{\infty}(-jt)^n x(t)e^{-j\omega t}\,dt\\ &=\mathcal{F}((-jt)^n x(t))\end{aligned} \tag{5.24}$$

5.3.7 Time Integration

If $X(\omega) = \mathcal{F}\left[x(t)\right]$ then, the integration of function $x(t)$ is given by

$$\boxed{\mathcal{F}\left[\int_{-\infty}^{t} x(t)\,dt\right] = \frac{X(\omega)}{j\omega} + \pi X(0)\delta(\omega)} \tag{5.25}$$

Integrating $x(t)$ in the time-domain leads to division of $X(\omega)$ by $j\omega$ plus an additive term $\pi X(0)\delta(\omega)$ to account for the possibility of the dc component that may appear in the integral of $x(t)$. If we replace ω by 0 in Equation 5.6

$$X(0) = \int_{-\infty}^{\infty} x(t)\,dt \tag{5.26}$$

indicating that the dc component is zero when the integral of $x(t)$ over all time vanishes.

5.3.8 Duality

Duality between the time and frequency domains is another important property of Fourier transforms. This property relates to the fact that the analysis equation and synthesis equation look almost identical except for a factor of $1/2\pi$ and the difference of a minus sign in the exponential in the integral. The duality property states that if $X(\omega)$ is the Fourier transform of $x(t)$, then the Fourier transform of $X(t)$ is $2\pi x(-\omega)$; that is,

$$\boxed{\mathcal{F}\left[x(t)\right] = X(\omega) \Rightarrow \mathcal{F}\left[X(t)\right] = 2\pi x(-\omega)} \tag{5.27}$$

This expresses the fact that the Fourier transform pairs are symmetric. To derive the property, from Equation 5.7

$$x(t) = \mathcal{F}^{-1}\left[X(\omega)\right] = \frac{1}{2\pi}\int_{-\infty}^{\infty} X(\omega)e^{j\omega t}\,d\omega \tag{5.28}$$

$$2\pi x(t) = \int_{-\infty}^{\infty} X(\omega)e^{j\omega t}\,d\omega \tag{5.29}$$

Replacing t with $-t$, in Equation 5.29, we obtain

$$2\pi x(-t) = \int_{-\infty}^{\infty} X(\omega)e^{-j\omega t}\,d\omega \tag{5.30}$$

By interchanging t and ω, and comparing the result with the Equation 5.28

$$2\pi x(-\omega) = \int_{-\infty}^{\infty} X(t)e^{-j\omega t}dt = \mathcal{F}\left[X(t)\right] \tag{5.31}$$

5.3.9 Convolution

According to the convolution property, the Fourier transform of the convolution of two time functions is the product of their corresponding Fourier transforms. If $x(t)$ and $h(t)$ are two signals, their convolution $y(t)$ is given by the convolution integral

$$y(t) = h(t) * x(t) = \int_{-\infty}^{\infty} h(\tau)x(t-\tau)d\tau \tag{5.32}$$

If $X(\omega)$, $H(\omega)$, and $Y(\omega)$ are the Fourier transforms of $x(t)$, $h(t)$, and $y(t)$ respectively, then

$$\boxed{Y(\omega) = \mathcal{F}[h(t) * x(t)] = H(\omega)X(\omega)} \tag{5.33}$$

which states that the Fourier transform of the convolution of two-time functions is the product of their corresponding Fourier transforms.

To derive the convolution property, we take the Fourier transform on both sides of Equation 5.32 we get

$$Y(\omega) = \int_{-\infty}^{\infty}\left[\int_{-\infty}^{\infty} h(\tau)x(t-\tau)d\tau\right]e^{-j\omega t}dt \tag{5.34}$$

$$Y(\omega) = \int_{-\infty}^{\infty} h(\tau)\left[\int_{-\infty}^{\infty} x(t-\tau)e^{-j\omega t}dt\right]d\tau$$

By taking $\lambda = t - \tau$ so that $t = \lambda + \tau$ and $dt = d\lambda$.

$$Y(\omega) = \int_{-\infty}^{\infty} h(\tau)\left[\int_{-\infty}^{\infty} x(\lambda)e^{-j\omega(\lambda+\tau)}d\lambda\right]d\tau$$

$$Y(\omega) = \int_{-\infty}^{\infty} h(\tau)e^{-j\omega\tau}d\tau\int_{-\infty}^{\infty} x(\lambda)e^{-j\omega\lambda}d\lambda = H(\omega)X(\omega) \tag{5.35}$$

Table 5.1 presents these and other properties of Fourier transform, while Table 5.2 lists the transform pairs of some common functions. Note the similarities between these tables on Fourier transform and Tables 3.1 and 3.2 on Laplace transform.

TABLE 5.1
Properties of the Fourier Transform

No.	Property	$x(t)$	$X(\omega)$
1.	Linearity	$a_1x_1(t) + a_2x_2(t)$	$a_1X_1(\omega) + a_2X_2(\omega)$
2.	Scaling	$x(at)$	$\frac{1}{\lvert a \rvert} X\left(\frac{\omega}{a}\right)$
3.	Time shift	$x(t - a)$	$e^{-j\omega a}X(\omega)$
4.	Frequency shift	$e^{j\omega_0 t}x(t)$	$X(\omega-\omega_0)$
5.	Modulation	$\cos(\omega_0 t)x(t)$	$\frac{1}{2}\left[X(\omega+\omega_0)+X(\omega-\omega_0)\right]$
		$\sin(\omega_0 t)x(t)$	$\frac{j}{2}\left[X(\omega+\omega_0)-X(\omega-\omega_0)\right]$
6.	Time differentiation	$\frac{dx}{dt}$	$j\omega X(\omega)$
		$\frac{d^n x}{dt^n}$	$(j\omega)^n X(\omega)$
7.	Frequency differentiation	$(-jt)^n x(t)$	$\frac{d^n}{d\omega^n} X(\omega)$
8.	Time integration	$\int_{-\infty}^{t} x(t)dt$	$\frac{X(\omega)}{j\omega} + \pi X(0)\delta(\omega)$
9.	Time reversal	$x(-t)$	$X(-\omega)$ or $X^*(\omega)$
10.	Duality	$X(t)$	$2\pi x(-\omega)$
11.	Time convolution	$x_1(t)* x_2(t)$	$X_1(\omega)X_2(\omega)$
12.	Frequency convolution	$x_1(t)x_2(t)$	$\frac{1}{2\pi} X_1(\omega) * X_2(\omega)$
13.	Parseval's relation	$\int_{-\infty}^{\infty} \lvert x(t) \rvert^2 \, dt$	$\frac{1}{2\pi}\int_{-\infty}^{\infty} \lvert X(\omega) \rvert^2 \, d\omega$

Example 5.4

A signal $x(t)$ has a Fourier transform given by

$$X(\omega) = \frac{5(1+j\omega)}{8-\omega^2+6j\omega}$$

Without finding $x(t)$, find the Fourier transform of the following:

(a) $x(t - 3)$
(b) $x(4t)$
(c) $e^{-j2t}x(t)$
(d) $x(-2t)$

TABLE 5.2
Fourier Transform Pairs

No.	$x(t)$	$X(\omega)$
1.	$\delta(t)$	1
2.	1	$2\pi\delta(\omega)$
3.	$u(t)$	$\pi\delta(\omega)+\frac{1}{j\omega}$
4.	$u(t+\tau)-u(t-\tau)$	$2\frac{\sin\omega\tau}{\omega}$
5.	$\lvert t\rvert$	$\frac{-2}{\omega^2}$
6.	$\mathrm{sgn}(t)$	$\frac{2}{j\omega}$
7.	$e^{-at}u(t)$	$\frac{1}{a+j\omega}$
8.	$e^{at}u(-t)$	$\frac{1}{a-j\omega}$
9.	$t^n e^{-at}u(t)$	$\frac{n!}{(a+j\omega)^{n+1}}$
10.	$e^{-a\lvert t\rvert}$	$\frac{2a}{a^2+\omega^2}$
11.	$\lvert t\rvert e^{-at}u(t)$	$\frac{4aj\omega}{a^2+\omega^2}$
12.	$e^{j\omega_0 t}$	$2\pi\delta(\omega-\omega_0)$
13.	$\sin\omega_0 t$	$j\pi[\delta(\omega+\omega_0)-\delta(\omega-\omega_0)]$
14.	$\cos\omega_0 t$	$\pi[\delta(\omega+\omega_0)+\delta(\omega-\omega_0)]$
15.	$(\sin\omega_0 t)u(t)$	ove
16.	$(\cos\omega_0 t)u(t)$	$\pi[\delta(\omega+\omega_0)-\delta(\omega-\omega_0)]+\frac{j\omega}{\omega_0^2-\omega^2}$
17.	$\sin(\omega_0 t+\theta)$	$j\pi[e{-}j\theta\delta(\omega+\omega_0)-ej\theta\delta(\omega-\omega_0)]$
18.	$\cos(\omega_0 t+\theta)$	$\pi[e{-}j\theta\delta(\omega+\omega_0)+ej\theta\delta(\omega-\omega_0)]$
19.	$e^{-at}\sin\omega_o t u(t)$	$\frac{\omega_0}{(a+j\omega)^2+\omega_0^2}$
20.	$e^{-at}\cos\omega_o t u(t)$	$\frac{a+j\omega}{(a+j\omega)^2+\omega_0^2}$

(Continued)

TABLE 5.2 (*Continued*)
Fourier Transform Pairs

No.	$x(t)$	$X(\omega)$
21.	$\Pi(t/\tau)$	$\tau \sin c\left(\frac{\omega\tau}{2}\right)$
22.	$\Lambda\left(\frac{t}{\tau}\right)$	$\tau \sin c^2\left(\frac{\omega\tau}{2}\right)$
23.	$\sum_{n=-\infty}^{\infty} \delta(t-nT)$	$\omega_0 \sum_{n=-\infty}^{\infty} \delta(\omega - n\omega_0)$

Solution

We apply the relevant property for each case:

(a) $\mathcal{F}[x(t-3)] = e^{-j\omega 3}X(\omega) = \dfrac{5(1+j\omega)e^{-j\omega 3}}{8-\omega^2+j6\omega}$

(b) $\mathcal{F}[x(4t)] = \dfrac{1}{4}X\left(\dfrac{\omega}{4}\right) = \dfrac{\frac{5}{4}(1+j\omega/4)}{8-\omega^2/16+j6\omega/4} = \dfrac{5(4+j\omega)}{128-\omega^2+j24\omega}$

(c) $\mathcal{F}[e^{-j2t}x(t)] = X(\omega+2) = \dfrac{5[1+j(\omega+2)]}{8-(\omega+2)^2+6j(\omega+2)} = \dfrac{5(1+j\omega+j2)}{4-\omega^2-4\omega+6j\omega+j12}$

(d) $\mathcal{F}[x(-2t)] = \dfrac{1}{2}X\left(\dfrac{\omega}{-2}\right) = \dfrac{\frac{5}{2}(1-j\omega/2)}{8-\frac{\omega^2}{4}-\frac{6j\omega}{2}} = \dfrac{5(2-j\omega)}{32-\omega^2-12j\omega}$

Practice Problem 5.4 A signal $x(t)$ has the Fourier transform

$$X(\omega) = \frac{9}{9+\omega^2}$$

Find the Fourier transform of the following signals:

(a) $y(t) = x(2t-1)$
(b) $z(t) = dx(2t)/dt$
(c) $h(t) = \int_{-\infty}^{t} x(\lambda)d\lambda$

Answer: (a) $\dfrac{18e^{-j\omega/4}}{36+\omega^2}$, (b) $\dfrac{18j\omega}{36+\omega^2}$, (c) $\pi\delta(\omega)+\dfrac{9}{j\omega(9+\omega^2)}$

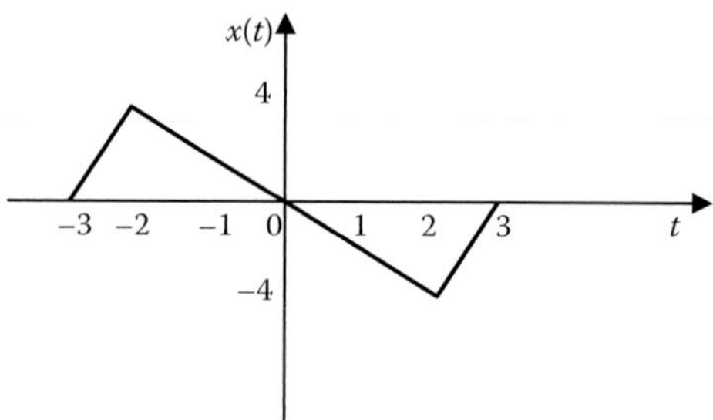

FIGURE 5.6 For Example 5.5.

Example 5.5

Determine the Fourier transform of the signal in Figure 5.6.

Solution

Although the Fourier transform of $x(t)$ can be found directly using Equation 5.6, it is much easier to find it using the derivative property. Taking the first derivative of $x(t)$ produces the signal in Figure 5.7a. Taking the second derivative gives us the signal in Figure 5.7b. From this,

$$x''(t) = 4\delta(t+3) - 6\delta(t+2) + 6\delta(t-2) - 4\delta(t-3)$$

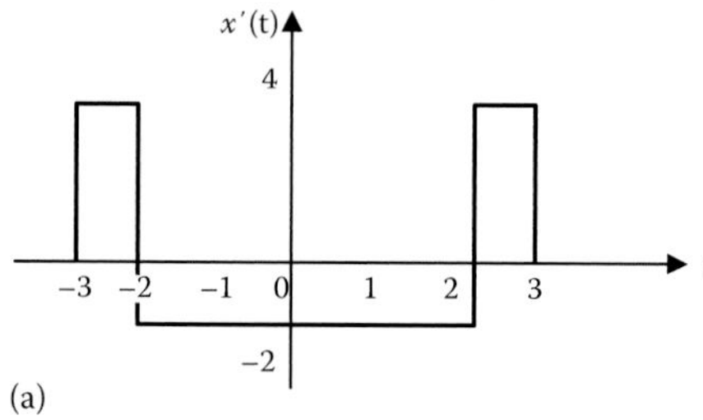

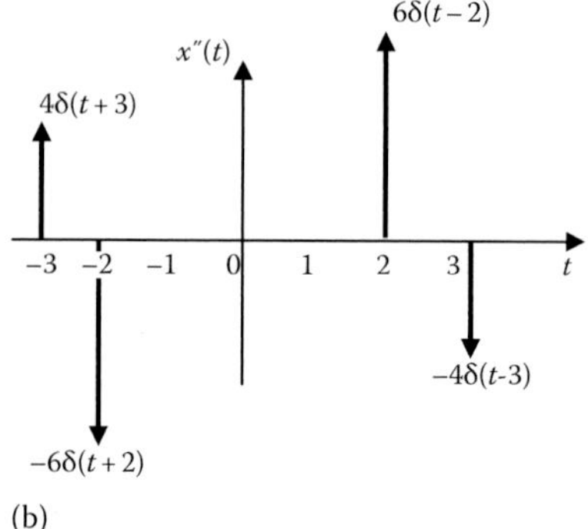

FIGURE 5.7 For Example 5.5: (a) first derivative of $x(t)$, (b) second derivative of $x(t)$.

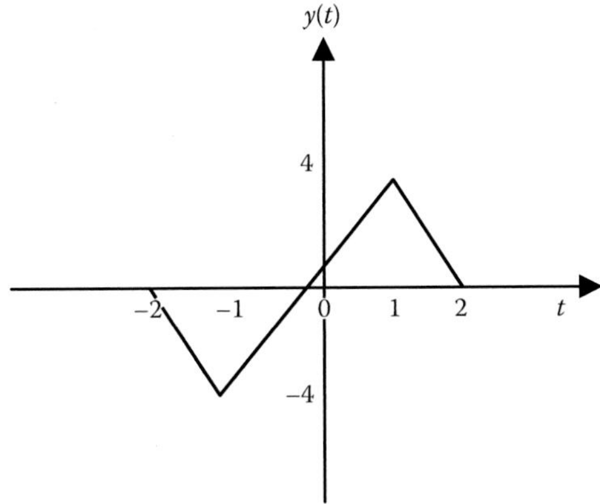

FIGURE 5.8 For Practice Problem 5.5.

Taking the Fourier transform of each term, we obtain

$$(j\omega)^2 X(\omega) = 4e^{j3\omega} - 6e^{j2\omega} + 6e^{-j2\omega} - 4e^{-j3\omega}$$

$$-\omega^2 X(\omega) = 4\left(e^{j3\omega} - e^{-j3\omega}\right) - 6\left(e^{j2\omega} - e^{-j2\omega}\right)$$

$$= j8\sin 3\omega - j12\sin 2\omega$$

$$X(\omega) = \frac{j}{\omega^2}\left(12\sin 2\omega - 8\sin 3\omega\right)$$

Practice Problem 5.5 Determine the Fourier transform of the function in Figure 5.8.

Answer: $Y(\omega) = \dfrac{j10}{\omega^2}(\sin 2\omega - 2\sin\omega)$

5.4 INVERSE FOURIER TRANSFORM

In general, we would use Equation 5.7 to find the inverse Fourier transform $x(t)$ of $X(\omega)$. Quite often, the Fourier transform is in the form of a rational function. In this case, we perform a partial fraction expansion of $X(\omega)$ and finding the corresponding time-domain signals using the Fourier transform pairs in Table 5.2. The procedure is similar to what we did in Section 3.4 for Laplace transform.

Example 5.6

Find the inverse Fourier transform of

(a) $G(\omega) = \dfrac{10j\omega}{(-j\omega + 2)(j\omega + 3)}$

(b) $Y(\omega) = \dfrac{\delta(\omega)}{(j\omega + 1)(j\omega + 2)}$

Solution

(a) To avoid complex algebra, let $s = j\omega$. Using partial fraction,

$$G(s) = \frac{10s}{(2-s)(3+s)} = \frac{-10s}{(s-2)(s+3)} = \frac{A}{s-2} + \frac{B}{s+3}, \quad s = j\omega$$

$$A = (s-2)G(s)\Big|_{s=2} = \frac{-10(2)}{2+3} = -4$$

$$B = (s+3)G(s)\Big|_{s=-3} = \frac{-10(-3)}{-3-2} = -6$$

$$G(\omega) = \frac{-4}{j\omega - 2} - \frac{6}{j\omega + 3}$$

Taking the inverse Fourier transform of each term,

$$g(t) = -4e^{2t}u(-t) - 6e^{-3t}u(t)$$

(b) Because of the delta function, we use Equation 5.7 to find the inverse.

$$y(t) = \frac{1}{2\pi}\int_{-\infty}^{\infty} \frac{\delta(\omega)e^{j\omega t}d\omega}{(2+j\omega)(j\omega+1)} = \frac{1}{2\pi}\frac{e^{j\omega t}}{(2+j\omega)(j\omega+1)}\Bigg|_{\omega=0} = \frac{1}{2\pi}\frac{1}{2} = \frac{1}{4\pi}$$

where the sifting property has been applied.

Practice Problem 5.6 Obtain the inverse Fourier transform of

(a) $\mathcal{F}(\omega) = \dfrac{e^{-j2\omega}}{1+j\omega}$

(b) $G(\omega) = \dfrac{\pi\delta(\omega)}{(5+j\omega)(2+j\omega)}$

Answer: (a) $f(t) = e^{-(t-2)}u(t-2)$, (b) $g(t) = 0.05$

5.5 APPLICATIONS

The Fourier transform is used extensively in a variety of fields such as optics, spectroscopy, acoustics, computer science, and electrical engineering. In electrical engineering, it is applied in circuit analysis, communications systems, and signal processing. Here we consider three application areas: circuit analysis, amplitude modulation (AM), and sampling. Filtering is another area of application, but we have already considered this in Section 4.6.3.

5.5.1 Circuit Analysis

We use Fourier transform to analyze circuits with nonsinusoidal excitations exactly in the same way we use phasor techniques to analyze circuits with sinusoidal excitations. For example, Ohm's law is written as

$$V(\omega) = Z(\omega)I(\omega) \tag{5.36}$$

where
$V(\omega)$ and $I(\omega)$ are the Fourier transforms of the voltage and current respectively
$Z(\omega)$ is the impedance

The impedances of resistors, inductors, and capacitors are the same as in phasor analysis, that is,

$$\begin{aligned} R &\Rightarrow R \\ L &\Rightarrow j\omega L \\ C &\Rightarrow \frac{1}{j\omega C} \end{aligned} \tag{5.37}$$

The Fourier approach to circuit analysis generalizes the phasor technique. It involves three steps. First, we transform the circuit elements into frequency-domain as in Equation 5.57 and take the Fourier transform of the excitations. Second, we apply circuit techniques such as Kirchhoff's voltage law, Kirchhoff's current law, voltage division, current division, source transformation, node or mesh analysis to find the unknown response (current or voltage). Third, we finally take the inverse Fourier transform to get the response in the time-domain. (We should note that Fourier analysis cannot handle circuits with initial conditions.)

As usual, the transfer function $H(\omega)$ is the ratio of the output response $Y(\omega)$ to the input excitation $X(\omega)$:

$$H(\omega) = \frac{Y(\omega)}{X(\omega)} \tag{5.38}$$

or

$$Y(\omega) = H(\omega)X(\omega) \tag{5.39}$$

Equation 5.39 states that if we know the transfer function $H(\omega)$ and the input $X(\omega)$, we can find the output $Y(\omega)$.

Example 5.7

Determine $v_o(t)$ in the circuit of Figure 5.9, where $i_s = 10e^{-2t}u(t)A$.

Solution

$$0.5\text{F} \rightarrow \frac{1}{j\omega C} = \frac{1}{j\omega 0.5} = \frac{2}{j\omega},$$

$$i_s = 10e^{-2t} \rightarrow I_s = \frac{10}{2 + j\omega}$$

Using current division, the current through the capacitor is

$$I_o = \frac{2}{2 + \frac{2}{j\omega}} I_s = \frac{j\omega}{1 + j\omega} \cdot \frac{10}{2 + j\omega}$$

$$V_o = I_o \frac{2}{j\omega} = \frac{20}{(1 + j\omega)(2 + j\omega)} = \frac{20}{(s+1)(s+2)}, s = j\omega$$

$$V_o = \frac{A}{s+1} + \frac{B}{s+2} = 20\left[\frac{1}{s+1} - \frac{1}{s+2}\right]$$

Taking the inverse Fourier transform produces

$$v_o(t) = 20(e^{-t} - e^{-2t})u(t)\ V$$

Practice Problem 5.7 Find $i_o(t)$ in the circuit of Figure 5.10. Let $v_s(t) = 4e^{-2t}u(t)$.

Answer: $i(t) = 4(e^{-0.5t} - e^{-t})u(t)$ A

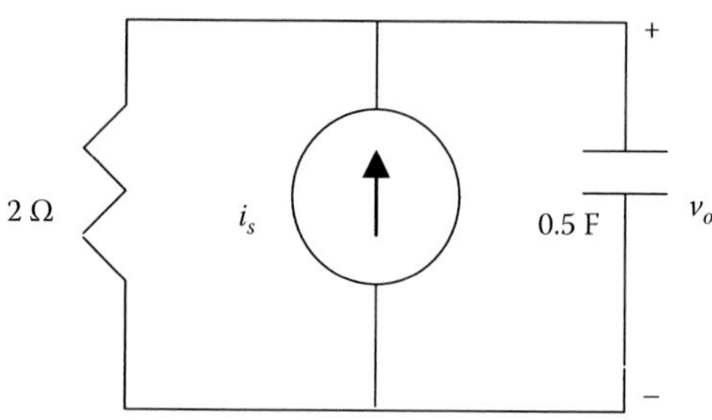

FIGURE 5.9 For Example 5.7.

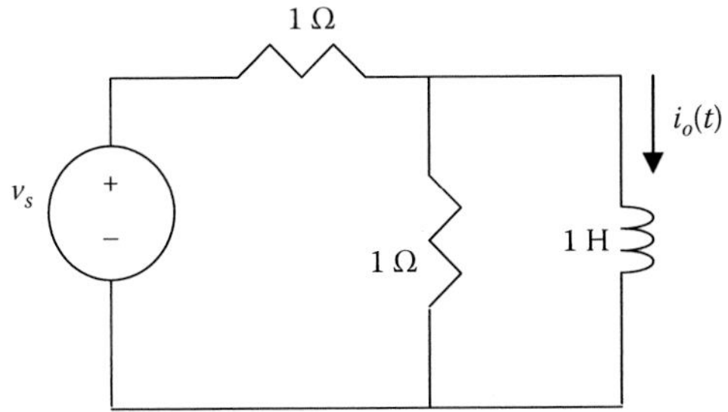

FIGURE 5.10 For Practice Problem 5.7.

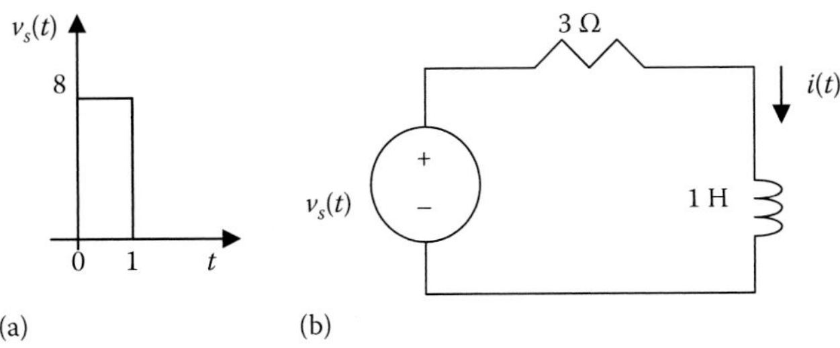

FIGURE 5.11 For Example 5.8.

Example 5.8

Given the circuit in Figure 5.11, with its excitation, determine the Fourier transform of $i(t)$.

Solution

One easy way to find the Fourier transform of v_s is to first find its derivative.

$$v_s{}'(t) = 8\delta(t) - 8\delta(t-1)$$

Taking the Fourier transform of each side,

$$j\omega V_s = 8(1 - e^{-j\omega}) \rightarrow V_s = \frac{8}{j\omega}\left(1 - e^{-j\omega}\right)$$

The current $I(\omega)$ is obtained as

$$I(\omega) = \frac{V_s}{3 + j\omega 1} = \frac{8(1 - e^{-j\omega})}{j\omega(3 + j\omega)}$$

Let $X(\omega) = \dfrac{8}{j\omega(3 + j\omega)}$, so that $I(\omega) = X(\omega)\left(1 - e^{-j\omega}\right)$

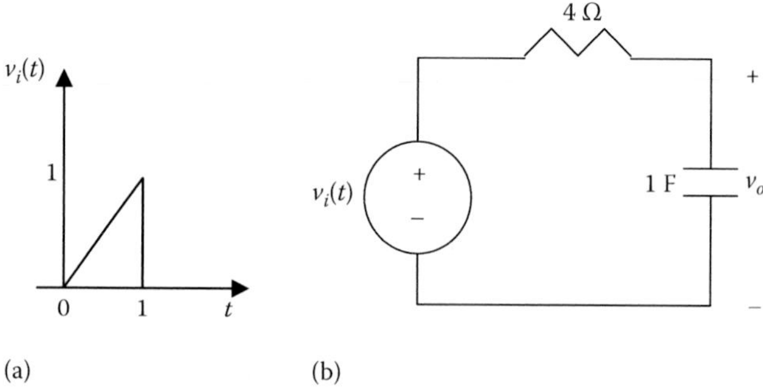

FIGURE 5.12 For Practice Problem 5.8.

$$X(s) = \frac{8}{s(s+3)} = \frac{A}{s} + \frac{B}{s+3'} \qquad s = j\omega$$

$$A = sX(s)\big|_{s=0} = \frac{8}{3}$$

$$B = (s+3)X(s)\big|_{s=-3} = -\frac{8}{3}$$

$$X(\omega) = \frac{8}{3}\left(\frac{1}{j\omega} - \frac{1}{3+j\omega}\right)$$

$$x(t) = \frac{8}{3}\left(\frac{1}{2}\text{sgn}(t) - e^{-3t}u(t)\right)$$

Since, $I(\omega) = X(\omega)(1-e^{-j\omega})$,

$$i(t) = x(t) - x(t-1)$$

$$= \frac{4}{3}\text{sgn}(t) - \frac{8}{3}e^{-3t}u(t) - \frac{4}{3}\text{sgn}(t-1) - \frac{8}{3}e^{-3(t-1)}u(t-1)$$

Practice Problem 5.8 Determine the voltage $V_o(\omega)$ in the circuit of Figure 5.12b, given the voltage source shown in Figure 5.12a.

Answer: $V_o(\omega) = \dfrac{1}{\omega^2(1+4j\omega)}\left(e^{-j\omega} + j\omega e^{-j\omega} - 1\right)$

5.5.2 Amplitude Modulation

AM is one of the applications of the Fourier transform. Transmission of information through space has become part of modern society, especially with the wireless telephony. However, transmitting intelligent signals—such as for speech and music—is not practical: it requires a lot of power and large antennas. One way of transmitting low-frequency audio information (50 Hz to 20 kHz) is to transmit it along with a high-frequency signal, called a *carrier*. Any of the three characteristics

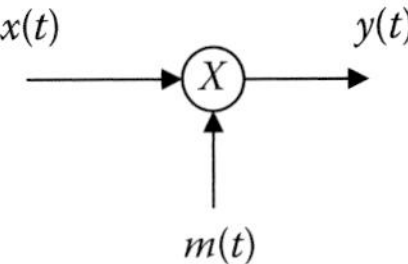

FIGURE 5.13 Modulated signal.

(amplitude, frequency, or phase) of a carrier can be controlled, to be able to carry the intelligent signal, called the *modulating signal*. Here, we are interested in the control of the carrier's amplitude. This is known as AM.

Amplitude modulation (AM) is a process whereby we let the modulating signal control the amplitude of the carrier.

Consider the signal multiplier shown in Figure 5.13. We focus only on the case when the modulating signal $m(t) = \cos\omega_o t$. The output of the multiplier is

$$y(t) = x(t)\cos\omega_o t \tag{5.40}$$

Since the multiplication in the time-domain always results in convolution in the frequency-domain,

$$\begin{aligned} Y(\omega) &= \frac{1}{2\pi}X(\omega) * \pi\left[\delta(\omega-\omega_o)+\delta(\omega+\omega_o)\right] \\ &= \frac{1}{2}\left[X(\omega-\omega_o)+X(\omega+\omega_o)\right] \end{aligned} \tag{5.41}$$

Thus, modulation results in shifting the spectrum of the signal as illustrated in Figure 5.14. The signal with frequency $\omega_c - \omega_o$ is known as the *lower sideband*, while the one with frequency $\omega_c + \omega_o$ is known as the *upper sideband*.

At the receiving end of the transmission, the audio information is received from the modulated carrier by a process known as **demodulation**. This shifts back the message spectrum to its original low frequency location. A synchronous demodulation multiplies the modulated signal $y(t)$ with $\cos\omega_o t$ as shown in Figure 5.15.

$$z(t) = y(t)\cos\omega_o t \tag{5.42}$$

Taking the Fourier transform as we did before,

$$\begin{aligned} Z(\omega) &= \frac{1}{2}\left[Y(\omega-\omega_o)+Y(\omega+\omega_o)\right] \\ &= \frac{1}{2}X(\omega)+\frac{1}{4}X(\omega-2\omega_o)+\frac{1}{4}X(\omega+2\omega_o) \end{aligned} \tag{5.43}$$

This is illustrated in Figure 5.16. The signal $z(t)$ is passed through a low-pass filter (see Figure 5.17a), which passes only the low-frequency components. Thus, $x(t)$ is extracted as shown in Figure 5.17b.

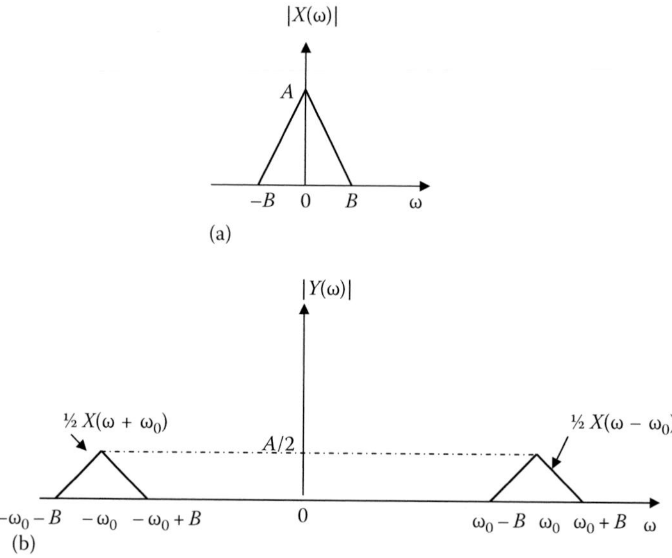

FIGURE 5.14 Amplitude spectra of: (a) signal $x(t)$, (b) modulated signal $y(t)$.

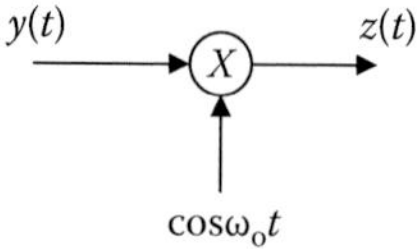

FIGURE 5.15 Demodulated signal.

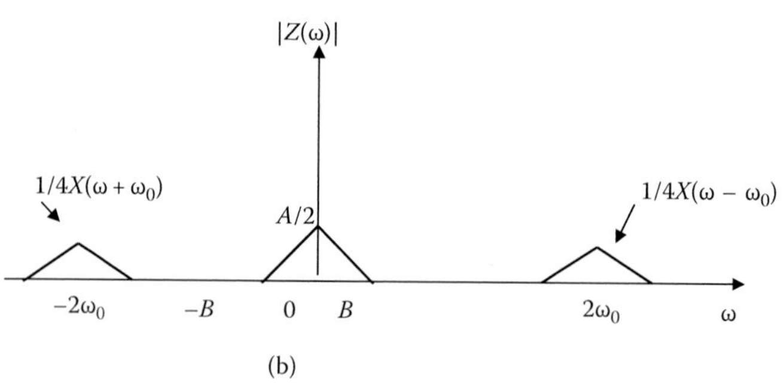

FIGURE 5.16 Amplitude spectra of $z(t)$.

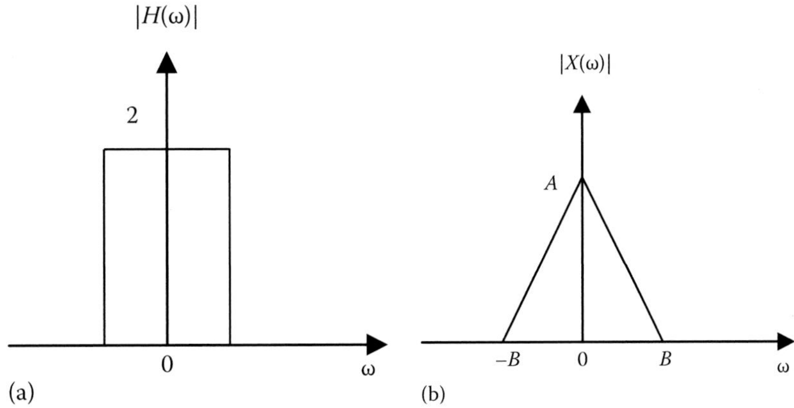

FIGURE 5.17 (a) Low-pass filter, (b) spectrum of the original signal.

Example 5.9

An AM signal is given by

$$x(t) = \cos 200\pi t \cos 10^4 t$$

Identify the upper sideband and the lower sideband.

Solution

$$\omega_o = 200\pi \rightarrow f_o = \frac{\omega_o}{2\pi} = 100$$

$$\omega_c = 10^4\pi \rightarrow f_c = \frac{\omega_c}{2\pi} = 5000$$

$$\text{USB} = f_c + f_o = 5100 = 5.1 \text{ kHz}$$

$$\text{LSB} = f_c - f_o = 4900 = 4.9 \text{ kHz}$$

Practice Problem 5.9 A music signal whose frequencies range from 80 Hz to 12 kHz is used to modulate a 2-MHz carrier. Find the range of frequencies for the lower and upper sidebands.

Answer: Lower sideband: 1,999,200–1,988,000 Hz
Upper sideband: 2,000,080–2,012,000 Hz

5.5.3 Sampling

Sampling is an important operation in signal processing. It may be regarded as a way of reducing analog signals to discrete signals. Thus sampling is the bridge from

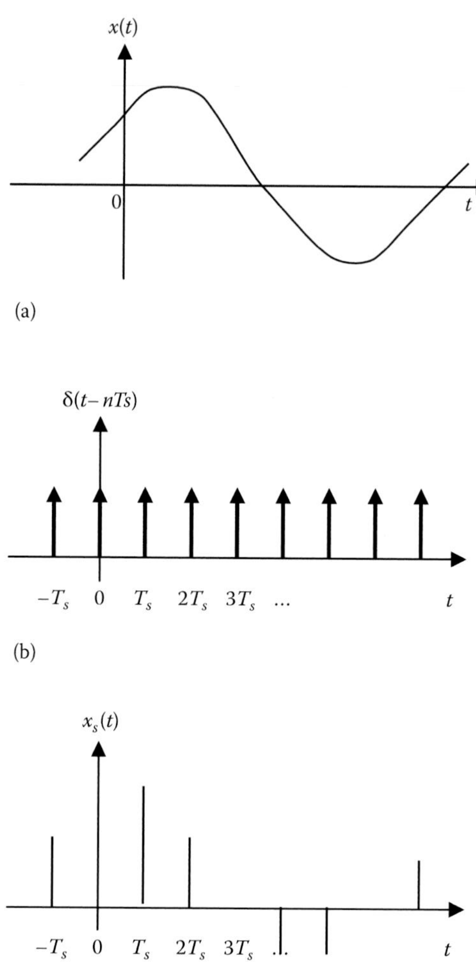

FIGURE 5.18 (a) Continuous (analog) signal to be sampled, (b) train of impulses, (c) sampled (digital) signal.

continuous to discrete signals. In analog systems we process entire signals, but in digital systems, only samples of signals are necessary. Sampling can be done by using a train of pulses or impulses. We will use impulse sampling here.

Consider the continuous-time signal $x(t)$ depicted in Figure 5.18a. We multiply by a train of impulses $\delta(t-nT_s)$ as shown in Figure 5.18b, where T_s is the *sampling interval* and $f_s = 1/T_s$ is the *sampling frequency* (or *rate*). The value of $x(t)$ at point nT_s is

$$x(t)\delta(t-nT_s) = x(nT_s)\delta(t-nT_s) \tag{5.44}$$

The sampled signal $x_s(t)$ in Figure 5.18c is, therefore,

$$x_s(t) = x(t)\sum_{n=-\infty}^{\infty} \delta(t - nT_s) \tag{5.45}$$

We take the Fourier transform of this and apply the frequency convolution property.

$$\begin{aligned} X_s(\omega) &= \frac{1}{2\pi}\left[X(\omega)\right] * \omega_s \sum_{n=-\infty}^{\infty} \delta(\omega - n\omega_s) \\ &= \frac{1}{T_s}\sum_{n=-\infty}^{\infty} X(\omega) * \delta(\omega - n\omega_s) \\ &= \frac{1}{T_s}\sum_{n=-\infty}^{\infty} X(\omega - n\omega_s) \end{aligned} \tag{5.46}$$

where $\omega_s = 2\pi/T_s$. We have used item 23 in Table 5.2. Equation 5.46 states that the Fourier transform $X_s(\omega)$ of the sampled signal is a sum of translates of the Fourier transform of the original signal $X(\omega)$ at a rate of $1/T_s$.

To ensure that the sampling process is successful, we must carefully select the sampling interval. This is stated in the following **sampling theorem**:

> A **band-limited** signal, with bandwidth W hertz, may be completely recovered from its samples if taken at a frequency at least twice as high as $2W$ samples per second.

(In practice, a signal is made band-limited by passing it through a low-pass filter before sampling it.) For a signal, with no frequency component higher than W hertz, there is no loss of information if the sampling frequency is at least twice the highest frequency in the signal, that is,

$$\frac{1}{T_s} = f_s \geq 2W \tag{5.47}$$

The minimum sampling frequency $f_s = 2W$ is known as the **Nyquist frequency** or rate, and the maximum spacing $1/f_s$ is the *Nyquist interval*, named after **Harry Nyquist** (1889–1976) for his paper published in 1928. The Nyquist rate is the minimum sampling rate necessary to preserve the information $x(t)$. A signal is said to be oversampled if it is sampled at a rate greater than its Nyquist rate. It is undersampled if it is sampled at less than the Nyquist rate.

Example 5.10

For frequencies above 4 kHz, the spectrum of a signal is zero. To sample the signal, find the maximum time spacing between samples.

Solution

$$W = 4\text{ kHz}$$

The Nyquist rate is

$$f_s = 2W = 8\text{ kHz}$$

The Nyquist interval is

$$T_s = \frac{1}{f_s} = \frac{1}{8\times 10^3} = 125\,\mu\text{s}$$

Practice Problem 5.10 A voice signal is band-limited to 2.5 kHz. Find the Nyquist rate and the Nyquist interval for the signal.

Answer: 5 kHz, 0.2 ms

5.6 PARSEVAL'S THEOREM

Parseval's theorem demonstrates another application of the Fourier transform. It states the relationship between energy in the time and frequency domains. The theorem is important in communications and signal processing.

If $p(t)$ is the power associated with the signal, the energy carried by the signal is

$$E = \int_{-\infty}^{\infty} p(t)dt \tag{5.48}$$

It is convenient to use a $1-\Omega$ resistor as the base for energy calculation. For a $1-\Omega$ resistor, $p(t) = v^2(t) = i^2(t) = x^2(t)$, where $x(t)$ stands for either voltage or current. Thus Equation 5.48 becomes

$$E_{1\Omega} = \int_{-\infty}^{\infty} x^2(t)dt \tag{5.49}$$

Parseval's theorem states that this same energy can be calculated in the frequency-domain as

$$E_{1\Omega} = \int_{-\infty}^{\infty} x^2(t)dt = \frac{1}{2\pi}\int_{-\infty}^{\infty} |X(\omega)|^2 \, d\omega \tag{5.50}$$

indicating that the Fourier transform is an energy-conserving relation.

Parseval's theorem states that the total energy delivered to a $1 - \Omega$ resistor equals the total area under the square $x^2(t)$ or $1/2\pi$ times the total area under the magnitude of the Fourier transform squared $|X(\omega)|^2$.

Parseval's theorem is named after the French mathematician Marc-Antoine Parseval des Chênes (1755–1836). It relates energy between the time-domain and the frequency-domain. It provides the physical significance of $X(\omega)$, namely, that $|X(\omega)|^2$ is the energy density (in joules per hertz) corresponding to $x(t)$. The plot of $|X(\omega)|^2$ versus ω is the **energy spectrum** of the signal $x(t)$.

To derive Equation 5.50, we begin with Equation 5.49 and substitute Equation 5.7 for one of the $x(t)$'s.

$$E_{1\Omega} = \int_{-\infty}^{\infty} x^2(t)dt = \int_{-\infty}^{\infty} x(t)\left[\frac{1}{2\pi}\int_{-\infty}^{\infty} X(\omega)e^{j\omega t} d\omega\right] dt \tag{5.51}$$

The function $x(t)$ can be moved inside the inner integral, since the integral does not involve time:

$$E_{1\Omega} = \frac{1}{2\pi}\int_{-\infty}^{\infty}\int_{-\infty}^{\infty} x(t)X(\omega)e^{j\omega t} d\omega dt \tag{5.52}$$

Reversing the order of integration,

$$\begin{aligned} E_{1\Omega} &= \frac{1}{2\pi}\int_{-\infty}^{\infty} X(\omega)\left[\int_{-\infty}^{\infty} x(t)e^{j\omega t} dt\right] d\omega \\ &= \frac{1}{2\pi}\int_{-\infty}^{\infty} X(\omega)X(-\omega)d\omega = \frac{1}{2\pi}\int_{-\infty}^{\infty} X(\omega)X^*(\omega)d\omega \end{aligned} \tag{5.53}$$

But $X(\omega)X^*(\omega) = |X(\omega)|^2$. Hence,

$$\boxed{E_{1\Omega} = \int_{-\infty}^{\infty} x^2(t)dt = \frac{1}{2\pi}\int_{-\infty}^{\infty} |X(\omega)|^2 \, d\omega} \tag{5.54}$$

as expected. Equation 5.54 shows the energy of $x(t)$ can be computed directly in the time-domain or indirectly from $X(\omega)$ in the frequency-domain.

Since $|X(\omega)|^2$ is an even function, we may integrate from 0 to ∞ and double the result; that is,

$$E_{1\Omega} = \int_{-\infty}^{\infty} x^2(t)dt = \frac{1}{\pi}\int_{0}^{\infty} |X(\omega)|^2 \, d\omega \tag{5.55}$$

We may calculate the energy in any frequency band $\omega_1 < \omega < \omega_2$ as

$$E_{1\Omega} = \frac{1}{2\pi}\int_{\omega_1}^{\omega_2} |X(\omega)|^2 \, d\omega \tag{5.56}$$

We should note that Parseval's theorem, as stated here, applies to nonperiodic signals. Parseval's theorem for periodic signals was covered in Section 4.4.6.

Example 5.11

A band-pass filter has its lower and upper cutoff frequencies as 10 and 20 Hz respectively. If the input signal is $V_i = 4e^{-t}u(t)$, calculate the $1-\Omega$ energy of the input and the percentage that appears at the output.

Solution

The energy of the input signal is

$$E_i = \int_{-\infty}^{\infty} v_i^2(t)dt = \int_{0}^{\infty} 16e^{-2t}dt = \frac{16}{-2}e^{-2t}\Big|_0^{\infty} = 8 \text{ J}$$

The Fourier transform of the input signal is

$$V_i = \frac{4}{1+j\omega}$$

The energy in the interval $10 < f < 20$ or $20\pi < \omega < 40\pi$ is

$$E_o = \frac{1}{2\pi}\int_{20\pi}^{40\pi} |V_i(\omega)|^2 d\omega = \frac{1}{2\pi}\int_{20\pi}^{40\pi} \frac{16}{(1+\omega^2)} d\omega = \frac{8}{\pi}\int_{20\pi}^{40\pi} \frac{d\omega}{1+\omega^2}$$

But $\int \frac{dx}{a^2+\omega^2} = \frac{1}{a}\tan^{-1}\frac{x}{a} + C$

$$E_o = \frac{8}{\pi}\tan^{-1}\omega\Big|_{20\pi}^{40\pi} = \frac{8}{\pi}(1.5628 - 1.5549) = 0.02012$$

$$\frac{E_o}{E_i}\times 100 = \frac{2.012}{8} = 0.252\%$$

Practice Problem 5.11 If $x(t) \Leftrightarrow X(\omega)$, and

$$X(\omega) = \begin{cases} 10, & |\omega| < 2 \\ 0, & |\omega| > 2 \end{cases}$$

Let $y(t) = \frac{d^2x(t)}{dt^2}$. Find $\int_{-\infty}^{\infty} |y(t)|^2\, dt$

Answer: 2010.62 J.

5.7 COMPARING THE FOURIER AND LAPLACE TRANSFORMS

The Fourier and Laplace transforms are closely related, as evident in their definitions, which are repeated here for convenience. The general (bilateral) Laplace transform of $x(t)$ is

$$X(s) = \int_{-\infty}^{\infty} x(t)e^{-st}dt \tag{5.57}$$

where $s = \sigma + j\omega$. If we let $s = j\omega$, Equation 5.57 becomes

$$X(\omega) = \int_{-\infty}^{\infty} x(t)e^{-j\omega t}dt \tag{5.58}$$

which is the Fourier transform of $x(t)$. For a causal signal $x(t)$ (i.e., positive-time signal, $x(t) = 0,\ t < 0$) and $\int_{-\infty}^{\infty} |x(t)|\, dt < \infty$, the two transforms are related by

$$X(\omega) = X(s)\Big|_{s=j\omega} \tag{5.59}$$

showing that the Fourier transform is a special case of the Laplace transform with $s = j\omega$. We can use Equation 5.59 to obtain the Fourier transform of signals for which the Fourier integral converges. This integral will converge when all the poles of $\mathcal{F}(s)$ lie in the left half of the s-plane and not on the $j\omega$ axis.

The Laplace transform defined and used in Chapter 4 is one-sided (or unilateral) in that the integral is over $0 < t < \infty$, restricting its usefulness to only positive-time signals, $x(t)$, $t > 0$. The Fourier transform accommodates signals existing for all time.

Although the Laplace transform has some advantages over the Fourier transform in circuit analysis, being an extension of Fourier analysis to nonperiodic signals, Fourier transform uses the same frequency-domain representation as phasor analysis. Fourier transform shares some operational properties with the Laplace transform but has some unique properties. For circuit analysis, Laplace transform is more widely used because the Laplace integral converges for a wider range of signals and automatically incorporates initial conditions.

5.8 COMPUTING WITH MATLAB®

MATLAB® function **fourier** can be used to find the Fourier transform of a signal $x(t)$, while the function **ifourier** can be used to find the inverse Fourier transform of $X(\omega)$. How to use them can be displayed with the **help Fourier** and **help ifourier** commands. Since transfer functions are already functions of frequency, it is straightforward and easy to plot them using MATLAB.

Example 5.12

Use MATLAB to find the Fourier transform of

$$x(t) = u(t-2) - e^{-2t}u(t)$$

Solution

The MATLAB script for finding the Fourier transform of $x(t)$ is as follows:

```
syms  x t
 x=heaviside(t-2)-exp(-2*t)*Heaviside(t) %define the signal
 X = fourier(x)
pretty(X) % simplify
```

This returns

```
x =
heaviside(t - 2) - heaviside(t)/exp(2*t)
 X =
- (1/exp(2*w*i))*(- pi*dirac(-w) + i/w) - 1/(2 + w*i)

          1      /                     i \      1
  - ----------- | - pi dirac(-w) + - | - -------
    exp(2 w i) \                       w /    2 + w i
```

From this, we obtain

$$X(\omega) = e^{-j2\omega}\left[-\pi\delta(-\omega) + \frac{j}{\omega}\right] - \frac{1}{2+j\omega}$$

Practice Problem 5.12 Let $y(t) = 4\delta(t) + 2e^{-t}u(t)$. Use MATLAB to find $Y(\omega)$.

Answer: $Y(\omega) = \dfrac{2}{1+j\omega} + 4$

Example 5.13

Use MATLAB to find the inverse Fourier transform of

$$X(\omega) = \frac{2j\omega}{1+j\omega}$$

Solution

The following MATLAB script is used to find the inverse of $X(\omega)$.

```
syms  X  x  w
X = 2*j*w/( 1 + j*w);
x = ifourier(X)
pretty(x)
```

This returns

```
x =
((4*pi*dirac(x))/exp(x) - (4*pi*heaviside(x))/exp(x))/(2*pi)
 4 PI exp(-x) dirac(x) - 4 PI exp(-x) heaviside(x)
 -------------------------------------------------
                        2 PI
```

From this, we get

$$x(t) = 2e^{-t}\delta(t) - 2e^{-t}u(t)$$

Practice Problem 5.13 Use MATLAB to find the inverse Fourier transform of

$$X(\omega) = \frac{\delta(\omega)}{1+j2\omega}$$

Answer: $1/(2\pi)$

Example 5.14

Consider a system with transfer function

$$H(s) = \frac{100s^2}{s^4 + 25s^3 + 50s^2 + 400s + 6000}, \qquad s = j\omega$$

Use MATLAB to obtain the plot of $H(\omega)$.

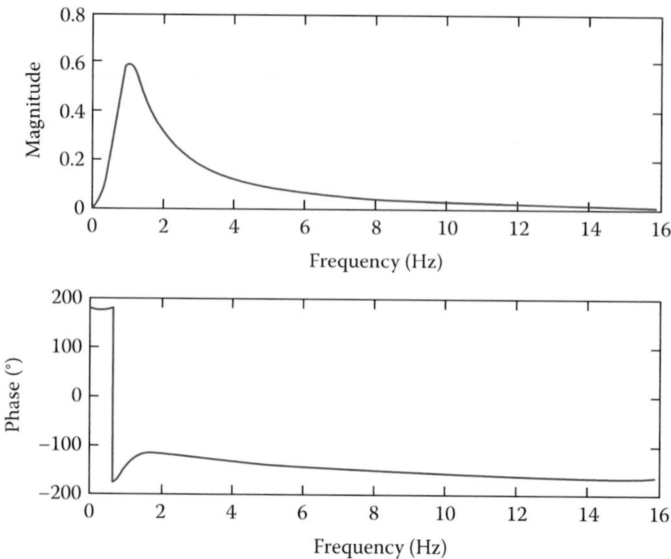

FIGURE 5.19 For Example 5.14.

Solution

The MATLAB script is as follows. Running the script results in the plot shown in Figure 5.19.

```
num = [100 0 0];
den = [ 1  25  50  400  6000];
w = 0.1:0.1:100
H = freqs(num, den,w);
mag =abs(H);
phase = angle(H)*180/pi   % converts phase to degrees
subplot(2,1,1)
 plot(w/(2*pi), mag)
  xlabel('Frequency (Hz)')
  ylabel('Magnitude')
subplot(2,1,2)
 plot(w/(2*pi), phase)
  xlabel('Frequency (Hz)')
ylabel('Phase (deg)')
```

Practice Problem 5.14 Use MATLAB to plot the frequency response to the transfer function

$$H(s) = \frac{1}{s^3 + 7s^2 + 14s + 8}$$

Answer: See Figure 5.20.

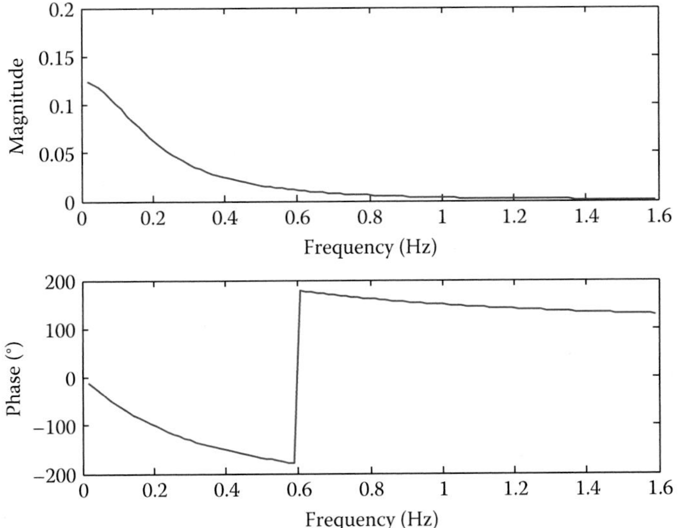

FIGURE 5.20 For Practice Problem 5.14.

5.9 SUMMARY

1. The Fourier transform is defined as

$$X(\omega) = \mathcal{F}\left[x(t)\right] = \int_{-\infty}^{\infty} x(t)e^{-j\omega t}dt$$

   It converts a nonperiodic signal $x(t)$ into a transform $X(\omega)$.

2. The inverse Fourier transform of $X(\omega)$ is defined as

$$x(t) = \mathcal{F}^{-1}[X(\omega)] = \frac{1}{2\pi}\int_{-\infty}^{\infty} X(\omega)e^{j\omega t}d\omega$$

3. The magnitude $|X(\omega)|$ plotted against ω is known as the magnitude spectrum of $x(t)$, while the plot of $|X(\omega)|^2$ against ω is known as the energy spectrum. The angle of $X(\omega)$ plotted against ω is known as the phase spectrum.
4. Common Fourier transform properties and pairs are in Tables 5.1 and 5.2 respectively.
5. If $H(\omega)$ is the transfer function of a system, the output $Y(\omega)$ can be obtained from the input $X(\omega)$ using

$$Y(\omega) = H(\omega)X(\omega)$$

6. Parseval's theorem provides the energy relationship between a signal $x(t)$ and its Fourier transform $X(\omega)$. The $1-\Omega$ energy is

$$E_{1\Omega} = \int_{-\infty}^{\infty} x^2(t)dt = \frac{1}{2\pi}\int_{-\infty}^{\infty} |X(\omega)|^2 \, d\omega$$

7. Applications of the Fourier transform include circuit analysis, AM, and sampling. Applying the Fourier transform approach to circuit analysis involves transforming the excitation and circuit elements into the frequency-domain, solving for the unknown response, and transforming the response to the time-domain using the inverse Fourier transform. For AM, the modulating signal is used to control the amplitude of the carrier. For sampling application, we found that no information is lost in sampling if the sampling frequency is equal to at least twice the Nyquist rate.
8. If $x(t) = 0$ for $t \leq 0$, we can obtain the Fourier transform of signal $x(t)$ from the one-sided Laplace transform $X(s)$ of $x(t)$ by replacing s with $j\omega$.
9. MATLAB commands can be used to find the Fourier transform of a signal or its inverse.

REVIEW QUESTIONS

5.1 The Fourier transform of $x(t)$ is called the spectrum of $x(t)$.
(a) True, (b) false

5.2 The Fourier transform and the Laplace transform of $\delta(t)$ are the same.
(a) True, (b) false.

5.3 Which of these functions is not Fourier transformable?
(a) $e^t u(-t)$, (b) $te^{-3t}u(t)$, (c) $1/t$, (d) $|t|u(t)$

5.4 If $x(t) = 2\delta(t)$ and $y(t) = dx/dt$, then $Y(\omega)$ is
(a) $2\delta'(\omega)$, (b) $2\delta(\omega)$, (c) $2j\omega$, (d) $2/j\omega$

5.5 The inverse Fourier transform of $\dfrac{e^{-j\omega}}{3+j\omega}$ is
(a) e^{-3t}, (b) $e^{-3t}\,u(t-1)$, (c) $e^{-3(t-1)}$, (d) $e^{-3(t-1)}u(t-1)$

5.6 The property that relates $e^{j\omega_0 t}X(t) \Leftrightarrow X(\omega-\omega_o)$ is
(a) Time shift, (b) frequency shift, (c) time differentiation, (d) frequency differentiation.

5.7 Evaluating the integral $\displaystyle\int_{-\infty}^{\infty} \frac{8\delta(\omega-2)}{4+\omega^2} d\omega$ gives
(a) 0, (b) 1 (c) 2, (d) ∞

5.8 A unit step current is applied through a 1 H inductor with no initial current. The voltage across the inductor is
(a) $u(t)$, (b) $\text{sgn}(t)$, (c) $e^{-t}u(t)$, (d) $\delta(t)$

5.9 The total energy of a signal is independent of its phase spectrum.
(a) True, (b) false

5.10 The MATLAB command for finding the inverse Fourier transform is
(a) invfourier, (b) ifourier, (c) fourierinv, (d) fourier

Answers: 5.1a, 5.2a, 5.3c,5.4 c,5.5d, 5.6b, 5.7b, 5.8d, 5.9b, 5.10b

PROBLEMS

SECTIONS 5.2 AND 5.3—DEFINITION OF THE FOURIER TRANSFORM AND PROPERTIES OF FOURIER TRANSFORM

5.1 Find the Fourier transform of

$$x(t) = \begin{cases} 1, & 0 < t < 1 \\ 0, & 1 < t < 2 \\ -1, & 2 < t < 3 \\ 0, & \text{otherwise} \end{cases}$$

5.2 Obtain the Fourier transforms of the following signals:

(a) $x_1(t) = \begin{cases} 1, & -2 < t < -1 \\ 0, & -1 < t < 1 \\ 1, & 1 < t < 2 \\ 0, & \text{otherwise} \end{cases}$

(b) $x_2(t) = \begin{cases} -1, & -2 < t < -1 \\ 0, & -1 < t < 1 \\ 1, & 1 < t < 2 \\ 0, & \text{otherwise} \end{cases}$

5.3 Calculate the Fourier transforms of the signals in Figure 5.21.
5.4 Compute the Fourier transform of the signal in Figure 5.22.
5.5 Find the Fourier transform of the signal shown in Figure 5.23.

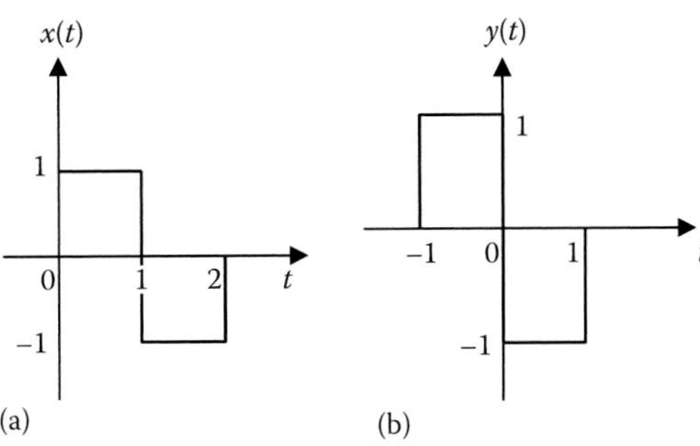

FIGURE 5.21 For Problem 5.3.

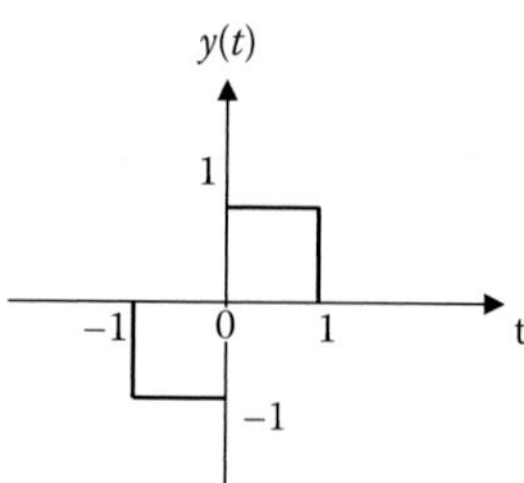

FIGURE 5.22 For Problem 5.4.

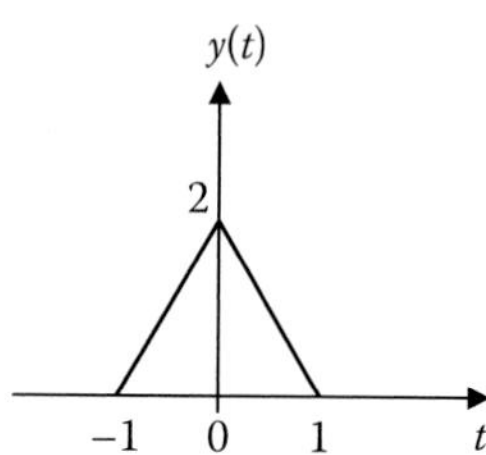

FIGURE 5.23 For Problem 5.5.

5.6 Using the properties of the Fourier transform, determine the Fourier transform of these signals:

(a) $x(t) = \sin(2t)u(t)$

(b) $y(t) = e^{-t}\cos(3t)u(t)$

(c) $z(t) = te^{-2t}u(t)$

5.7 Verify that

(a) $$\int_{-\infty}^{\infty} \cos\omega t d\omega = 2\pi\delta(t)$$

(b) $$\frac{j}{2}\int_{-\infty}^{\infty} \left[\delta(\omega+\omega_o) - \delta(\omega-\omega_o)\right] e^{j\omega t} d\omega = \sin(\omega_o t)$$

5.8 Find the Fourier transform of the following signals:

(a) $x_1(t) = e^{-3t}\sin 10tu(t)$

(b) $x_2(t) = e^{-4t}\cos 10tu(t)$

5.9 (a) Find the Fourier transform of $x(t) = 2\cos(\omega_o t + \theta)$

(b) Use the result in part (a) to derive the Fourier transform of $y(t) = 2\cos(100\pi t - \pi/4)$.

5.10 Given that $\mathcal{F}(\omega) = \mathcal{F}[f(t)]$, prove the following results using the definition of Fourier transform

(a) $$\mathcal{F}\left[f(t-t_o)\right] = e^{-j\omega t_o}\mathcal{F}(\omega)$$

(b) $$\mathcal{F}\left[\frac{df(t)}{dt}\right] = j\omega\mathcal{F}(\omega)$$

(c) $\mathcal{F}[f(-t)] = \mathcal{F}(-\omega)$

(d) $\mathcal{F}\left[tf(t)\right] = j\frac{d}{d\omega}\mathcal{F}(\omega)$

5.11 A signal has Fourier transform

$$X(\omega) = \frac{5 + j\omega}{-\omega^2 + j2\omega + 1}$$

Determine the Fourier transform of these signals:

(a) $y(t) = x(2t - 5)$

(b) $y(t) = tx(t)$

(c) $y(t) = x(t)\sin 2t$

(d) $y(t) = \frac{d^2x(t)}{dt^2}$

5.12 A signal $f(t)$ has a Fourier transform given by

$$\mathcal{F}(\omega) = \frac{5(1 + j\omega)}{8 - \omega^2 + 6j\omega}$$

Without finding $f(t)$, find the Fourier transform of the following:

(a) $f(t - 3)$

(b) $f(4t)$

(c) $e^{-j2t}f(t)$

(d) $f(-2t)$

5.13 The Fourier transform of a signal $x(t)$ is

$$X(\omega) = \frac{2}{j\omega} + \pi\delta(\omega)$$

Derive the Fourier transform of the following signals:

(a) $x_1(t) = 4x(2t-1)$

(b) $x_2(t) = x(t-2)\sin t$

(c) $x_3(t) = \frac{d}{dt}x(t)$

(d) $x_4(t) = tx(t)$

5.14 Obtain the Fourier transform of each of the following signals:

(a) $x(t) = 4\cos 2t + 2\delta(t)$

(b) $y(t) = 4\sin 2t - 2\sin 2(t-1) - 2\sin 2(t + 1)$

5.15 If $X(\omega)$ is the Fourier transform of $x(t)$, show that

$$x(t + a) + x(t - a) \Leftrightarrow 2X(\omega)\cos a\omega$$

5.16 Use the convolution property to derive the Fourier transform of the time-domain product $y(t) = \sin 2t \cdot u(t)$

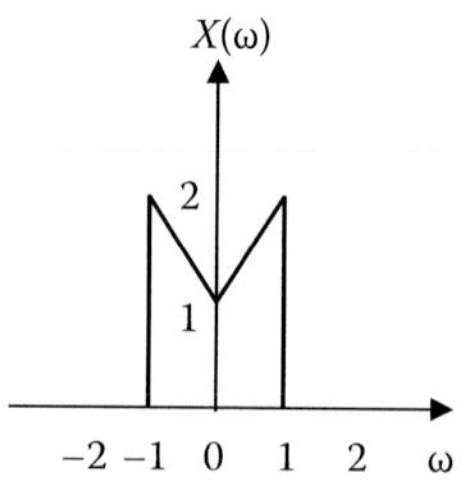

FIGURE 5.24 For Problem 5.17.

5.17 Let a signal $x(t)$ have the Fourier transform $X(\omega)$ shown in Figure 5.24. Without computing $x(t)$, sketch $Y(\omega)$ when
(a) $y(t) = x(2t)$
(b) $y(t) = x(t/4)$
(c) $y(t) = x(t)\cos 2t$
(d) $y(t) = x(t)\sin 2t$

5.18 Find the Fourier transform of the following signals:
(a) $x(t) = \delta(t-2) + \delta(t-1) + \delta(t+1) + \delta(t+2)$
(b) $y(t) = 4te^{-t}u(t)$
(c) $z(t) = 2e^{-t}\sin(4t)u(t)$

5.19 Verify the Fourier transform pairs:

$$te^{-at}u(t) \Rightarrow \frac{1}{(a+j\omega)^2}, \quad a > 0$$

5.20 Verify the following identities:

(a) $X(0) = \int_{-\infty}^{\infty} x(t)dt$

(b) $x(0) = \frac{1}{2\pi}\int_{-\infty}^{\infty} X(\omega)d\omega$

5.21 (a) A signal $x(t)$ has the equivalent bandwidth B and equivalent duration T defined as

$$T = \frac{\int_{-\infty}^{\infty} x(t)dt}{x(0)}, \quad B = \frac{\int_{-\infty}^{\infty} X(\omega)d\omega}{X(0)}$$

Show that $BT = 2\pi$.

(b) Determine the equivalent bandwidth of (i) $\Pi(t/\tau)$, (ii) $\Lambda(t/\tau)$.

5.22 If $X(\omega) = \mathcal{F}[x(t)]$ and a and b are constants, prove the following identities:

(a) $\mathcal{F}\left[x\left(\dfrac{t}{a}+b\right)\right] = |a| e^{jab\omega} X(\omega)$

(b) $\mathcal{F}[x(t)\delta(t-a)] = X(a)e^{-j\omega a}$

(c) $\mathcal{F}\left[x(t) * \delta\left(\dfrac{t}{a}-b\right)\right] = |a| e^{-jab\omega} X(\omega)$

5.23 Let $x(t)$ be given by $x(t) = \begin{cases} t, & 0 \le t \le 1 \\ 0, & \text{otherwise} \end{cases}$

(a) Sketch $x(t)$. (b) Obtain $X(\omega)$ by using differentiation property.

5.24 (a) If $x(t)$ is real and even, show that $X(\omega)$ is real and even.
(b) Give an example for part (a).

5.25 If $x(t)$ and $y(t)$ are two signals with Fourier transforms $X(\omega)$ and $Y(\omega)$, show that

$$\int_{-\infty}^{\infty} x(t)y^*(t)dt = \frac{1}{2\pi}\int_{-\infty}^{\infty} X(\omega)Y^*(\omega)d\omega$$

5.26 Consider $X(\omega)$ as the Fourier transform of $x(t)$ in Figure 5.25. Evaluate

(a) $X(0)$

(b) $\displaystyle\int_{-\infty}^{\infty} X(\omega)d\omega$

(c) $\displaystyle\int_{-\infty}^{\infty} |X(\omega)|^2 d\omega$

5.27 Evaluate the Fourier transform of

$$x(t) = \begin{cases} \sin t, & 0 < t < 2\pi \\ 0, & \text{otherwise} \end{cases}$$

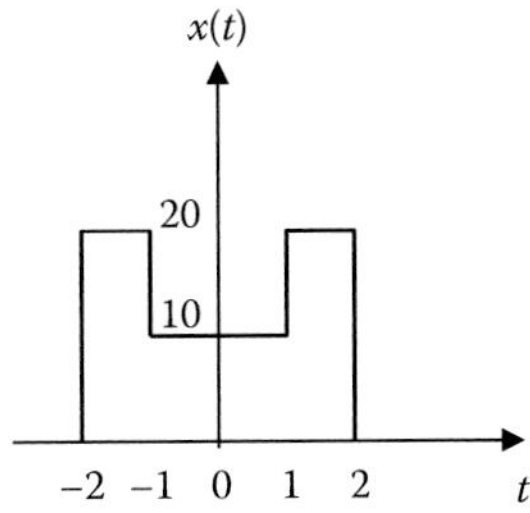

FIGURE 5.25 For Problem 5.26.

Section 5.4—Inverse Fourier Transform

5.28 Find $x(t)$ for each of the following:

(a) $X(\omega) = \dfrac{4\pi\delta(\omega - 1)}{2 + j\omega}$

(b) $X(\omega) = 4\pi\delta(\omega) - \dfrac{2}{j\omega}$

(c) $X(\omega) = 10\text{sinc}(4\omega)$

5.29 Find the signal corresponding to each of the following Fourier transforms:

(a) $X(\omega) = \dfrac{2j\omega}{1 + j\omega}$

(b) $Y(\omega) = 10e^{-|\omega|}$

(c) $Z(\omega) = \dfrac{j\omega}{-\omega^2 + j5\omega + 6}$

5.30 Determine the inverse Fourier transform of the signal in Figure 5.26.

Section 5.5—Applications

5.31 The transfer function of a system is

$$H(\omega) = \frac{10}{2 + j\omega}$$

Determine the response $y(t)$ for the following inputs:

(a) $x(t) = 5\cos(2t)$

(b) $x(t) = \delta(t)$

(c) $x(t) = e^{-4t}u(t)$

5.32 A signal $x(t)$ is passed through a system as shown in Figure 5.27.

(a) Find the transfer function $H(\omega)$.

(b) If $x(t) = \cos(t)$, find $y(t)$.

(c) If $x(t) = e^{-3t}u(t)$, find $y(t)$.

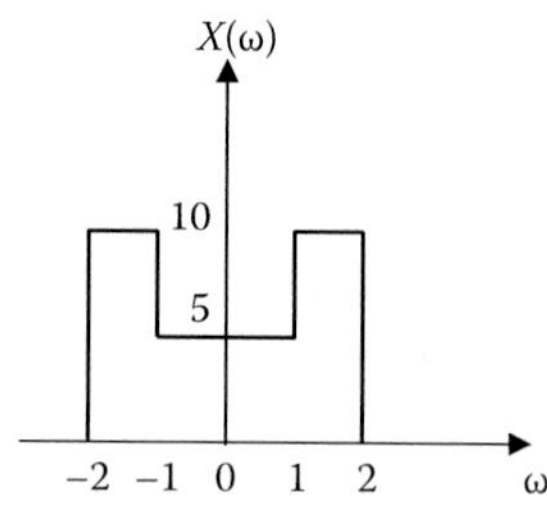

FIGURE 5.26 For Problem 5.30.

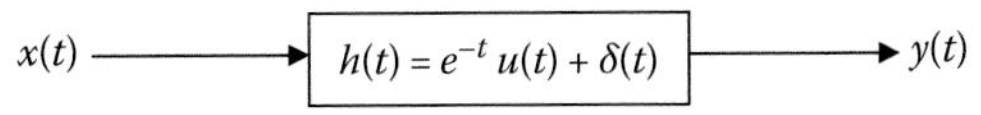

$x(t)$ → $h(t) = e^{-t}u(t) + \delta(t)$ → $y(t)$

FIGURE 5.27 For Problem 5.32.

5.33 The transfer function of a circuit is

$$H(\omega) = \frac{2}{j\omega + 2}$$

If the input signal to the circuit is $v_s(t) = e^{-4t}u(t)$, find the output signal. Assume all initial conditions are zero.

5.34 Suppose $v_s(t) = u(t)$ for $t > 0$. Determine $i(t)$ in the circuit of Figure 5.28 using Fourier transform.

5.35 Use Fourier transform to find $i(t)$ in the circuit of Figure 5.29 if $v_s(t) = 10e^{-2t}u(t)$.

5.36 Find the voltage $v_o(t)$ in the circuit of Figure 5.30. Let $i_s(t) = 8e^{-t}u(t)$ A.

5.37 For the circuit in Figure 5.31, find the transfer function $H(\omega) = V_o(\omega)/V_i(\omega)$.

5.38 Obtain the response $v_o(t)$ of the circuit in Figure 5.32, where $v_s(t) = e^{-2t}u(t)$ V.

5.39 Refer to the filter shown in Figure 5.33. Find the transfer function and determine the type of filter.

5.40 A continuous-time linear system is described by

$$\frac{dy(t)}{dt} + 3y(t) = x(t)$$

Find the output $y(t)$ due to the input $x(t) = e^{-2t}u(t)$.

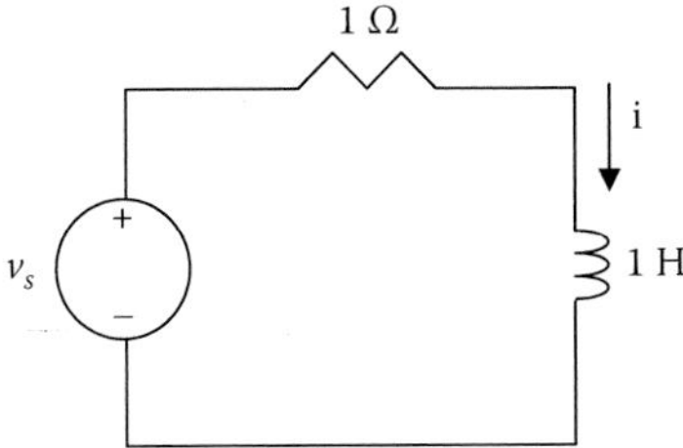

FIGURE 5.28 For Problem 5.34.

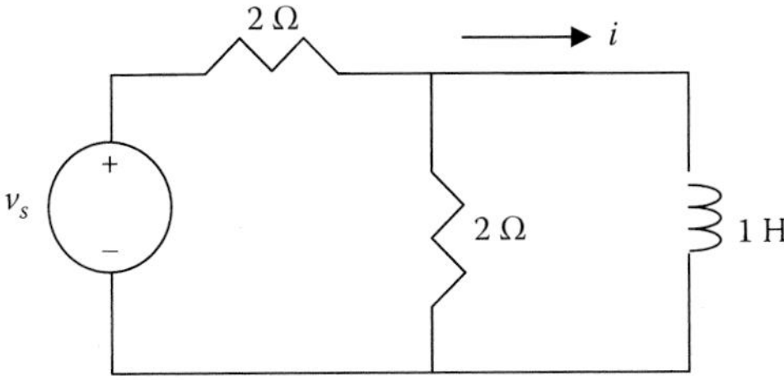

FIGURE 5.29 For Problem 5.35.

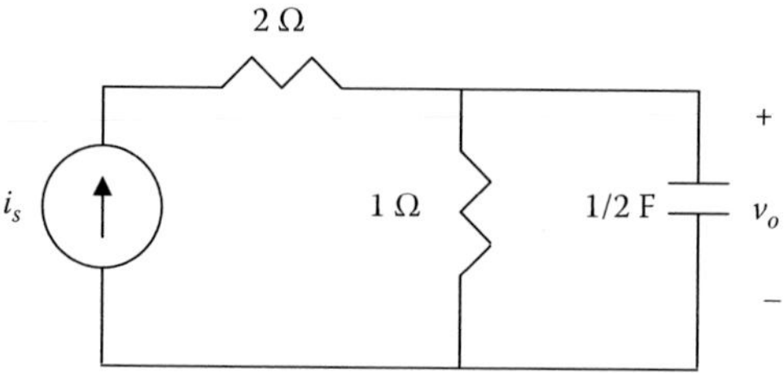

FIGURE 5.30 For Problem 5.36.

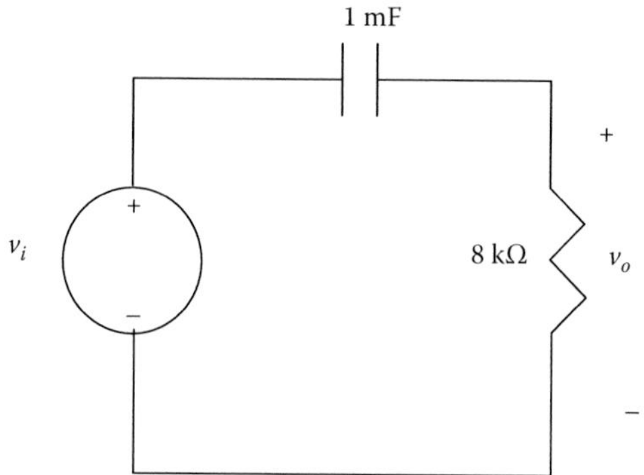

FIGURE 5.31 For Problem 5.37.

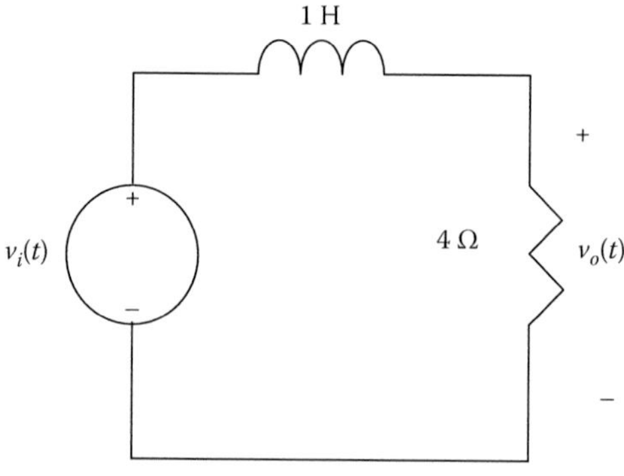

FIGURE 5.32 For Problem 5.38.

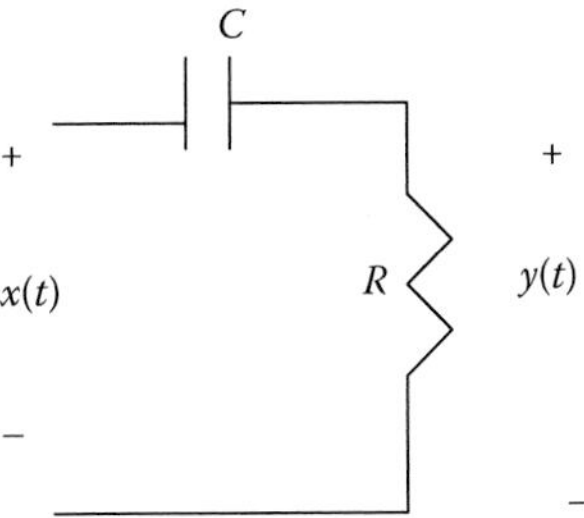

FIGURE 5.33 For Problem 5.39.

5.41 Determine the transfer function of the linear system characterized by

$$\frac{d^2y(t)}{dt^3}+5\frac{dy(t)}{dt}+6y(t)=\frac{dx(t)}{dt}+x(t)$$

and find its impulse response.

5.42 Find the response of a system described by

$$\frac{dy(t)}{dt}+2y(t)=x(t), \quad x(t)=e^{-t}u(t)$$

using the Fourier transform.

5.43 In a system, the input signal $x(t)$ is amplitude modulated by $m(t) = 2 + \cos\omega_o t$. The response $y(t) = m(t)x(t)$. Find $Y(\omega)$ in terms of $X(\omega)$.

5.44 In a communication system, two signals

$$x_1(t)=\cos(100\pi t) \quad \text{and} \quad x_2(t)=2\cos(400\pi t)$$

are multiplied to give $x_3(t) = x_1(t)x_2(t)$. Sketch the frequency spectra of $x_1(t)$, $x_2(t)$, and $x_3(t)$.

5.45 Determine the Nyquist rates for these signals:
(a) $x(t) = 10\cos(60\pi t)$
(b) $x(t) = 3\cos(40\pi t) + 2\sin(50\pi t)$

5.46 A signal $x(t)$ is passed through a low-pass filter. If the filter blocks frequencies above 6 Hz, find the output $y(t)$ for the following cases:
(a) $x(t) = 2–3\cos(2\pi t) + 4\sin(13\pi t) + 10\cos(15\pi t)$
(b) $x(t) = 4–2\sin^2(4\pi t)$.

Section 5.6—Parseval's Theorem

5.47 Find the energy dissipated by the resistor in the circuit of Figure 5.34.

5.48 Calculate the total energy of the signal whose magnitude spectrum is shown in Figure 5.35.

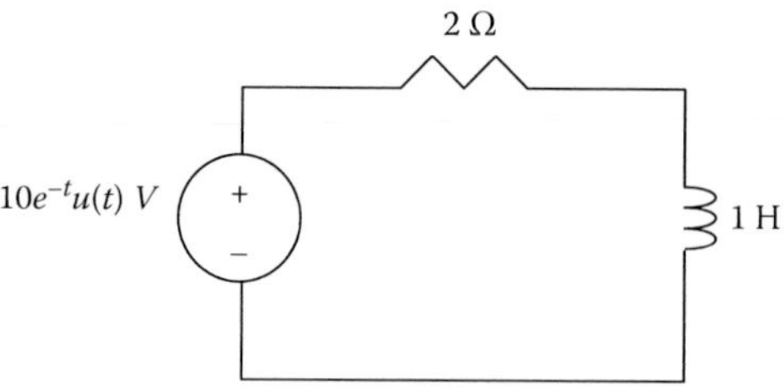

FIGURE 5.34 For Problem 5.47.

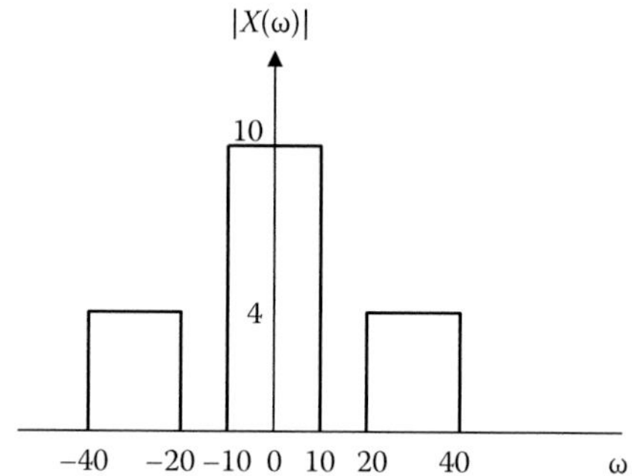

FIGURE 5.35 For Problem 5.48.

5.49 If $x(t)$ and $X(\omega)$ are Fourier transform pair and

$$X(\omega) = \begin{cases} 10, & |\omega| < 2 \\ 0, & |\omega| > 2 \end{cases}$$

Let $y(t) = \dfrac{d^2x(t)}{dt^2}$. Find $\displaystyle\int_{-\infty}^{\infty} |y(t)|^2\, dt$.

Section 5.8—Computing with MATLAB®

5.50 Use MATLAB to obtain the Fourier transform of the following signals:

(a) $x(t) = t\exp(-t)u(t)$

(b) $y(t) = (e^{-2t} - e^{-t})u(t)$

(c) $z(t) = \delta(t + 1/2) + e^{-1}u(t)$

5.51 Use MATLAB to computer the Fourier transform of

$$x(t) = 10e^{-2t}(4t)u(t)$$

Plot the magnitude $|X(\omega)|$.

5.52 Use MATLAB to find the inverse Fourier transform of

(a) $X_1(\omega) = \dfrac{e^{j\omega}}{1 - j\omega}$

(b) $X_2(\omega) = \dfrac{\delta(\omega)}{1 + j2\omega}$

(c) $X_3(\omega) = \dfrac{\pi\delta(\omega)}{(2 + j\omega)(5 + j\omega)}$

5.53 A system has the transfer function

$$H(s) = \frac{s^2 + s + 1}{2s^3 + 10s^2 + 16s + 8} \qquad s = j\omega$$

Use MATLAB to plot the frequency response.

5.54 Use MATLAB to plot the transfer function

$$H(\omega) = \frac{100(1 + j0.01\omega)}{(1 + j0.02\omega)(1 + j0.05\omega)}$$

5.55 The three-pole Butterworth filter has the transfer function

$$H(s) = \frac{\omega_c^3}{s^3 + 2\omega_c s^2 + 2\omega_c^2 s + \omega_c^3}, \qquad s = j\omega$$

Use MATLAB to sketch $H(s)$ for $\omega_c = 2\pi$.

5.56 Use MATLAB to plot the transfer function

$$H(\omega) = \frac{1}{1 + j0.2\omega - 0.05\omega^2}$$

6 Discrete Fourier Transform

To accomplish great things, we must not only act, but also dream; not only plan, but also believe.

—Anatole France

CAREER IN COMMUNICATIONS SYSTEMS

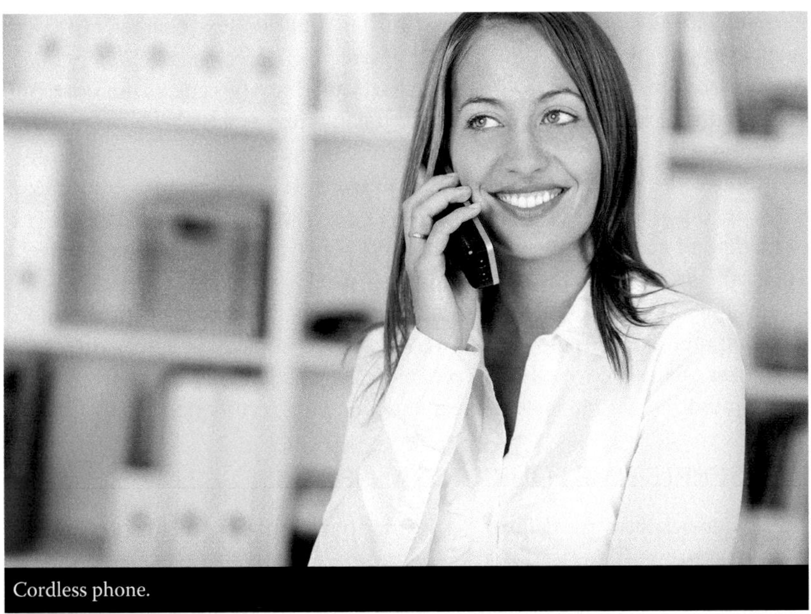

Cordless phone.

Communications systems apply the principles of signals and systems. A communication system is designed to convey information from a source (the transmitter) to a destination (a receiver) via a channel (the propagation medium). Communication engineers design systems for transmitting and receiving information. The information can be in the form of voice, data, or video.

We live in the information age—news, weather, sports, shopping, financial, business inventory, and other sources make information available to us almost instantly via communications systems. Some obvious examples of communication systems are the telephone network, mobile cellular telephones, radios, cable TV, fax, and radar.

The field of communications is perhaps the fastest growing area in electrical engineering. The merging of the communications field with computer technology in recent years has led to digital data communication networks such as local area networks, metropolitan area networks, and broadband integrated services digital networks. For example, the Internet (the "information superhighway") allows educators, business people, and others to send electronic mail from their computers worldwide, log onto remote databases, and transfer files.

A communication engineer designs systems that provide high-quality information services. The systems include hardware for generating, transmitting, and receiving information signals. More and more government agencies, academic departments, and businesses are demanding faster and more accurate transmission of information. To meet these needs, communication engineers are in high demand.

6.1 INTRODUCTION

In the previous chapters, we have discussed applying frequency-domain techniques to continuous-time signals and systems. In this and the next chapter, we consider frequency-domain analysis of discrete-time systems. Here, we specifically consider the Fourier analysis of discrete-time systems. Fourier analysis plays the same important role in discrete-time systems as in continuous-time systems.

Our approach is similar to that used for continuous-time signals and systems. We first introduce the discrete-time Fourier transform (DTFT) and its properties. We then discuss discrete Fourier transform (DFT), and the fast Fourier transform (FFT), a fast method of computing the DFT. The DFT plays a vital role in signal processing for spectral, while FFT is the most commonly used Fourier analysis technique. The main advantage of DFT and FFT is that they are well suited for computation by a digital computer. We learn how to use MATLAB® to implement numerically the discrete Fourier analysis covered in this chapter. We finally apply the concepts learned in the chapter to touch-tone telephone and windowing technique.

6.2 DISCRETE-TIME FOURIER TRANSFORM

In the previous chapter, we defined the Fourier transform of a continuous-time signal $x(t)$ as

$$X(\omega) = \int_{-\infty}^{\infty} x(t)e^{-j\omega t}dt \tag{6.1}$$

By replacing the integral with summation, we obtain the *discrete-time Fourier transform* (DTFT) of $x[n]$ as

$$X(\Omega) = \sum_{n=-\infty}^{\infty} x[n]e^{-j\Omega n} \tag{6.2}$$

where we have used the uppercase omega (Ω) as the frequency variable to distinguish the continuous-time and discrete-time cases. $X(\Omega)$ is referred as the Fourier spectrum of $x[n]$.

If we introduce $\Omega + 2\pi$ in Equation 6.2, we find that

$$X(\Omega+2\pi) = X(\Omega), \quad -\infty < \Omega < \infty \tag{6.3}$$

since $e^{-j2\pi} = 1$. This shows that $X(\Omega)$ is periodic. For this reason, $X(\Omega)$ can be expanded in a Fourier series. If we multiply both sides of Equation 6.2 by $e^{j\Omega m}d\Omega$ and integrate, we obtain

$$\int_{-\pi}^{\pi} X(\Omega)e^{j\Omega m}d\Omega = \int_{-\pi}^{\pi}\left[\sum_{n=-\infty}^{\infty} x[n]e^{-j\Omega n}\right]e^{j\Omega m}d\Omega \tag{6.4}$$

But we know that the integral of any sinusoidal signal is zero when it is integrated over a period. Hence,

$$\sum_{n=-\infty}^{\infty} x[n]\int_{-\pi}^{\pi} e^{j\Omega(m-n)}d\Omega = \begin{cases} 2\pi x[n], & m = n \\ 0, & m \neq n \end{cases} \tag{6.5}$$

That is, only one term in the summation in Equation 6.4 remains. Thus, Equation 6.4 becomes

$$x[n] = \frac{1}{2\pi}\int_{-\pi}^{\pi} X(\Omega)e^{j\Omega n}d\Omega \tag{6.6}$$

which is the inverse discrete-time Fourier transform (IDTFT).

From Equations 6.2 and 6.6, we have the discrete-time Fourier transform (DTFT) pair as

$$\boxed{\text{DFT: } X(\Omega) = F\{x[n]\} = \sum_{n=-\infty}^{\infty} x[n]e^{-j\Omega n}} \tag{6.7}$$

$$\boxed{\text{IDFT: } x[n] = F^{-1}\{X(\Omega)\} = \frac{1}{2\pi}\int_{-\pi}^{\pi} X(\Omega)e^{j\Omega n}d\Omega} \tag{6.8}$$

where $F\{x[n]\}$ denotes the DTFT of $x[n]$. The one-to-one relationship between $x[n]$ and $X(\Omega)$ is symbolized as

$$x[n] \xleftrightarrow{\text{DTFT}} X(\Omega) \tag{6.9}$$

Since $X(\Omega)$ is periodic, the integral in Equation 6.8 can be evaluated over any interval of length 2π. For example,

$$x[n]=\frac{1}{2\pi}\int_{0}^{2\pi} X(\Omega)e^{j\Omega n}d\Omega \tag{6.10}$$

The sufficient condition for $X(\Omega)$ to exist (or to converge) is that $x[n]$ is absolutely summable, that is,

$$\sum_{n=-\infty}^{\infty}|x[n]|<\infty \tag{6.11}$$

We should keep in mind that although ω has units of radians/second, Ω has units of radians. Although $x[n]$ is discrete, $X(\Omega)$ is a function of the continuous variable Ω.

Example 6.1

Given the discrete-time signal

$$x[n]=\begin{cases} 1, & n=-1 \\ -2, & n=0 \\ 0, & n=1 \\ 3, & n=2 \\ 0, & \text{otherwise} \end{cases}$$

Find the discrete-time Fourier transform.

Solution

Since $x[n]$ has few nonzero terms, the finite sum becomes a sum of few terms.

$$\begin{aligned} X(\Omega) &= \sum_{n=-\infty}^{\infty} x[n]e^{-j\Omega n} \\ &= e^{j\Omega} - 2e^{j0} + 0e^{j\Omega} + 3e^{-j2\Omega} \\ &= e^{j\Omega} - 2 + 3e^{-j2\Omega} \end{aligned}$$

Practice Problem 6.1 Find the DTFT of the signal

$$x[n]=\begin{cases} 2, & n=-2 \\ 3, & n=0 \\ 1, & n=1 \\ -1, & n=2 \\ 0, & \text{otherwise} \end{cases}$$

Answer: $2e^{j2\Omega} + 3 + e^{-j\Omega} - e^{-j2\Omega}$

Example 6.2

Find the DTFT of the discrete-time signal

(a) $x[n] = \delta[n]$

(b) $x[n] = \begin{cases} a^n, & 0 \le n \le m \\ 0, & \text{otherwise} \end{cases}$

where a is a constant.

Solution

To find the DTFT of $x[n]$, we apply Equation 6.7.

(a) $$X(\Omega) = \sum_{n=-\infty}^{\infty} x[n]e^{-j\Omega n} = \sum_{n=0}^{m} \delta[n]e^{-j\Omega n} = e^{-j\Omega n}\Big|_{n=0} = 1$$

Thus,

$$\delta[n] \xleftrightarrow{\text{DTFT}} 1$$

(b) $$X(\Omega) = \sum_{n=-\infty}^{\infty} x[n]e^{-j\Omega n} = \sum_{n=0}^{m} a^n e^{-j\Omega n} = \sum_{n=0}^{m} (ae^{-j\Omega})^n \tag{6.2.1}$$

But the geometric series sums as

$$\sum_{n=0}^{m} q^n = \frac{1-q^{m+1}}{1-q}, \quad |q| < 1 \tag{6.2.2}$$

with common factor q. By letting $q = ae^{-j\Omega}$, we combine Equations 6.2.1 and 6.2.2 to get

$$X(\Omega) = \frac{1-(ae^{-j\Omega})^{m+1}}{1-ae^{-j\Omega}}, \quad |a| < 1$$

Practice Problem 6.2 Determine the DTFT of the discrete-time signal $x[n] = a^n u[n]$, that is,

$$x[n] = \begin{cases} a^n, & n \ge 0 \\ 0, & n < 0 \end{cases}$$

Answer: $\dfrac{1}{1-ae^{-j\Omega}}, |a| < 1$

Example 6.3

Obtain the DTFT of the signal $x[n] = a^{|n|}$, $|a| < 1$.

Solution

The Fourier transform is

$$X(\Omega) = \sum_{n=-\infty}^{\infty} x[n]e^{-j\Omega n} = \sum_{n=-\infty}^{\infty} a^{|n|}e^{-j\Omega n}$$

$$= \sum_{n=0}^{\infty} a^{n}e^{-j\Omega n} + \sum_{n=-\infty}^{-1} a^{-n}e^{-j\Omega n} \tag{6.3.1}$$

If we make the substitution $m = -n$ in the second summation, we get

$$X(\Omega) = \sum_{n=0}^{\infty} (ae^{-j\Omega})^{n} + \sum_{m=1}^{\infty} (ae^{j\Omega})^{m} \tag{6.3.2}$$

Both summations are infinite geometric series

$$\sum_{n=0}^{\infty} q^{n} = 1 + q + q^{2} + \cdots = \frac{1}{1-q}, \quad |q| < 1 \tag{6.3.3}$$

The first series in Equation 6.3.2 starts from $n = 0$ and its summation is given by Equation 6.3.3. However, the second series starts from $m = 1$. We need to subtract 1 from both sides of Equation 6.3.3 so that

$$\sum_{n=1}^{\infty} q^{n} = q + q^{2} + \cdots = \frac{1}{1-q} - 1 = \frac{q}{1-q}, \quad |q| < 1 \tag{6.3.4}$$

Hence, applying Equations 6.3.3 and 6.3.4 into Equation 6.3.2 gives

$$X(\Omega) = \frac{1}{1-ae^{-j\Omega}} + \frac{ae^{j\Omega}}{1-ae^{j\Omega}} = \frac{1-a^{2}}{1-2a\cos\Omega + a^{2}} \tag{6.3.5}$$

Practice Problem 6.3 Find the DTFT of $a^{n}\, u[-(n + 1)]$.

Answer: $\dfrac{1}{ae^{-j\Omega}-1}, \quad |a| < 1$

Example 6.4

Obtain the DTFT of the constant Signal $x[n]$ for all n. That is,

$$x[n] = 1, \quad n = 0, \pm 1, \pm 2, \ldots$$

Solution

$$X(\Omega) = \sum_{n=-\infty}^{\infty} x[n]e^{-jn\Omega} = \sum_{n=-\infty}^{\infty} (1)e^{-jn\Omega} \tag{6.4.1}$$

By expanding the frequency-domain train of impulse delta function into Fourier series, we can show that

$$\sum_{n=-\infty}^{\infty} \delta(\omega - n\omega_0) = \frac{1}{\omega_0}\sum_{n=-\infty}^{\infty} e^{-jnT_0\omega}, \quad T_0 = \frac{2\pi}{\omega_0}$$

or

$$\sum_{n=-\infty}^{\infty} e^{-jn\Omega} = 2\pi\sum_{n=-\infty}^{\infty} \delta(\Omega - n2\pi) \tag{6.4.2}$$

where we have used $\Omega = T_0\omega$ and the property $\delta(at) = \frac{1}{|a|}\delta(t)$. Substituting Equation 6.4.2 in Equation 6.4.1 leads the DTFT pair

$$x[n] = 1 \quad \text{for all } n \leftrightarrow 2\pi\sum_{n=-\infty}^{\infty} \delta(\Omega - 2\pi n)$$

Practice Problem 6.4 Obtain the DTFT of

$$x[n] = e^{j\Omega_0 n}$$

Answer: $2\pi\sum_{k=-\infty}^{\infty} \delta(\Omega - \Omega_o - 2\pi k)$

6.3 PROPERTIES OF DTFT

The discrete-time Fourier transform has some properties similar to those of the continuous-time Fourier transform. For each property, we shall state it and derive it, and then illustrate with some examples.

6.3.1 Linearity

The linearity property of the DTFT states that if

$$x_1[n] \xleftrightarrow{\text{DTFT}} X_1(\Omega)$$

and

$$x_2[n] \xleftrightarrow{\text{DTFT}} X_2(\Omega)$$

then

$$\boxed{ax_1[n] + bx_2[n] \xleftrightarrow{\text{DTFT}} aX_1(\Omega) + bX_2(\Omega)} \tag{6.12}$$

where a and b are constants. The proof is trivial.

Let us apply the linearity property to find the DTFT of the sinusoidal sequence

$$x[n] = \cos\Omega_0 n, \quad |\Omega_0| \leq \pi \tag{6.13}$$

Using Euler's formula,

$$\cos\Omega_0 n = \frac{1}{2}\left(e^{j\Omega_0 n} + e^{-j\Omega_0 n}\right)$$

Applying the result in Practice Problem 6.4 and the linearity property,

$$X(\Omega) = 2\pi \sum_{k=-\infty}^{\infty} \delta(\Omega - \Omega_0 - 2\pi k) + \delta(\Omega + \Omega_0 - 2\pi k) \tag{6.14}$$

that is,

$$\cos\Omega_0 n \xleftrightarrow{\text{DTFT}} 2\pi \sum_{k=-\infty}^{\infty} \delta(\Omega - \Omega_0 - 2\pi k) + \delta(\Omega + \Omega_0 - 2\pi k) \tag{6.15}$$

6.3.2 Time Shifting and Frequency Shifting

A delay or advance in the time domain results in a phase shift in the frequency domain. That is, if

$$x[n] \xleftrightarrow{\text{DTFT}} X(\Omega)$$

then

$$\boxed{x[n - n_0] \xleftrightarrow{\text{DTFT}} e^{-j\Omega n_0} X(\Omega)} \tag{6.16}$$

This property can be proved using Equation 6.7.

$$F\{x[n - n_0]\} = \sum_{n=-\infty}^{\infty} x[n - n_0] e^{-j\Omega n} \tag{6.17}$$

Let $n - n_0 = k$ on the right side of this equation so that $n = k + n_0$.

$$F\{x[n-n_0]\} = \sum_{n=-\infty}^{\infty} x[k]e^{-j\Omega(k+n_0)} = e^{-j\Omega n_0}\sum_{n=-\infty}^{\infty} x[k]e^{-j\Omega k}$$

$$= e^{-j\Omega n_0}X(\Omega) \tag{6.18}$$

For example, suppose we want to obtain the DTFT of the sequence $x[n] - x[n-1]$. Using the time-shifting property,

$$x[n]-x[n-1] \xleftrightarrow{\text{DTFT}} (1-e^{-j\Omega})X(\Omega) \tag{6.19}$$

The dual property is the frequency shifting property, which states that

$$\boxed{x[n]e^{j\Omega_0 n} \xleftrightarrow{\text{DTFT}} X(\Omega-\Omega_0)} \tag{6.20}$$

This can be derived by using Equation 6.7.

$$x[n]e^{j\Omega_0 n} \overset{DTFT}{\leftrightarrow} \sum_{n=-\infty}^{\infty} x[n]e^{j\Omega_0 n}e^{-j\Omega n} = \sum_{n=-\infty}^{\infty} x[n]e^{-j(\Omega-\Omega_0)} = X(\Omega-\Omega_0) \tag{6.21}$$

From this, it follows that

$$x[n]e^{-j\Omega_0 n} \xleftrightarrow{\text{DTFT}} X(\Omega+\Omega_0) \tag{6.22}$$

Combining Equations 6.19 and 6.21, we get

$$x[n]\cos\Omega_0 n \xleftrightarrow{\text{DTFT}} \frac{1}{2}\left[X(\Omega-\Omega_0)+X(\Omega+\Omega_0)\right] \tag{6.23}$$

which is the modulation property.

6.3.3 Time Reversal and Conjugation

If $x[n] \xleftrightarrow{\text{DTFT}} X(\Omega)$, then

$$\boxed{x[-n] \xleftrightarrow{\text{DTFT}} X(-\Omega)} \tag{6.24}$$

We can prove this using Equation 6.7.

$$F\{x[-n]\} = \sum_{n=-\infty}^{\infty} x[-n]e^{-j\Omega n}$$

If we let $-n = k$, then

$$F\{x[-n]\} = \sum_{n=-\infty}^{\infty} x[k]e^{j\Omega k} = \sum_{n=-\infty}^{\infty} x[k]e^{-j(-\Omega)k} = X(-\Omega) \tag{6.25}$$

A related property is the conjugation of $x[n]$. That is,

$$\boxed{x^*[n] \xleftrightarrow{\text{DTFT}} X^*(-\Omega)} \tag{6.26}$$

where * denotes the complex conjugate.

6.3.4 Time Scaling

Time scaling for discrete-time signal is different from time scaling for continuous-time signal. We recall if $X(\omega)$ is the Fourier transform of continuous-time signal $x(t)$, then

$$F[x(at)] = \frac{1}{|a|}X\left(\frac{\omega}{a}\right) \tag{6.27}$$

In the case of discrete-time signals, $x[an]$ is not a sequence if a is not an integer. To derive a result that closely parallels Equation 6.27, let k be a positive integer. We define the signal

$$x_{(k)}[n] = \begin{cases} x[n/k] = x[m], & \text{if } n = km,\ m = \text{integer} \\ 0, & \text{if } n \text{ is not a multiple of } k \end{cases} \tag{6.28}$$

Since $x_{(k)}[n]$ is zero unless n is a multiple of k, the DTFT of $x_{(k)}[n]$ is given by

$$X_{(k)}(\Omega) = \sum_{n=-\infty}^{\infty} x_{(k)}[n]e^{-j\Omega n} = \sum_{m=-\infty}^{\infty} x_{(k)}[mk]e^{-j\Omega mk}$$

Replacing $x_{(k)}[n]$ by $x[m]$ gives

$$X_{(k)}(\Omega) = \sum_{m=-\infty}^{\infty} x[m]e^{-j(k\Omega)m} = X(k\Omega) \tag{6.29}$$

That is,

$$\boxed{x_{(k)}[n] \xleftrightarrow{\text{DTFT}} X(k\Omega)} \tag{6.30}$$

This shows that as the signal spreads out in time ($k > 1$), the corresponding Fourier transform is compressed. Since $X(\Omega)$ is periodic with period 2π, $X(k\Omega)$ is periodic with period $2\pi/k$.

6.3.5 Frequency Differentiation

This property states that if $x[n] \xleftrightarrow{\text{DTFT}} X(\Omega)$, then

$$\boxed{nx[n] \xleftrightarrow{\text{DTFT}} j\frac{dX(\Omega)}{d\Omega}} \tag{6.31}$$

This can be proved by differentiating both sides of Equation 6.7 with respect to Ω. We obtain

$$\frac{dX(\Omega)}{d\Omega} = \sum_{n=-\infty}^{\infty} -jnx[n]e^{-j\Omega n}$$

Multiplying both sides by j gives

$$j\frac{dX(\Omega)}{d\Omega} = \sum_{n=-\infty}^{\infty} nx[n]e^{-j\Omega n} \tag{6.32}$$

The right-hand side of this equation is the Fourier transform of $nx[n]$. Thus, Equation 6.31 follows.

For example, suppose we know that

$$F\left\{a^n u[n]\right\} = \frac{e^{j\Omega}}{e^{j\Omega} - a} \qquad |a| < 1 \tag{6.33}$$

Using Equation 6.31, we obtain

$$F\left\{na^n u[n]\right\} = j\frac{d}{d\Omega}\left(\frac{e^{j\Omega}}{e^{j\Omega} - a}\right) = \frac{ae^{j\Omega}}{(e^{j\Omega} - a)^2} \tag{6.34}$$

6.3.6 Time and Frequency Convolution

If $x_1[n] \xleftrightarrow{\text{DTFT}} X_1(\Omega)$ and $x_2[n] \xleftrightarrow{\text{DTFT}} X_2(\Omega)$, then

$$\boxed{x_1[n] * x_2[n] \xleftrightarrow{\text{DTFT}} X_1(\Omega)X_2(\Omega)} \tag{6.35}$$

To prove this, we substitute the definition of discrete-time convolution in Equation 6.7.

$$F\{x_1[n] * x_2[n]\} = \sum_{n=-\infty}^{\infty} \left(\sum_{k=-\infty}^{\infty} x_1[k]x_2[n-k] \right) e^{-j\Omega n} \tag{6.36}$$

If we change the order of the summations, we obtain

$$F\{x_1[n] * x_2[n]\} = \sum_{k=-\infty}^{\infty} x_1[k] \left(\sum_{n=-\infty}^{\infty} x_2[n-k]e^{-j\Omega n} \right) \tag{6.37}$$

Applying the time-shifting property,

$$\sum_{n=-\infty}^{\infty} x_2[n-k]e^{-j\Omega n} = e^{-i\Omega k} X_2(\Omega) \tag{6.38}$$

Substituting this in Equation 6.37 yields

$$F\{x_1[n] * x_2[n]\} = \sum_{k=-\infty}^{\infty} x_1[k]e^{-j\Omega k} X_2(\Omega) = X_1(\Omega)X_2(\Omega) \tag{6.39}$$

This confirms Equation 6.35 for time-convolution.

For example, for a discrete-time system with input $x[n]$, output $y[n]$, and impulse response $h[n]$, we obtain

$$Y(\Omega) = H(\Omega)X(\Omega) \tag{6.40}$$

For frequency-convolution,

$$\boxed{x_1[n]x_2[n] \xleftrightarrow{\text{DTFT}} \frac{1}{2\pi} X_1(\Omega) \otimes X_2(\Omega)} \tag{6.41}$$

This is also called the *multiplication property* since multiplication is involved in the time domain. This property states the multiplication of two signals in the time domain corresponds to periodic convolution in the frequency domain, scaled by the inverse of the period. Periodic convolution is defined by

$$X_1(\Omega)\otimes X_2(\Omega)=\int_0^{2\pi} X_1(\theta)X_2(\Omega-\theta)d\theta \tag{6.42}$$

To prove Equation 6.41, we have by definition

$$F\{x_1[n]x_2[n]\}=\sum_{n=-\infty}^{\infty} x_1[n]x_2[n]e^{-j\Omega n}=\sum_{n=-\infty}^{\infty} x_2[n]\left[\frac{1}{2\pi}\int_0^{2\pi} X_1(\tau)e^{-jn\tau}d\tau\right]e^{-j\Omega n} \tag{6.43}$$

Interchanging the order of integration and summation, we have

$$F\{x_1[n]x_2[n]\}=\frac{1}{2\pi}\int_0^{2\pi} X_1(\tau)\left[\sum_{n=-\infty}^{\infty} x_2[n]e^{-j(\Omega-\tau)n}\right]d\tau$$

$$=\frac{1}{2\pi}\int_0^{2\pi} X_1(\tau)X_2(\Omega-\tau)d\tau=\frac{1}{2\pi}X_1(\Omega)\otimes X_2(\Omega) \tag{6.44}$$

For example, we can derive the modulation property as a special case of the frequency-convolution.

$$F\left[e^{j\Omega_0 n}x[n]\right]=\frac{1}{2\pi}\left[2\pi\delta(\Omega-\Omega_0)\right]\otimes X(\Omega)=X(\Omega-\Omega_0) \tag{6.45}$$

6.3.7 Accumulation

Accumulation (or summation) is the discrete-time counterpart of integration. The accumulation property can be stated as

$$\boxed{\sum_{k=-\infty}^{\infty} x[k]\xleftrightarrow{\text{DTFT}}\frac{1}{1-e^{-j\Omega}}X(\Omega)+\pi X(0)\sum_{n=-\infty}^{\infty}\delta(\Omega-2n\pi)} \tag{6.46}$$

The accumulation operation is defined by

$$y[n]=\sum_{k=-\infty}^{\infty} x[k] \tag{6.47}$$

This is equivalent to convolution according to Equation 2.25,

$$y[n]=\sum_{k=-\infty}^{\infty}x[k]=x[n]*u[n] \tag{6.48}$$

Applying the frequency-convolution property,

$$Y(\Omega)=X(\Omega)U(\Omega)=X(\Omega)\left[\frac{1}{1-e^{-j\Omega}}+\pi\sum_{n=-\infty}^{\infty}\delta(\Omega-2n\pi)\right]$$

$$=\pi X(0)\sum_{n=-\infty}^{\infty}\delta(\Omega-2n\pi)+\frac{1}{1-e^{-j\Omega}}X(\Omega) \tag{6.49}$$

6.3.8 Parseval's Relation

This can be derived from the convolution property. The energy of a discrete-time signal $x[n]$ is

$$E_x=\sum_{n=-\infty}^{\infty}|x[n]|^2 \tag{6.50}$$

We may regard this as the value of DTFT of the signal

$$g[n]=\left|x[n]\right|^2=x[n]x^*[n] \tag{6.51}$$

Hence,

$$G(\Omega)=\frac{1}{2\pi}X(\Omega)X^*(-\Omega) \tag{6.52}$$

Thus,

$$E_x=\sum_{n=-\infty}^{\infty}|x[n]|^2=\frac{1}{2\pi}\int_0^{2\pi}|X(\Omega)|^2\,d\Omega \tag{6.53}$$

This is Parseval's relation. $|X(\Omega)|^2$ is the energy spectrum of the signal.

For convenience, Table 6.1 summarizes the properties of discrete-time Fourier transform, while Table 6.2 presents the transforms of some common transform pairs.

TABLE 6.1
Properties of Discrete-Time Fourier Transform (DTFT)

Property	Time Domain	Frequency Domain
1. Periodicity	$x[n]$	$X(\Omega + 2\pi) = X(\Omega)$
2. Linearity	$ax_1[n] + bx_2[n]$	$aX_1(\Omega) + bX_2(\Omega)$
3. Time-shifting	$x[n - k]$	$e^{-j\Omega k}X(\Omega)$
4. Frequency-shifting (modulation)	$e^{j\Omega_0 n}x[n]$	$X(\Omega - \Omega_0)$
5. Time reversal	$x[-n]$	$X(-\Omega)$
6. Conjugation	$x^*[n]$	$X^*(-\Omega)$
7. Time scaling	$x_{(k)}[i]$	$X(k\Omega)$
8. Differentiation	$nx[n]$	$j\dfrac{dX\Omega}{\Omega}$
9. Time-convolution	$x_1[n]*x_2[n]$	$X_1(\Omega)X_2(\Omega)$
10. Frequency-convolution	$x_1[n]x_2[n]$	$\dfrac{1}{2\pi}X_1(\Omega)\otimes X_2(\Omega)$
11. Accumulation	$\sum_{n=-\infty}^{\infty} x[k]$	$\dfrac{1}{1-e^{-j\Omega}}X(\Omega)+\pi X(0)\sum_{n=-\infty}^{\infty}\delta(\Omega-2n\pi)$
12. Parseval's relation	$E_x=\sum_{n=-\infty}^{\infty}\lvert x[n]\rvert^2$	$E_x=\dfrac{1}{2\pi}\int_0^{2\pi}\lvert X(\Omega)\rvert^2\,d\Omega$

Example 6.5

Determine the DTFT of $x[n] = a^{|n|}$.

Solution

$x[n]$ can be expressed in terms of $a^n u[n]$ and $a^{-n}u[-n]$, that is,

$$a^{|n|} = a^n u[n] + a^{-n}u[-n] - \delta[n] \tag{6.5.1}$$

We know from Table 6.2 that

$$F\left\{a^n u[n]\right\} = \frac{1}{1-ae^{-j\Omega}} = \frac{e^{j\Omega}}{e^{j\Omega}-a} \qquad |a| \le 1 \tag{6.5.2}$$

Using the time reversal property,

$$F\left\{a^{-n}u[-n]\right\} = \frac{e^{-j\Omega}}{e^{-j\Omega}-a} \qquad |a| \le 1 \tag{6.5.3}$$

TABLE 6.2
Common DTFT Pairs

Signal	Type	$x[n]$	$X(\Omega)$
1.	Impulse	$\delta[n]$	1
2.	Shifted impulse	$\delta[n-k]$	$e^{-j\Omega k}$
3.	Unit step	$u[n]$	$\frac{1}{1-e^{-j\Omega}}+\pi\sum_{n=-\infty}^{\infty}\delta(\Omega-2n\pi),\quad \lvert\Omega\rvert\le\pi$
4.	Shifted unit step	$-u[-n-1]$	$\frac{1}{1-e^{-j\Omega}}-\pi\sum_{n=-\infty}^{\infty}\delta(\Omega-2n\pi),\quad \lvert\Omega\rvert\le\pi$
5.	DC signal	1, for all n	$2\pi\sum_{k=-\infty}^{\infty}\delta(\Omega-2\pi k)$
6.	Gated function	$u[n]-u[n-k]$	$\frac{\sin(k\Omega/2)}{\sin(\Omega/2)}e^{-j\Omega(k-1)/2}$
7.	Exponential	$a^n u[n]$	$\frac{1}{1-ae^{-j\Omega}}\quad \lvert a\rvert\le 1$
8.	Weighted exponential	$na^n u[n]$	$\frac{ae^{-j\Omega}}{(1-ae^{-j\Omega})^2}\quad \lvert a\rvert\le 1$
		$(n+1)a^n u[n]$	$\frac{1}{(1-ae^{-j\Omega})^2}\quad \lvert a\rvert\le 1$
9.	Two-sided exponential	$a^{\lvert n\rvert}$	$\frac{1-a^2}{1+a^2-2a\cos\Omega}\quad \lvert a\rvert<1$
10.	Complex sinusoid	$e^{j\Omega_0 n}$	$2\pi\sum_{k=-\infty}^{\infty}\delta(\Omega-\Omega_0-2\pi k)$
11.	Cosine wave	$\cos\Omega_0 n$	$2\pi\sum_{k=-\infty}^{\infty}\delta(\Omega-\Omega_0-2\pi k)+\delta(\Omega+\Omega_0-2\pi k)$
12.	Sine wave	$\sin\Omega_0 n$	$\frac{\pi}{j}\sum_{k=-\infty}^{\infty}\delta(\Omega-\Omega_0-2\pi k)-\delta(\Omega+\Omega_0-2\pi k)$

Taking the DTFT of each term in Equation 6.5.1 and substituting Equations 6.5.2 and 6.5.3 gives

$$F\left\{a^{\lvert n\rvert}\right\}=\frac{e^{j\Omega}}{e^{j\Omega}-a}+\frac{e^{-j\Omega}}{e^{-j\Omega}-a}-1=\frac{1-a^2}{1-2a\cos\Omega+a^2}\quad \lvert a\rvert\le 1 \tag{6.5.4}$$

Practice Problem 6.5 Determine the DTFT of the composite signal

$$x[n]=a^n u[n]-\sin\Omega_0 n,\quad \lvert a\rvert<1$$

Answer: $\frac{1}{1-ae^{-j\Omega}}+j\pi\sum_{k=-\infty}^{\infty}\delta(\Omega-\Omega_0-2\pi k)-\delta(\Omega+\Omega_0-2\pi k)$

Example 6.6

Given the signal

$$x[n] = \begin{cases} 1, & |n| \leq 2 \\ 0, & \text{otherwise} \end{cases} \tag{6.6.1}$$

(a) Sketch $x[n]$ and its DTFT $X(\Omega)$.
(b) Sketch the time-scaled signal $x_{(2)}[n]$ and its DTFT $X_{(2)}(\Omega)$.

Solution

(a) The rectangular pulse is shown in Figure 6.1a. It is also known as 5-point rectangular window function. Its Fourier transform is

$$X(\Omega) = \sum_{n=-\infty}^{\infty} x[n]e^{-j\Omega n} = \sum_{n=-(N-1)/2}^{(N-1)/2} (1)(e^{-j\Omega})^n, \quad N = 5 \tag{6.6.2}$$

This is a geometric sequence with factor $e^{-j\Omega}$. From Appendix A.10,

$$\sum_{k=M}^{N} a^k = \frac{a^{N+1} - a^M}{a-1} \quad a \neq 1 \tag{6.6.3}$$

Applying this to Equation 6.6.2 yields

$$\begin{aligned} X(\Omega) &= \frac{e^{-j[(N+1)/2]\Omega} - e^{j[(N-1)/2]\Omega}}{e^{-j\Omega} - 1} = \frac{e^{-j\Omega/2}\left(e^{-j(N/2)\Omega} - e^{j(N/2)\Omega}\right)}{e^{-j\Omega/2}(e^{-j\Omega/2} - e^{j\Omega/2})} \\ &= \frac{\sin(N\Omega/2)}{\sin(0.5\Omega)}, \quad N = 5 \\ &= \frac{\sin(2.5\Omega)}{\sin(0.5\Omega)} \end{aligned} \tag{6.6.4}$$

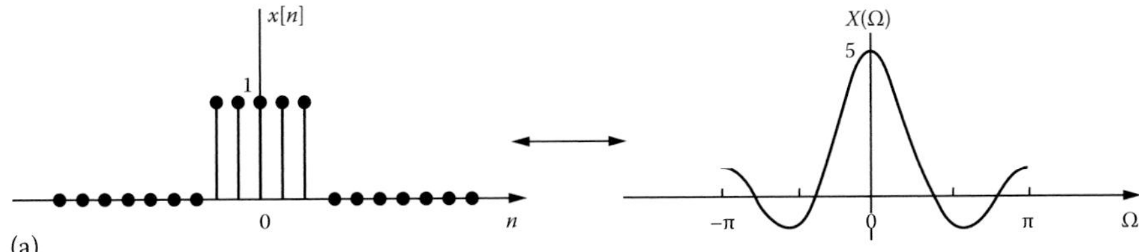

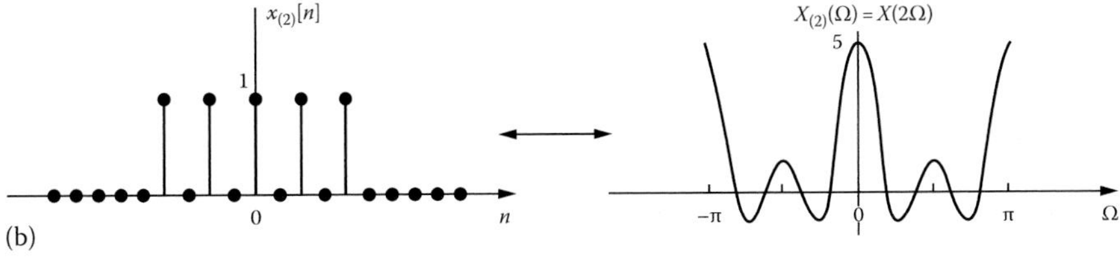

FIGURE 6.1 For Example 6.6.

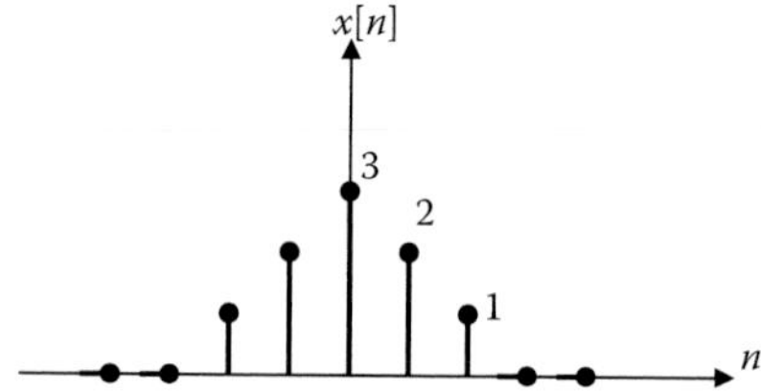

FIGURE 6.2 For Practice Problem 6.5.

The sequence in Equation 6.6.1 and its Fourier transform in Equation 6.6.4 are sketched in Figure 6.1a. Notice that although $x[n]$ is discrete, $X(\Omega)$ is continuous.

(b) Using the time-scaling property in Equation 6.30 and Equation 6.6.4, we obtain

$$X_{(2)}(\Omega) = X(2\Omega) = \frac{\sin(5\Omega)}{\sin(\Omega)}$$

Both the signal $x_{(2)}[n]$ and its DTFT $X_{(2)}(\Omega)$ are sketched in Figure 6.1b. Notice that $x_{(2)}[n]$ is obtained from $x[n]$ by placing $k - 1 = 2 - 1$ zeros between successive values of the original signal $x[n]$.

Practice Problem 6.6 Find the DTFT of the sequence shown in Figure 6.2.

Answer: $3 + 2\cos 2\Omega + 4\cos \Omega$.

Example 6.7

Determine the Inverse of the Function

$$X(\Omega) = \frac{1}{(1 - ae^{-j\Omega})^2}, \quad |a| < 1$$

Solution

From Table 6.2,

$$a^n u[n] \xleftrightarrow{\text{DTFT}} \frac{1}{1 - ae^{-j\Omega}}, \quad |a| < 1$$

The given function can be expressed as

$$X(\Omega) = \frac{1}{(1 - ae^{-j\Omega})^2} = \left(\frac{1}{1 - ae^{-j\Omega}}\right)\left(\frac{1}{1 - ae^{-j\Omega}}\right)$$

By the convolution theorem,

$$x[n] = a^n u[n] * a^n u[n] = \sum_{k=-\infty}^{\infty} a^k u[k] a^{n-k} u[n-k]$$

$$= a^n \sum_{k=0}^{n} 1 = (n+1)a^n u[n] \quad |a| < 1$$

Practice Problem 6.7 Given that $h[n] = \left(\frac{1}{2}\right)^n u[n]$ and $x[n] = \left(\frac{1}{3}\right)^n u[n]$, find

$$y[n] = h[n] * u[n].$$

Answer: $3\left(\frac{1}{2}\right)^n u[n] - 2\left(\frac{1}{3}\right)^n u[n]$

6.4 DISCRETE FOURIER TRANSFORM

The discrete Fourier transform (DFT) may be regarded as a logical extension of the discrete-time Fourier transform (DTFT) covered earlier. By sampling the DTFT $X(\Omega)$ at uniformly spaced frequencies $\Omega = 2\pi k/N$, where $k = 0, 1, 2,\ldots, N-1$, we can define the DFT of $x[n]$. Let $x[n]$, $n = 0, 1, 2,\ldots, N-1$, be an N-point sequence. We define the DFT of $x[n]$ as

$$X[k] = \sum_{n=0}^{N-1} x[n] e^{-j2\pi nk/N}, \qquad k = 0,1,2,\ldots, N-1 \tag{6.54}$$

The discrete Fourier transform relates discrete-time sequence $x[n]$, $n = 0, 1, 2,\ldots, N-1$, to the discrete-frequency sequence $X[k]$, $k = 0, 1, 2,\ldots, N-1$. The discrete Fourier transform is defined by the pair

$$\boxed{\text{DFT: } X[k] = F[x[n]] = \sum_{n=0}^{N-1} x[n] e^{-j2\pi nk/N}} \tag{6.55}$$

$$\boxed{\text{IDFT: } x[n] = F^{-1}\left[X[k]\right] = \frac{1}{N}\sum_{k=0}^{N-1} X[k] e^{j2\pi nk/N}} \tag{6.56}$$

Equation 6.56 specifies the inverse DFT or IDFT. We use the notation

$$x[n] \xleftrightarrow{\text{DFT}} X[k] \tag{6.57}$$

Note that the DFT $X[k]$ as given in Equation 6.55 is a periodic sequence with period N. This should be expected since $X[k]$ is a sampled version of the periodic DTFT $X(\Omega)$. Also, note that both DFT and the inverse DFT (IDFT) begin the summation from zero and end at $N - 1$ so that both are finite. Since DFT can be derived from DTFT, we would expect their properties to be similar. The properties of DFT are listed in Table 6.3.

Notice that items 6 and 7 in Table 6.3 involve circular convolution. The convolution of two discrete-time signals covered in Chapter 2

$$y[n] = x[n] * h[n] = \sum_{k=-\infty}^{\infty} x[k]h[n-k] \tag{6.58}$$

may be regarded as *linear convolution*. Consider two sequences, $x[n]$ and $h[n]$, defined for $n = 0, 1,\ldots, N - 1$, that is, they both have length N and zero values of $n < 0$ and $n \geq N$. The *periodic or circular convolution* is defined as

$$y[n] = x[n] \otimes h[n] = \sum_{k=0}^{N-1} x[k]h[n-k] \tag{6.59}$$

TABLE 6.3
Properties of the DFT

Property	Time Domain	Frequency Domain
1. Linearity	$ax_1[n] + bx_2[n]$	$aX_1[k] + bX_2[k]$
2. Time-shifting	$x[n - m]$	$e^{-j2\pi km}X(k)$
3. Frequency-shifting (modulation)	$e^{-j2\pi k_0 n/N}x[n]$	$X(k - k_0)$
4. Time reversal	$x[-n]$	$X(-k)$
5. Conjugation	$x^*[n]$	$X^*(-k)$
6. Time-convolution	$x_1[n] \otimes x_2[n]$	$X_1[k]X_2[k]$
7. Frequency-convolution	$x_1[n]x_2[n]$	$\frac{1}{N}X_1[k] \otimes X_2[k]$
8. Parseval's relation	$E_x = \sum_{n=0}^{N-1} \lvert x[n] \rvert^2$	$E_x = \frac{1}{N}\sum_{k=0}^{N-1} \lvert X[k] \rvert^2$

It is often regarded as the evaluation of two signals around two concentric circles. Equation 6.59 is evaluated by writing the N values of $x[n]$ equally spaced around an outer circle in a counterclockwise direction, while the N values of $h[n]$ are equally spaced in a clockwise direction on an inner circle.

In general, $X[k]$ is complex and it may be easier to express it in terms of its real part $R[k]$ and imaginary part $I[k]$, that is,

$$X\left[k\right]=R\left[k\right]+jX\left[k\right] \tag{6.60}$$

where

$$R[k]=x[0]+\sum_{n=1}^{N-1}x[n]\cos\left(\frac{2\pi kn}{N}\right) \tag{6.61}$$

$$I[k]=-\sum_{n=1}^{N-1}x[n]\sin\left(\frac{2\pi kn}{N}\right) \tag{6.62}$$

Example 6.8

Find the DFT of the sequence $x[n] = \{1, -2, 1, 3\}$.

Solution

We can find the DFT of $x[n]$ by either directly using Equation 6.55 or indirectly using Equations 6.61 and 6.62.
$x[0] = 1$, $x[1] = -2$, $x[2] = 1$, and $x[3] = 3$. With $N = 4$, Equation 6.55 becomes

$$\begin{aligned} X[k] &= \sum_{n=0}^{N-1} x[n]e^{-j2\pi nk/N} = \sum_{n=0}^{3} x[n]e^{-j\pi nk/2}, && k=0,1,2,3 \\ &= x[0]+x[1]e^{-j\pi k/2}+x[2]e^{-j\pi k}+x[3]e^{-j\pi 3k/2}, && k=0,1,2,3 \\ &= 1-2e^{-j\pi k/2}+e^{-j\pi k}+3e^{-j\pi 3k/2}, && k=0,1,2,3 \end{aligned} \tag{6.8.1}$$

The real part of this is

$$\begin{aligned} R[k] &= 1-2\cos(-\pi k/2)+\cos(-\pi k)+3\cos(-3\pi k/2), && k=0,1,2,3 \\ &= 1-2\cos(\pi k/2)+\cos(\pi k)+3\cos(3\pi k/2), && k=0,1,2,3 \end{aligned}$$

Evaluating this for each k leads to

$$R[k]=\begin{cases} 3, & k=0 \\ 0, & k=1 \\ 1, & k=2 \\ 0, & k=3 \end{cases} \tag{6.8.2}$$

The imaginary part of $X[k]$ is

$$\begin{aligned} I[k] &= -2\sin(-\pi k/2) + \sin(-\pi k) + 3\sin(-\pi 3k/2) \\ &= 2\sin(\pi k/2) - \sin(\pi k) - 3\sin(3\pi k/2), \quad k = 0,1,2,3 \end{aligned}$$

Evaluating this for each value of k, we get

$$I[k] = \begin{cases} 0, & k=0 \\ 5, & k=1 \\ 0, & k=2 \\ -5, & k=3 \end{cases} \tag{6.8.3}$$

From Equations 6.8.2 and 6.8.3, we obtain $X[k] = R[k] + jI[k]$, that is,

$$X[k] = \begin{cases} 3, & k=0 \\ j5, & k=1 \\ 1, & k=2 \\ -j5, & k=3 \end{cases} \tag{6.8.4}$$

Practice Problem 6.8 Determine the IDFT of the $X[k]$ obtained in Example 6.8. *Hint*: Use Equation 6.56.

Answer: $x[n] = [1,-2, 1, 3]$.

Example 6.9

Find the periodic convolution of the following two sequences:

$$x[n] = [1,0,-2,3] \quad \text{and} \quad h[n] = [3,1,2,-1]$$

Solution
$N = 4$ and we have a case similar to the previous example. Instead of using real and imaginary parts, we apply Equation 6.55 directly. From Equation 6.8.1,

$$\begin{aligned} X[k] &= x[0] + x[1]e^{-j\pi k/2} + x[2]e^{-j\pi k} + [3]e^{-j\pi 3k/2}, \quad k = 0,1,2,3 \\ X[0] &= x[0] + x[1] + x[2] + x[3] = 2 \\ X[1] &= x[0] + x[1]e^{-j\pi/2} + x[2]e^{-j\pi} + x[3]e^{-j3\pi/2} = 3 + j3 \\ X[2] &= x[0] + x[1]e^{-j\pi} + x[2]e^{-j2\pi} + x[3]e^{-j3\pi} = -4 \\ X[3] &= x[0] + x[1]e^{-j3\pi/2} + x[2]e^{-3j\pi} + x[3]e^{-j9\pi/2} = 3 - j3 \end{aligned}$$

Similarly,

$$H[0] = h[0] + h[1] + h[2] + h[3] = 5$$
$$H[1] = h[0] + h[1]e^{-j\pi/2} + h[2]e^{-j\pi} + h[3]e^{-j3\pi/2} = 1 - j2$$
$$H[2] = h[0] + h[1]e^{-j\pi} + h[2]e^{-j2\pi} + h[3]e^{-j3\pi} = 5$$
$$H[3] = h[0] + h[1]e^{-j3\pi/2} + h[2]e^{-3j\pi} + h[3]e^{-j9\pi/2} = 1 + j2$$

Convolution in the time domain produces multiplication in the frequency domain, that is,

$$y[n] = h[n] \otimes x[n] \leftrightarrow Y[k] = H(k)X(k)$$

$$Y[0] = H[0]X[0] = 10$$
$$Y[1] = H[1]X[1] = 9 - j3$$
$$Y[2] = H[2]X[2] = -20$$
$$Y[3] = H[3]X[3] = 9 + j3$$

We now use inverse discrete Fourier transform (or IDFT) to find $y[n]$ from $Y[k]$. From Equation 6.56,

$$y[n] = F^{-1}[Y[k]] = \frac{1}{N}\sum_{k=0}^{N-1} Y[k]e^{j2\pi nk/N}$$

Since $N = 4$, we have

$$y[0] = \frac{1}{4}[Y[0] + Y[1] + Y[2] + Y[3]] = 2$$

$$y[1] = \frac{1}{4}\left[Y[0] + Y[1]e^{j\pi/2} + Y[2]e^{j\pi} + Y[3]e^{j3\pi/2}\right] = 9$$

$$y[2] = \frac{1}{4}\left[Y[0] + Y[1]e^{j\pi} + Y[2]e^{j2\pi} + Y[3]e^{j3\pi}\right] = -7$$

$$y[3] = \frac{1}{4}\left[Y[0] + Y[1]e^{j3\pi/2} + Y[2]e^{3j\pi} + Y[3]e^{j9\pi/2}\right] = 6$$

From this, we have $y[n] = [2, 9, -7, 6]$.

Practice Problem 6.9 Repeat Example 6.9 for

$$x[n] = [0, 1, 2, -3] \quad \text{and} \quad h[n] = \left(\frac{1}{2}\right)^n, \quad n = 0, 1, 2, 3$$

Answer: $y[n] = [-0.875, 0.5, 2.125, -1.75]$.

Example 6.10

Verify Parseval's relation using the sequence $x[n] = \{1, -2, 1, 3\}$.

Solution

Using the given sequence directly, we get

$$E_x = \sum_{n=0}^{N-1} |\, x[n] \,|^2 = 1 + 4 + 1 + 9 = 15$$

In Example 6.8, we obtained the DFT of the sequence as

$$X[k] = \begin{cases} 3, & k = 0 \\ j5, & k = 1 \\ 1, & k = 2 \\ -j5, & k = 3 \end{cases}$$

$$E_x = \frac{1}{N} \sum_{k=0}^{N-1} |\, X[k] \,|^2 = \frac{1}{4}(9 + 25 + 1 + 25) = 15$$

confirming Parseval's relation.

Practice Problem 6.10 Repeat Example 6.10 for the sequence $x[n] = [0, 1, 2, -3]$.

Answer: $E_x = 14$.

6.5 FAST FOURIER TRANSFORM

The number of complex computations required in calculating DFT is drastically reduced by an algorithm known as the **fast Fourier transform** (FFT). The development of FFT, which dramatically reduces the number of computational operations, made DFT a very useful transform in many areas of applied science and technology. FFT was developed by James Cooley and John Tukey in 1965. Keep in mind that FFT is not a transform by itself; it is an algorithm—a fast computation of DFT.

For a discrete-time signal, we defined the N-point DFT and IDFT as

$$X[k] = \sum_{n=0}^{N-1} x[n] e^{-j2\pi nk/N} \quad k = 0, 1, \ldots, N-1 \tag{6.63}$$

$$x[n] = \frac{1}{N} \sum_{k=0}^{N-1} X[k] e^{j2\pi nk/N} \quad n = 0, 1, 2, \ldots, N-1 \tag{6.64}$$

It is evident from Equation 6.63 that to compute each $X[k]$ requires N complex multiplications and $N - 1$ complex additions. Since we have to evaluate $X[k]$ for $k = 0, 1, 2, \ldots, N - 1$, the entire computation of DFT requires N^2 complex multiplications and $N(N - 1)$ complex additions. This becomes prohibitively time consuming for large N.

TABLE 6.4
Number of Multiplications Required in DFT and FFT

M	$N = 2^m$	DFT	FFT
1	2	4	1
2	4	16	4
3	8	64	12
4	16	256	32
5	32	1,024	80
6	64	4,096	192
7	128	16,384	448
8	256	65,536	1,024
9	512	261,144	2,304
10	1,024	1,048,576	5,120

The numbers of complex multiplications required for *N*-point DFT (N^2) and FFT $\left(\frac{N}{2}\log_2 N\right)$ are compared in Table 6.4. We can see from the table that as *N* increases, the speed advantage of FFT becomes obvious. (For FFT computations, *N* need not be a power of 2.) The FFT algorithm is very useful in communications and digital signal processing.

Like inverse DFT (or IDFT), we also have inverse FTT (or IFFT). Although there are several variations of Tukey–Cooley algorithm on FFT, they can be grouped into two types: decimation in time and decimation in frequency. Discussion of these algorithms is beyond the scope of this book. All that is needed as this stage is using FFT and IFFT in MATLAB.

6.6 COMPUTING WITH MATLAB®

MATLAB has in-built commands **fft** and **ifft** for finding respectively the fast Fourier transform and inverse fast Fourier transform of a discrete-time sequence. As we know, finding DFT is the same as finding FFT.

Example 6.11

A square wave is represented by the sequence $x[n]$ defined as

$$x[0] = x[1] = x[2] = x[3] = 1 \text{ and } x[4] = x[5] = x[6] = x[7] = -1$$

Use MATLAB to find the FFT.

Solution

We use the following MATLAB script to find $X[k]$.

```
x = [ 1 1 1 1 -1 -1 -1 -1];
X = fft(x)
stem(abs(X))
```

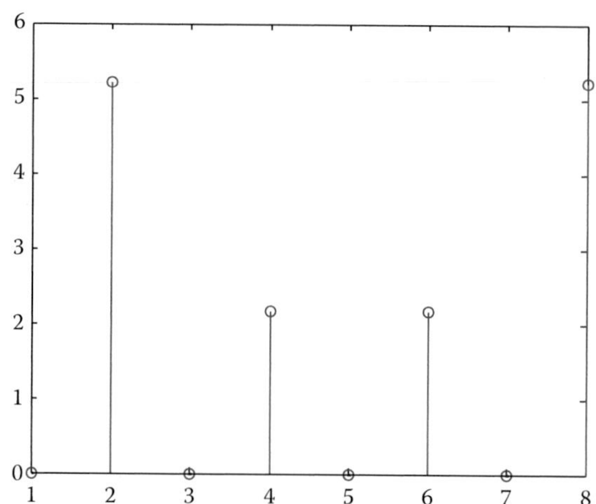

FIGURE 6.3 For Example 6.10.

This results in

$X = [0\ \ 2 - j4.8284\ \ 0\ \ 2 - j0.8284\ \ 0\ \ 2 + j0.8284\ \ 0\ \ 2 + j4.8284]$

The MATLAB command stem is used to plot the absolute value of the result, as shown in Figure 6.3. To ensure that the result is correct, we can find the IFFT of X.

```
X = [0 2-j*4.8284 0 2-j*0.8284 0 2+j*0.8284 0 2+j*4.8284]
x = ifft(X)
```

The result is the same as the given $x[n]$.

Practice Problem 6.11 Find the FTT of the sequence $x[n] = \{1, -2, 1, 3\}$.

Answer: $[3, j5, 1, -j5]$.

Example 6.12

Repeat Example 6.9 using MATLAB.

Solution

We find the FFT of sequences $x[n]$ and $h[n]$. We multiply the results together and then find IFFT to get $y[n]$. The MATLAB for doing the convolution follows.

```
x = [ 1  0 -2 3];
h = [ 3  1  2 -1];
X = fft(x);
H = fft(h);
Y = X.*H;
y = ifft(Y)
```

This produces $y[n] = [2, 9, -7, 6]$, which is what we got the hard way in Example 6.9.

Practice Problem 6.12 Repeat Practice Problem 6.9 using MATLAB.

Answer: $y[n] = [-0.875, 0.5, 2.125, -1.75]$.

Example 6.13

Although MATLAB does not have a function for DTFT, we can use it to plot the Fourier spectrum $X(\Omega)$. In Practice Problem 6.2, the DTFT of the discrete-time signal $x[n] = a^n u[n]$ is

$$X(\Omega) = \frac{1}{1 - ae^{-j\Omega}}$$

Plot this for $a = 0.8$,

Solution

The necessary MATLAB script is shown below. The resulting plot of $X(\Omega)$ is in Figure 6.4.

```
a=0.8;
Omega= 0:0.01:2.0;
X =1./(1- a*exp(-j*Omega));
plot(Omega,abs(X));
xlabel('\Omega')
ylabel('X(\Omega)')
```

Practice Problem 6.13 Use MATLAB to plot the Fourier spectrum

$$X(\Omega) = \frac{ae^{j\Omega}}{(e^{j\Omega} - a)^2}, \qquad a = 0.9$$

Answer: See Figure 6.5.

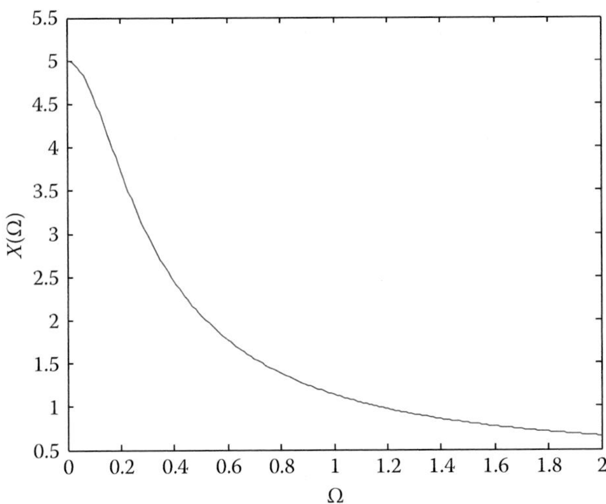

FIGURE 6.4 For Example 6.13.

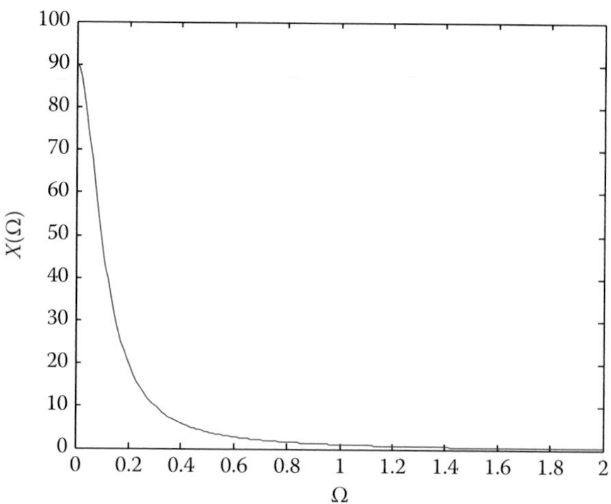

FIGURE 6.5 For Practice Problem 6.13.

6.7 APPLICATIONS

We will consider some applications of the discrete Fourier transforms. Specifically, we consider touch-tone telephone and windowing technique.

6.7.1 Touch-Tone Telephone

The touch-tone telephone uses signals of different frequencies to specify which button is being pushed. The keypad has 12 buttons arranged in four rows and three columns. The DTFT of a sampled signal can be used to identify these frequencies. Pushing of a button generates a sum of two sinusoids corresponding to its unique pair of frequencies. The higher frequency sinusoid indicates the column of the key, while the lower frequency sinusoid indicates the row of the key. Table 6.5 shows the DTFT frequencies for touch-tone signals sampled at 8192 Hz. For example, the digit 8 is represented by the signal

$$x_8[n] = \sin(0.6535n) + \sin(1.0247n) \tag{6.65}$$

TABLE 6.5
DTFT Frequencies for Touch-Tone Signals

	ω_{column}		
ω_{row}	**0.9273**	**1.0247**	**1.1328**
0.5346	1	2	3
0.5906	4	5	6
0.6535	7	8	9
0.7217		0	

6.7.2 Windowing

Sampling a signal for storage in a computer requires that we limit the signal to a finite length. This truncation causes discontinuities at either end of the sampled signal. In other to minimize this effect called *spectral leakage*, the data samples can be multiplied by windows which approach zero smoothly at the beginning and end of the signal.

Windowing is a fundamental operation in digital signal processing and is implied in DFT development. It refers to the operation of multiplying a signal $x[n]$ by a finite-duration sequence $w[n]$ to produce a finite-duration signal $x_w[n]$, that is,

$$x_w[n] = x[n]w[n] \tag{6.66}$$

where

$w[n] = 0$ for $n < 0$ and $n > N$

This makes both the window sequence $w[n]$ and the windowed sequence $x_w[n]$ to be of length N. A few examples of window sequences are as follows:

1. *Rectangular*:

$$w[n] = \begin{cases} 1, & 0 \le n \le N-1 \\ 0, & \text{otherwise} \end{cases} \tag{6.67}$$

2. *Hanning*:

$$w[n] = \frac{1}{2}\left[1 - \cos\left(\frac{2\pi n}{N-1}\right)\right], \quad 0 \le n < \frac{N-1}{2} \tag{6.68}$$

3. *Hamming*:

$$w[n] = 0.54 - 0.46\cos\left(\frac{2\pi n}{N-1}\right), \quad 0 \le n < N-1 \tag{6.69}$$

4. *Barlett*:

$$w[n] = 1 - \frac{2n}{N-1}, \quad 0 \le n < \frac{N-1}{2} \tag{6.70}$$

5. *Blackman*:

$$w[n] = 0.4 - 0.5\cos\left(\frac{2\pi n}{N-1}\right) + 0.08\cos\left(\frac{4\pi n}{N-1}\right), \quad 0 \le n < N \tag{6.71}$$

The windowing technique is used in many applications. For example, it is used in Fourier analysis to decrease the size of the Gibbs overshoot and ripples. Windowing is also used in filter design. By using the windowing technique, we can obtain a filter that satisfies the design specs with fewer elements and less processing delay.

Example 6.14

Compute the DTFT of the discrete sequence $x[n] = 0.8^n u[n]$. Use $N = 32$ and compare the exact spectrum with the one obtained using the Hamming window.

Solution

From Table 6.2, the exact DTFT of $u[n]$ is

$$X(\Omega) = \frac{1}{1 - ae^{-j\Omega}} \qquad |a| \le 1$$

where $a = 0.8$. We use the following MATLAB code for computing $X(\Omega)$ and Hamming window.

```
a=0.8; N = 32;
n = 0:N-1
w = 0:0.01:pi; % values of Omega
fe = abs(1./(1 - a*exp(-j*w))); % exact DTFT
subplot(2,1,1)
  plot(w,fe)
  title('Exact spectrum');
  xlabel('\Omega'); ylabel('Magnitude');
   wh = 0.54 - 0.46*cos(2*pi*n'/(N-1)) %Hamming window
fh=abs((a.^n'.*wh)'*exp(-j*n'*w));
fhs = sum(fh,1); %sums columns of N x length(w) matrix fh
subplot(2,1,2)
   plot(w,fhs)
   title('Spectrum using Hamming window');
  xlabel('\Omega'); ylabel('Magnitude');
```

The exact spectrum and the spectrum obtained using Hamming window are shown in Figure 6.6.

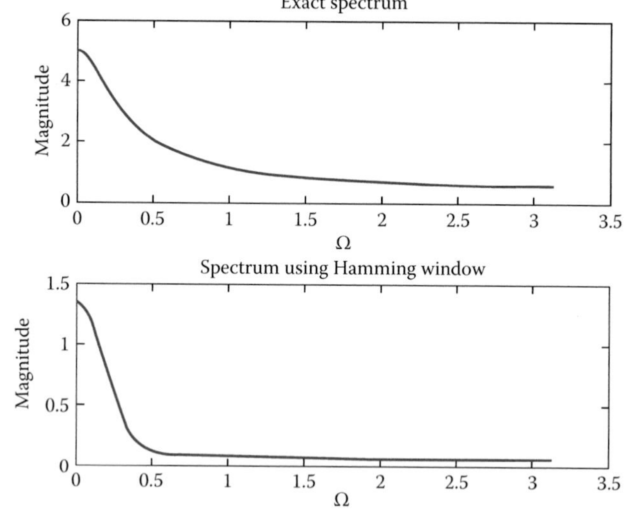

FIGURE 6.6 For Example 6.14.

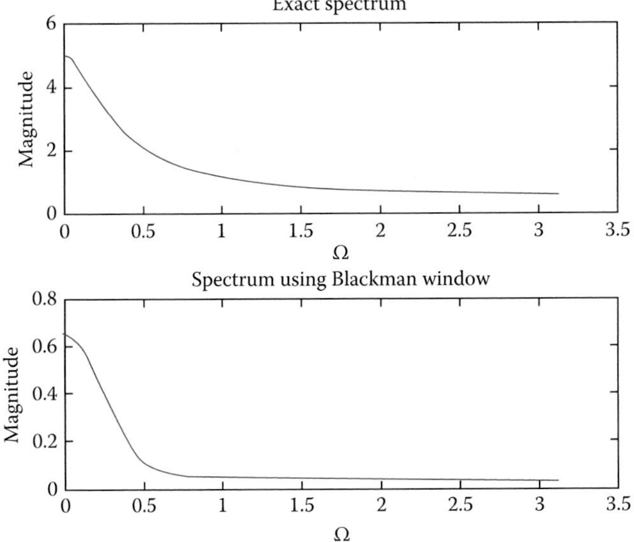

FIGURE 6.7 For Practice Example 6.14.

Practice Problem 6.14 Repeat Example 6.14 using Blackman window.

Answer: See Figure 6.7.

6.8 SUMMARY

1. The discrete-time Fourier transform (DTFT) of a signal $x[n]$ is given by

$$X(\Omega) = \sum_{n=-\infty}^{\infty} x[n]e^{-j\Omega n}$$

   where the variable Ω has the units of radians.

2. The inverse discrete-time Fourier transform (IDTFT) of $X(\Omega)$ is

$$x[n] = \frac{1}{2\pi}\int_{-\pi}^{\pi} X(\Omega)e^{j\Omega n}\, d\Omega$$

3. The properties of discrete-time Fourier transform are listed in Table 6.1, while Table 6.2 presents the transforms of some common transform pairs.

4. The discrete Fourier transform (DFT) of a sequence $x[n]$ of length N is given by

$$X[k] = \sum_{n=0}^{N-1} x[n] e^{-j2\pi nk/N}$$

The properties of DFT are listed in Table 6.3.

5. The inverse discrete Fourier transform (IDFT) is given by

$$x[n] = \frac{1}{N} \sum_{k=0}^{N-1} X[k] e^{j2\pi nk/N}$$

6. The fast Fourier transform (FTT) is an algorithm designed for efficient and fast machine computation of the DFT.
7. MATLAB commands **fft** and **ifft** can be used respectively to find the fast Fourier transform and inverse fast Fourier transform of a discrete-time sequence.
8. The two applications considered in this chapter are touch-tone telephone and windowing technique. The touch-tone telephone uses signals of different frequencies to specify which button is being pushed. When a discrete-time signal has an infinite length, we can terminate it by multiplying it by a window function.

REVIEW QUESTIONS

6.1 The condition for signal $x[n]$ to have DTFT is that $x[n]$ is
(a) Integratable, (b) differentiable, (c) summable, (d) compressible.

6.2 The DTFT of the discrete-time signal $x[n] = 1$, $n = 0, \pm 1, \pm 2, \ldots$, does not exist.
(a) True, (b) False.

6.3 Which of these is not convolution?

(a) $x[n]*h[n]$, (b) $x[n]h[n]$, (c) $\sum_{k=-\infty}^{\infty} x[k]h[n-k]$, (d) $\sum_{k=-\infty}^{\infty} h[k]x[n-k]$

6.4 If $X(\Omega)$ is the DTFT of $x[n]$, then the Fourier transform of $x[-n]$ is
(a) $X(\Omega)e^{-j\Omega}$, (b) $X(\Omega)e^{j\Omega}$, (c) $X(\Omega - 1)$, (d) $X(-\Omega)$

6.5 The DFT is discrete but not periodic.
(a) True, (b) False.

6.6 Given the following discrete-time sequence $x[0] = 1$, $x[1] = -1$, $x[0] = 3$, $x[3] = 2$, the value of $X[0]$ is
(a) $2 + j3$, (b) $2 - j4$, (c) 5, (d) none of the above.

6.7 For a 4-point computation of DFT, how many complex multiplications are involved?
(a) 4, (b) 8, (c) 16, (d) 32.

6.8 An N-point computation of DFT can be done by FFT provided N is even.
(a) True, (b) False.

6.9 The MATLAB command for finding IDFT is
(a) fft, (b) ifft, (c) stem, (d) idft, (e) none of the above.

6.10 Which of the following MATLAB command is not used for windowing technique?
(a) Hanning, (b) Hamming, (c) Blackman, (d) Bartlett.

Answers: 6.1c, 6.2a, 6.3b, 6.4d, 6.5b, 6.6c, 6.7c, 6.8b, 6.9b, 6.10a.

PROBLEMS

SECTIONS 6.2 AND 6.3—DISCRETE-TIME FOURIER TRANSFORM AND PROPERTIES OF DTFT

6.1 (a) Show that $X(\Omega)$ is periodic with period 2π, that is, $X(\Omega + 2\pi) = X(\Omega)$.
(b) Specifically show that $X(\Omega) = \dfrac{1}{1 - ae^{-j\Omega}}$ is periodic.

6.2 Find the DTFT of the signal shown in Figure 6.8.

6.3 Determine the DTFT of the sequence $x[n]$ shown in Figure 6.9.

6.4 Compute the DTFT of the signals in Figure 6.10.

6.5 Determine the DTFT $X(\Omega)$ for each of the following signals:
(a) $x[n] = 0.8^{|n|}$
(b) $x[n] = d[n + k] - d[n - k]$
(c) $x[n] = 0.4^n \cos 2n u[n]$

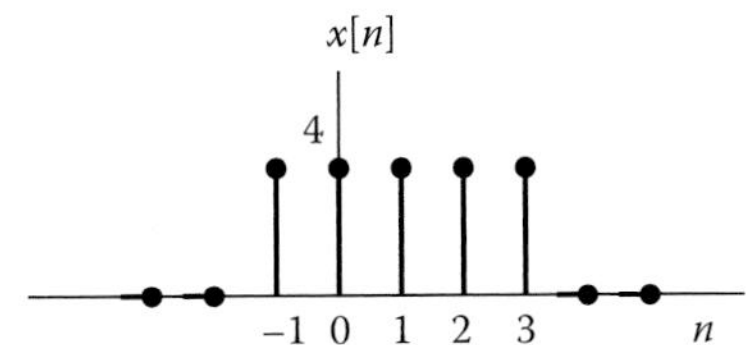

FIGURE 6.8 For Problem 6.2.

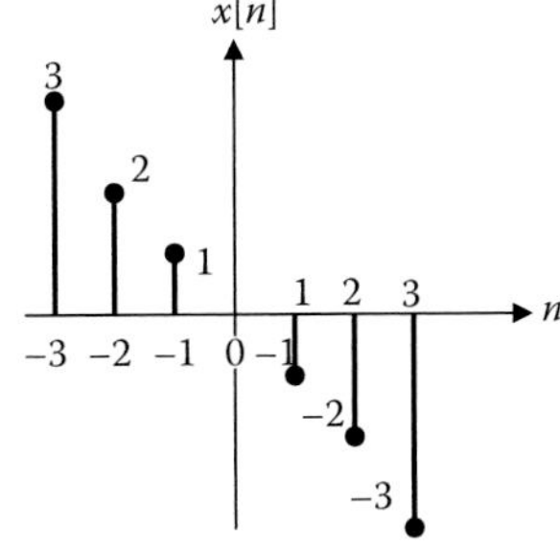

FIGURE 6.9 For Problem 6.3.

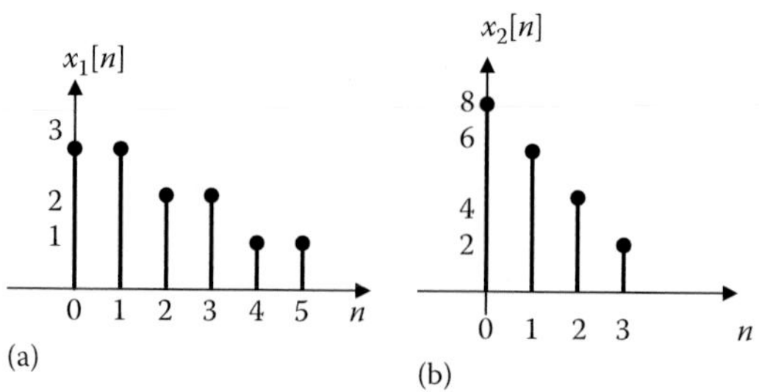

FIGURE 6.10 For Problems 6.4 and 6.24.

6.6 Find the DTFT of each of the following discrete-time signals:
(a) $(0.5)^n u[-n]$
(b) $\delta[n-2] - \delta[n+2]$
(c) $(0.8)^n u[n+2]$

6.7 Obtain the DTFT of the following composite signals:
(a) $x[n] = (0.6)^n u[n] - \sin 2\alpha n$
(b) $y[n] = 2^n u[-n] + \cos 2\alpha n$

6.8 Use Tables 6.1 and 6.2 to find the DTFT of

$$x[n] = a^n u[n] - a^{-n} u[-n]$$

6.9 A certain signal $x[n]$ has DTFT given by

$$X(\Omega) = \frac{e^{j\Omega}}{e^{j\Omega} + a}$$

where a is a constant. Find DTFT $Y(\Omega)$ of the following:
(a) $y[n] = x[n-2]$
(b) $y[n] = x[n]e^{j4n}$
(c) $y[n] = x[n]\sin 2n$
(d) $y[n] = x[n] * x[n]$

6.10 The DTFT of a signal $x[n]$ is

$$X(\Omega) = \frac{2}{3 + e^{-j\Omega}}$$

Find the DTFT of the following signals:
(a) $y[n] = x[-n]$
(b) $z[n] = nx[n]$
(c) $w[n] = x[n] + x[n-1]$
(d) $v[n] = x[n]\cos(n\pi)$

6.11 Let $X(\Omega)$ be the DTFT of $x[n] = (0.8)^n u[n]$. Find the signal corresponding to the following spectra:

(a) $Y(\Omega) = X(-\Omega)$

(b) $Z(\Omega) = X(\Omega-\pi)$

(c) $W(\Omega) = jX'(\Omega)$

6.12 Obtain the IDTFT of each of the following Fourier transforms:

(a) $X(\Omega) = 10\sin(4\Omega)$

(b) $X(\Omega) = 2\cos^2\Omega$

6.13 Determine the signal $x[n]$ corresponding to each of the following Fourier transforms:

(a) $X(\Omega) = 1 + 3e^{-j2\Omega} - 2e^{j4\Omega} + e^{j5\Omega}$

(b) $X(\Omega) = \dfrac{e^{-j\Omega} - \frac{1}{2}}{1 - \frac{1}{2}e^{-j\Omega}}$

(c) $X(\Omega) = 3\pi[\delta(\Omega-2) + \delta(\Omega + 2)]$

6.14 Obtain the IDTFT $x[n]$ of the rectangular pulse spectrum $X(\Omega)$ shown in Figure 6.11.

6.15 Let $X(\Omega) = \begin{cases} 1, & -\pi < \Omega < \pi \\ 0, & \text{otherwise} \end{cases}$. Use this function to verify Parseval's relation.

6.16 Use Equation 6.8 to determine the IDTFT of each of the spectra in Figure 6.12.

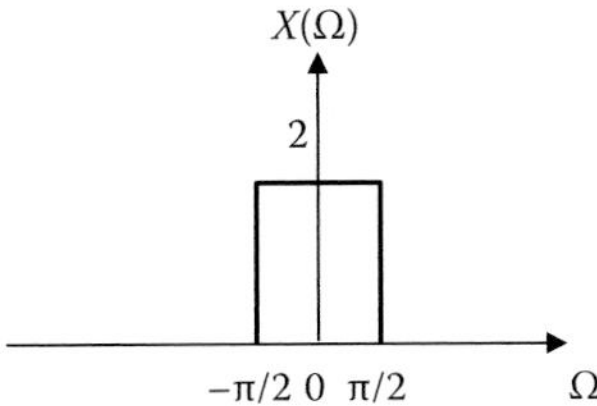

FIGURE 6.11 For Problem 6.14.

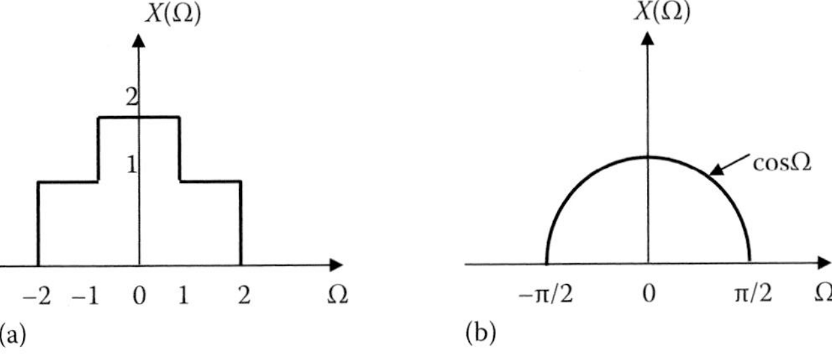

FIGURE 6.12 For Problem 6.16.

6.17 Show that if $x[n] \leftrightarrow X(\Omega)$, then

$$x[n+k]-x[n-k] \leftrightarrow 2jX(\Omega)\sin k\Omega$$

6.18 Find the convolution $y[n] = h[n] * x[n]$ of the following pairs of signals:

(a) $x[n]=\left(\frac{1}{4}\right)^n u[n], \quad h[n]=1$

(b) $x[n]=\left(\frac{1}{3}\right)^n u[n], \quad h[n]=\delta[n]+\delta[n-1]$

(c) $x[n]=\left(\frac{1}{2}\right)^n u[n], \quad h[n]=\left(\frac{1}{3}\right)^n u[n]$

Sections 6.4 and 6.5—Discrete Fourier Transform and Fast Fourier Transform

6.19 Find the DFT of the sequence $x[n] = a^n$.

6.20 Prove the following DFT properties:

(a) $X[0]=\sum_{n=0}^{N-1} x[n]$

(b) $X[N/2]=\sum_{n=0}^{N-1} (-1)^n x[n]$

6.21 Find the DFT of the following sequences:

(a) $x[n] = \{0, 1, 2, 3\}$

(b) $y[n] = \{1, 1, -1, -1, 1, 1, -1, -1\}$

6.22 Calculate the DFT of the following discrete-time signals:

(a) $x[0] = -1$, $x[1] = 1$, $x[2] = 0$, $x[3] = 2$

(b) $x[0] = 1$, $x[1] = 2$, $x[2] = 3$, $x[3] = -1$

6.23 Show that

(a) $x[n]\cos\left(\frac{2\pi km}{N}\right) \longleftrightarrow \frac{1}{2}\left[X(m-k)+X(m+k)\right]$

(b) $x[n]\sin\left(\frac{2\pi km}{N}\right) \longleftrightarrow \frac{1}{2j}\left[X(m-k)-X(m+k)\right]$

6.24 Determine the DFT of the signals in Figure 6.10.

6.25 The DFT of a signal $x[n]$ is

$$X(0)=1, \quad X(1)=1+j2, \quad X(2)=1-j, \quad X(3)=1+j, \quad X(4)=1-j2$$

Compute $x[n]$.

6.26 Given that $X[k] = \text{DFT}\{x[n]\}$, show that

$$x[n] = \frac{1}{N}\left(\text{DFT}\{X^*[k]\}\right)^*$$

where * denotes complex conjugation. This shows that the same algorithm used for computing DFT can be used for computing IDFT.

6.27 How many multiplications are required in DFT and FFT when $N = 2048$?

Section 6.6—Computing with MATLAB®

6.28 Use MATLAB to compute the FFT of the following signals. For each signal, plot $|X(k)|$.

(a) $x[n] = 1,\ 0 \le n \le 12$

(b) $x[n] = n,\ 0 \le n \le 10$

(c) $x[n] = \begin{cases} 1, & n = 0 \\ 1/n, & n = 1,2,\ldots,10 \\ 0, & \text{otherwise} \end{cases}$

(d) $x[n] = n(0.8)^n,\quad 0 \le n \le 10$

6.29 In Example 6.3, the DTFT of the signal $x[n] = a^{|n|}$ is

$$X(\Omega) = \frac{1-a^2}{1-2a\cos\Omega + a^2}$$

For $a = 0.75$ and $0 < \Omega < 2\pi$, plot $|X(\Omega)|$.

6.30 Use MATLAB to find the DFT of the discrete signal

$$x[n] = \{1, 2, 0, -1, -2, 1, 5, 4\}$$

6.31 Consider the 16-point sequence

$$x[n] = \begin{cases} 1, & n = 0,1,\ldots,16 \\ 0, & \text{otherwise} \end{cases}$$

Using MATLAB, plot the absolute value of the DFT of $x[n]$.

Section 6.7—Applications

6.32 Use MATLAB to plot Blackman windowing for $N = 15$.

6.33 Modify the window analysis program of Example 6.14 by replacing Hamming window with Hanning window.

7 z-Transform

There is no end to education. We are all in the kindergarten of God.

—Elbert Hubbard

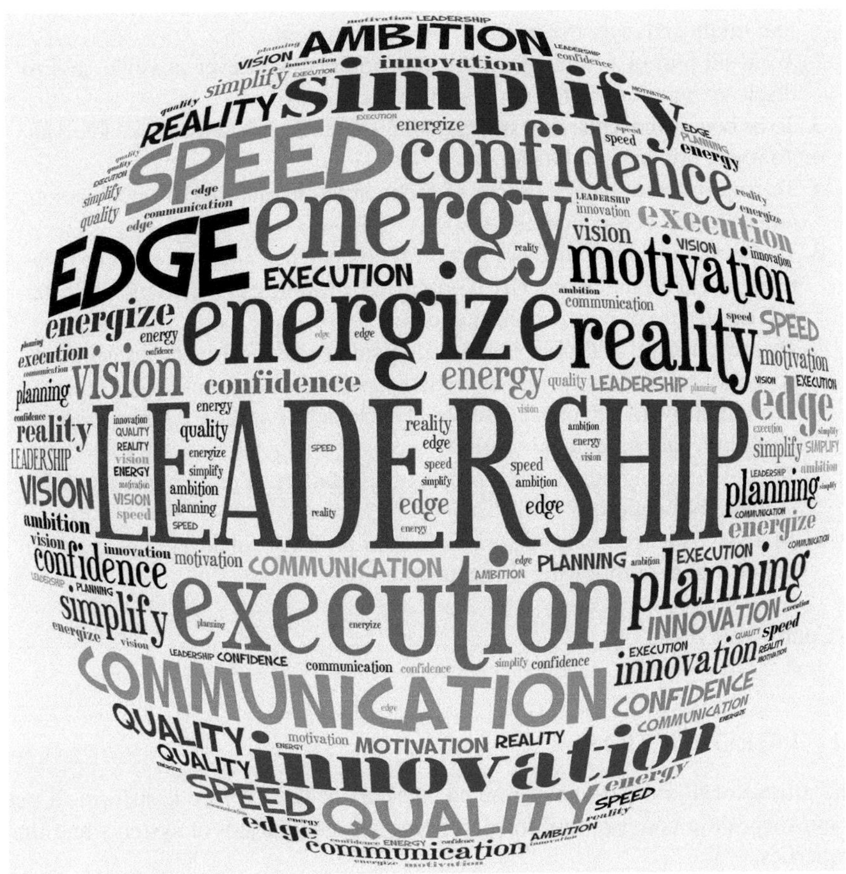

CODES OF ETHICS

Engineering is a profession that makes significant contributions to the economic and social well-being of people all over the world. As members of this important profession, engineers are expected to exhibit the highest standards of honesty and integrity. Unfortunately, the engineering curriculum is so crowded that there is no room for a course on ethics in most schools. The Institute of Electrical and Electronics

Engineers (or IEEE) Code of ethics is presented here to acquaint students with ethical behavior in engineering professions.

We, the members of the IEEE, in recognition of the importance of our technologies in affecting the quality of life throughout the world, and in accepting a personal obligation to our profession, its members and the communities we serve, do hereby commit ourselves to the highest ethical and professional conduct and agree:

1. To accept responsibility in making engineering decisions consistent with the safety, health and welfare of the public, and to disclose promptly factors that might endanger the public or the environment.
2. To avoid real or perceived conflicts of interest whenever possible, and to disclose them to affected parties when they do exist.
3. To be honest and realistic in stating claims or estimates based on available data.
4. To reject bribery in all its forms.
5. To improve the understanding of technology, its appropriate application, and potential consequences.
6. To maintain and improve our technical competence and to undertake technological tasks for others only if qualified by training or experience, or after full disclosure of pertinent limitations.
7. To seek, accept, and offer honest criticism of technical work, to acknowledge and correct errors, and to credit properly the contributions of others.
8. To treat fairly all persons regardless of such factors as race, religion, gender, disability, age, or national origin.
9. To avoid injuring others, their property, reputation, or employment by false or malicious action.
10. To assist colleagues and coworkers in their professional development and to support them in following this code of ethics.

—Courtesy of IEEE

7.1 INTRODUCTION

The z-transform is the discrete-time counterpart of the Laplace transform. It puts at our disposal a compact tool for describing a broad variety of systems and their properties.

Just as Laplace transform is useful in handling signals that do not have Fourier transform, the z-transform enables us to treat discrete-time signals that do not have discrete-time Fourier transforms (DTFT). (For example, the unit sequence $u[n]$ does not have a DTFT.) Also, just as the Laplace transform converts integrodifferential equations into algebraic equations, the z-transform converts difference equations into algebraic equations that are easier to manipulate and solve. Therefore, techniques in this chapter parallel the techniques used in Chapter 3 on Laplace transform. Although the properties of the z-transform are similar to those of the Laplace transform, there are some differences. Like Laplace transform, the z-transform is applicable to systems with initial conditions.

We begin by defining the z-properties and studying its important properties. The z-transform of some common discrete-time signals is derived. We present methods for finding the inverse z-transform. Two important applications of the z-transform are discussed. Finally, we consider using MATLAB® to find z-transform and its inverse.

7.2 DEFINITION OF THE z-TRANSFORM

The z-transform is the generalization of the discrete-time Fourier transform (DTFT). We recall that from Chapter 6, the DTFT of a signal $x[n]$ is given by

$$X(\Omega) = \sum_{n=-\infty}^{\infty} x[n]e^{-j\Omega n} \tag{7.1}$$

Inserting a factor ρ^{-n} in Equation 7.1 leads to

$$X(\Omega) = \sum_{n=-\infty}^{\infty} x[n]\rho^{-n}e^{-j\Omega n} = \sum_{n=-\infty}^{\infty} x[n](\rho e^{j\Omega})^{-n} \tag{7.2}$$

If we let $z = \rho e^{j\Omega}$. Then, Equation 7.2 becomes

$$X(z) = \sum_{n=-\infty}^{\infty} x[n]z^{-n} \tag{7.3}$$

This is the two-sided (or bilateral) z-transform. The one-sided (or unilateral) z-transform is

$$\boxed{X(z) = \mathcal{Z}\{x[n]\} = \sum_{n=0}^{\infty} x[n]z^{-n}} \tag{7.4}$$

Only the one-sided z-transform will be discussed in this chapter. $X(z)$ and $x[n]$ constitute a z-transform pair.

$$x[n] \xrightarrow{Z} X(z) \tag{7.5}$$

Note the following:

1. From Equation 7.4, we notice that the z-transform of $x[n]$ is a power series of z^{-1} whose coefficients are $x[n]$, that is,

$$X(z) = \sum_{n=0}^{\infty} x[n]z^{-n} = x[0] + x[1]z^{-1} + x[2]z^{-2} + x[3]z^{-3} + \cdots \tag{7.6}$$

   In this power series, z^{-n} can be interpreted as indicating the nth sampling instant.

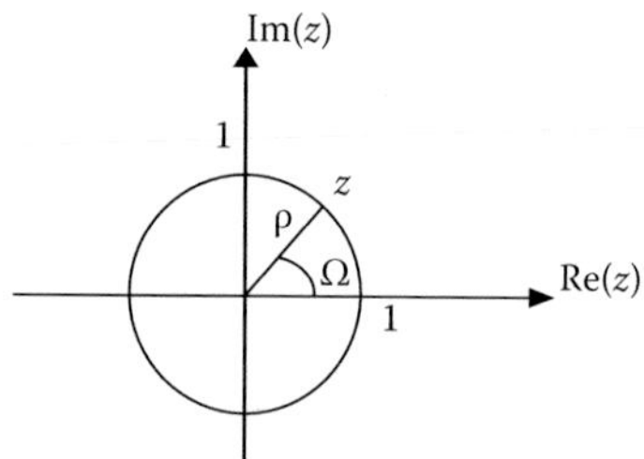

FIGURE 7.1 The unit circle in the z-plane.

2. We also notice that the sum in Equation 7.4 may not converge for all values of z. Each z-transform is associated with a region in the complex plane for which the transform exists. We will discuss this further in the next section.
3. Generally, z is a complex variable with $z = \Sigma + j\Omega$ just as the Laplace transform variable $s = \sigma + j\omega$. z as a complex number $z = \rho e^{j\Omega}$ is illustrated in Figure 7.1.
4. In our development of z-transforms, we will be using the formula for the geometric series

$$1 + a + a^2 + a^3 + \cdots = \sum_{n=0}^{\infty} a^n = \frac{1}{1-a} \tag{7.7}$$

which is convergent if and only if $|a| < 1$.

Example 7.1

Let $x[n]$ by defined by

$$x[n] = \begin{cases} 1, & n = 1 \\ 3, & n = 4 \\ 0, & \text{otherwise} \end{cases}$$

Find the z-transform.

Solution

Using Equation 7.6,

$$X(z) = 1z^{-1} + 3z^{-4} = \frac{1}{z} + \frac{3}{z^4} = \frac{z^3 + 3}{z^4}$$

Practice Problem 7.1 Find the z-transform of the discrete-time signal

$$x[n] = \begin{cases} 2, & n = 11 \\ 4, & n = 42 \\ 0, & \text{otherwise} \end{cases}$$

Answer: $\dfrac{4 + 2z^{31}}{z^{42}}$

Example 7.2

Find the z-transform of the impulse sequence $\delta[n - m]$, $m \geq 0$.

Solution

The given sequence can be written as

$$\delta[n-m] = \begin{cases} 1, & n = m \\ 0, & \text{otherwise} \end{cases}$$

that is, the sequence is zero for all n except when $n = m$. Hence, the z-transform is

$$\mathcal{Z}\{\delta[n-m]\} = \sum_{n=0}^{\infty} \delta[n-m]z^{-n} = z^{-m} \quad (7.2.1)$$

Notice that if $m = 0$, then

$$\mathcal{Z}\{\delta[n]\} = 1 \quad (7.2.2)$$

Practice Problem 7.2 Find the z-transform of the unit-sample sequence

$$x[n] = \begin{cases} 1, & n = 0 \\ 0, & n \neq 0 \end{cases}$$

Answer: 1.

7.3 REGION OF CONVERGENCE

Just like Laplace transform, the z-transform $X(z)$ converges only for certain range of z known as the **region of convergence** (ROC). In order for the z-transform to be complete, its ROC must be specified. The ROC specifies the region where the z-transform converges. The ROC may be within a circle (such as $|z| < |a|$, as shown typically in Figure 7.2a), outside a circle (such as $|z| > |a|$, as typically shown in Figure 7.2b), or within an annulus (such as $|a| < |z| < |b|$, as shown typically in Figure 7.2c). An ROC does not include a pole because the z-transform does not converge at a pole.

To illustrate how to find the ROC, let us consider the following examples.

Example 7.3

Determine the z-transform of an exponential discrete-time signal represented by $x[n] = a^n u[n]$.

Solution

$$X(z) = \sum_{n=0}^{\infty} a^n z^{-n} = 1 + \frac{a}{z} + \frac{a^2}{z^2} + \frac{a^3}{z^3} + \cdots$$

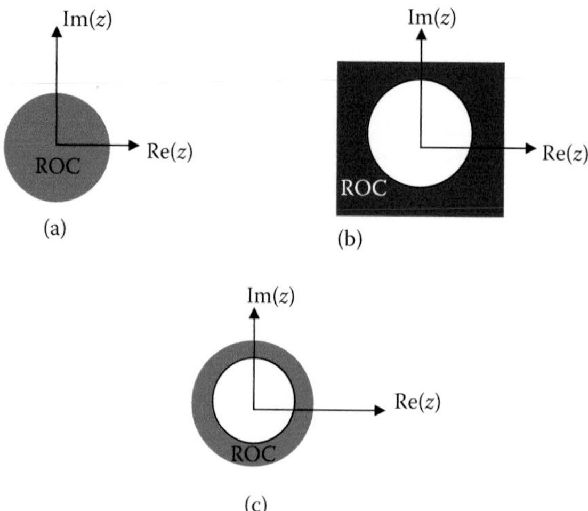

FIGURE 7.2 The regions of convergence (ROCs) of the z-transforms of various sequences.

This is a geometric series which has the form

$$\sum_{n=0}^{\infty} q^n = \frac{1}{1-q}, \quad |q| < 1$$

In our case, $q = a/z$ so that

$$X(z) = \frac{1}{1-\frac{a}{z}} = \frac{z}{z-a}$$

The region of convergence is $\left|\frac{a}{z}\right| < 1$ or $|z| > |a|$. This is similar to ROC in Figure 7.2b.

Practice Problem 7.3 Obtain the z-transform of the discrete-time unit step signal $x[n] = u[n]$. Specify the ROC.

Answer: $\frac{z}{z-1}, \quad |z| > 1$

Example 7.4

Find the z-transform of the following composite signal and determine its region of convergence.

$$x[n] = \left(\frac{1}{2}\right)^n u[n] + 4\left(-\frac{1}{3}\right)^{2n} u[n]$$

Solution

$x[n]$ consists of two signals. We may find the z-transform of each separately and add them.

$$x_1[n] = \left(\frac{1}{2}\right)^n u[n]$$

From Example 7.3,

$$X_1(z) = \frac{z}{z - \frac{1}{2}}, \quad |z| > \frac{1}{2} \tag{7.4.1}$$

Similarly,

$$x_2[n] = 4\left(-\frac{1}{3}\right)^{2n} u[n] = 4\left(\frac{1}{9}\right)^n u[n]$$

Using the result from Example 7.3 again,

$$X_2(z) = \frac{4z}{z - \frac{1}{9}}, \quad |z| > \frac{1}{9} \tag{7.4.2}$$

Combining Equations 7.4.1 and 7.4.2 gives

$$X(z) = X_1(z) + X_2(z) = \frac{z}{z - \frac{1}{2}} + \frac{4z}{z - \frac{1}{9}} = \frac{2z}{2z-1} + \frac{36z}{9z-1}$$

To determine the ROC, we realize that $X(z)$ exists if and only if both $X_1(z)$ and $X_2(z)$ exist. Hence, the region of converge is $|z| > ½$, which covers the ROC of both $X_1(z)$ and $X_2(z)$.

Practice Problem 7.4 Determine the z-transform and the ROC for

$$y[n] = 4\left(\frac{1}{3}\right)^n u[n] - 5\left(\frac{2}{3}\right)^n u[n]$$

Answer: $Y(z) = \frac{12z}{3z-1} - \frac{15z}{3z-2}, \quad |z| > \frac{2}{3}$

7.4 PROPERTIES OF THE z-TRANSFORM

We now consider some basic properties of the z-transform. We use those properties to derive the z-transform of some signals and develop a z-transform table. We first state each property, prove it, and then illustrate it with examples.

7.4.1 Linearity

Let $x[n]$ and $y[n]$ be two discrete-time signals with transforms $X(z)$ and $Y(z)$, respectively. The z-transform of their linear combination is

$$\boxed{\mathcal{Z}\{ax[n]+by[n]\} = aX(z)+bY(z)} \tag{7.8}$$

where a and b are constants.

For example, from Example 7.3 and Practice Problem 7.3, we know that

$$a^n u[n] \leftrightarrow \frac{z}{z-a}, \quad |z|<|a| \quad \text{and} \quad u[n] \leftrightarrow \frac{z}{z-1}, \quad |z|>1 \tag{7.9}$$

By the linearity property,

$$2a^n u[n] + 3u[n] = \frac{2z}{z-a} + \frac{3z}{z-1}, \quad |z|>|a| \quad \text{and} \quad |z|>1 \tag{7.10}$$

7.4.2 Time-Shifting

This properties states that if $X(z)$ is the z-transform of $x[n]$, then

$$\boxed{x[n-m] \leftrightarrow z^{-m}X(z)} \tag{7.11}$$

assuming zero initial conditions. To prove this, we use Equation 7.4.

$$\mathcal{Z}\{x[n-m]\} = \sum_{n=0}^{\infty} x[n-m]z^{-n}$$

By changing variables $k = n - m$, we get

$$\mathcal{Z}\{x[n-m]\} = \sum_{k=0}^{\infty} x[k]z^{-(m+k)} = z^{-m}\sum_{k=0}^{\infty} x[k]z^{-k} = z^{-m}X(z) \tag{7.12}$$

This assumes that the signal is causal ($x[n] = 0$ for $n < 0$).

For example, when $m = 1$ or -1, we have

$$\begin{aligned} x[n-1] &\leftrightarrow z^{-1}X(z) \\ x[n+1] &\leftrightarrow zX(z) \end{aligned} \tag{7.13}$$

For nonzero initial conditions, the time delay property becomes

$$x[n-m] \leftrightarrow z^{-m}X(z)+z^{-m+1}x[-1]+\cdots+z^{-1}x[-m+1]+x[-m] \tag{7.14}$$

7.4.3 Frequency Scaling

If $X(z)$ is the z-transform of $x[n]$, then

$$\boxed{a^n x[n] \leftrightarrow X\left(\frac{z}{a}\right)} \tag{7.15}$$

This implies that multiplication by a^n in the time domain corresponds to scaling in the frequency or z domain.

By definition, the z-transform of the sequence $a^n x[n]$ is

$$\mathcal{Z}\left\{a^n x[n]\right\} = \sum_{n=0}^{\infty} a^n x[n]z^{-n} = \sum_{n=0}^{\infty} x[n]\left(\frac{z}{a}\right)^{-n}$$

$$= X\left(\frac{z}{a}\right) \tag{7.16}$$

This property can be used to find the z-transform of signals multiplied by an exponential signal. For example,

$$e^{j\Omega_0 n}x[n] \leftrightarrow X(e^{-j\Omega_0}z) \tag{7.17}$$

We know that

$$\cos(\Omega n)u[n] = \frac{1}{2}\left(e^{j\Omega n} + e^{-j\Omega n}\right)u[n] \tag{7.18}$$

and that

$$u[n] \leftrightarrow \frac{z}{z-1} \tag{7.19}$$

$$e^{j\Omega n}u[n] \leftrightarrow \frac{ze^{j\Omega}}{ze^{j\Omega}-1} = \frac{z}{z-e^{-j\Omega}} \tag{7.20}$$

Similarly,

$$e^{-j\Omega n}u[n] \leftrightarrow \frac{z}{z-e^{j\Omega}} \tag{7.21}$$

Combining Equations 7.20 and 7.21 gives

$$Z\{\cos\Omega n u[n]\} = \frac{1}{2}\left[\frac{z}{z-e^{-j\Omega}} + \frac{z}{z-e^{j\Omega}}\right] = \frac{z(z-\cos\Omega)}{z^2 - 2z\cos\Omega + 1} \tag{7.22}$$

7.4.4 Time Reversal

If $x[n]\leftrightarrow X(z)$, then

$$\boxed{x[-n] \leftrightarrow X\left(\frac{1}{z}\right)} \tag{7.23}$$

To prove this, we use Equation 7.4

$$\mathcal{Z}\{x[-n]\} = \sum_{n=0}^{\infty} x[-n]z^{-n}$$

Replacing $-n$ with k,

$$\mathcal{Z}\{x[-n]\} = \sum x[k]z^{k} = \sum x[k]\left(\frac{1}{z}\right)^{-k} = X\left(\frac{1}{z}\right) \tag{7.24}$$

For example, from Equation 7.17,

$$u[n] \leftrightarrow \frac{z}{z-1}$$

Applying Equation 7.24,

$$u[-n] \leftrightarrow \frac{1/z}{1/z-1} = \frac{1}{1-z} \tag{7.25}$$

7.4.5 Modulation

The modulation property is very important in communication theory and digital signal processing. If $x[n]\leftrightarrow X(z)$, then

$$\boxed{(\cos\Omega n)x[n] \leftrightarrow \frac{1}{2}\left[X(e^{j\Omega}z) + X(e^{-j\Omega}z)\right]} \tag{7.26}$$

$$\boxed{(\sin\Omega n)x[n] \leftrightarrow \frac{j}{2}\left[X(e^{j\Omega}z) - X(e^{-j\Omega}z)\right]} \tag{7.27}$$

To prove Equations 7.26 and 7.27, we use Euler's identity

$$\cos\Omega n x[n] = \frac{1}{2}\left(e^{-j\Omega n}x[n] + e^{j\Omega n}x[n]\right) \tag{7.28}$$

$$\sin\Omega n x[n] = \frac{j}{2}\left(e^{-j\Omega n}x[n] - e^{j\Omega n}x[n]\right) \tag{7.29}$$

From Equation 7.17,

$$e^{j\Omega_0 n}x[n] \leftrightarrow X(e^{-j\Omega_0}z) \tag{7.30}$$

$$e^{-j\Omega_0 n}x[n] \leftrightarrow X(e^{j\Omega_0}z) \tag{7.31}$$

Applying Equation 7.30 and 7.31 on Equations 7.28 and 7.29 produces Equations 7.26 and 7.27. We can use this property to find the z-transforms of cosine and sine signals. We already got the z-transform of cosine in Equation 7.22.

7.4.6 Accumulation

Let $y[n]$ be the summation or accumulation of the discrete-time signal $x[n]$, with $x[n] = 0$ for $n = -1, -2, -3, \ldots$. Then

$$y[n] = \sum_{k=0}^{n} x[k] \tag{7.32}$$

The accumulation property states that

$$\boxed{\sum_{k=0}^{n} x[k] \leftrightarrow \frac{z}{z-1}X(z)} \tag{7.33}$$

It states that the z-transform of the sum of $x[n]$ is equal to the $z/(z-1)$ times the z-transform of $x[n]$.

To prove Equation 7.33, we write

$$y[n] = \sum_{k=0}^{n-1} x[k] + x[n] \tag{7.34}$$

This may be written as

$$y[n] = y[n-1] + x[n] \tag{7.35}$$

Taking the z-transform of both sides and applying the time-shifting property, we obtain

$$Y(z) = \frac{1}{z}Y(z) + X(z) \tag{7.36}$$

Solving for $Y(z)$ yields

$$Y(z) = \frac{1}{1-z^{-1}} X(z) = \frac{z}{z-1} X(z) \tag{7.37}$$

7.4.7 Convolution

This property plays a very useful role in linear system theory. It states that if

$$x[n] \leftrightarrow X(z) \quad \text{and} \quad h[n] \leftrightarrow H(z) \tag{7.38}$$

then their time convolution in the time domain is equivalent to their product in the frequency or z domain, that is,

$$\boxed{y[n] = x[n] * h[n] \leftrightarrow Y(z) = X(z)H(z)} \tag{7.39}$$

Since the property applies to causal as well as noncausal signals, we prove it to the general case of noncausal signals, where the convolution sum ranges from $-\infty$ to ∞. By definition,

$$\begin{aligned} Y(z) &= \sum_{n=-\infty}^{\infty} x[n] * h[n] z^{-n} = \sum_{n=-\infty}^{\infty} \left(\sum_{k=-\infty}^{\infty} x[n] h[n-k] \right) z^{-n} \\ &= \sum_{k=-\infty}^{\infty} x[k] \left(\sum_{n=-\infty}^{\infty} h[n-k] z^{-n} \right) \end{aligned} \tag{7.40}$$

We note that the term in brackets is the z-transform of the shifted signal $h[n-k]$. By applying the time-shifting property, we get

$$Y(z) = \sum_{k=-\infty}^{\infty} x[k] \left(z^{-k} H(z) \right) = \left(\sum_{k=-\infty}^{\infty} x[k] z^{-k} \right) H(z) = X(z)H(z) \tag{7.41}$$

7.4.8 Initial and Final Values

The initial $x[0]$ and the final value $x[\infty]$ are given by the initial and value theorems, which are stated as

$$\boxed{x[0] = \lim_{z \to \infty} X(z)} \tag{7.42}$$

$$\boxed{x[\infty] = \lim_{z \to 1} (1 - z^{-1}) X(z)} \tag{7.43}$$

assuming that $x[\infty]$ exists.

To prove Equation 7.42, we write Equation 7.4 explicitly as

$$X(z) = x[0] + x[1]z^{-1} + x[2]z^{-2} + \cdots + x[k]z^{-k} + \cdots \tag{7.44}$$

As $z \to \infty$, the terms $z^{-k} \to 0$ for all $k > 0$. Hence, Equation 7.44 becomes

$$\lim_{z\to\infty} X(z) = x[0]$$

To prove Equation 7.43, we apply the time-shifting property

$$\mathcal{Z}\{x[n] - x[n-1]\} = (1 - z^{-1})X(z) \tag{7.45}$$

The left-hand side of Equation 7.45 can be written as

$$\sum_{n=0}^{\infty} \{x[n] - x[n-1]\}z^{-n} = \lim_{N\to\infty} \sum_{n=0}^{N} \{x[n] - x[n-1]\}z^{-n} \tag{7.46}$$

If we take the limit as $z \to 1$, combining Equations 7.45 and 7.46 produces

$$\lim_{z\to 1}(1 - z^{-1})X(z) = \lim_{N\to\infty} \sum_{n=0}^{\infty} \{x[n] - x[n-1]\}$$

$$= \lim_{N\to\infty} x(N) = x(\infty) \tag{7.47}$$

The initial value theorem can be used to check the z-transform of a given signal. The final value theorem gives the correct result only when $x[\infty]$ exists. It can be shown that if there are poles on or outside the unit circle, except for a single pole at $z = 1$, the **final value theorem** does not apply.

For example, let us find the initial and final values of a unit step function, $u[n]$. We know that $u[0] = 1 = u[\infty]$. By the initial value theorem,

$$u[0] = \lim_{z\to\infty} U(z) = \lim_{z\to\infty}\left(\frac{z}{z-1}\right) = 1 \tag{7.48}$$

By the final value theorem,

$$x[\infty] = \lim_{z\to 1}(1 - z^{-1})X(z) = \lim_{z\to 1}\left(\frac{z-1}{z}\frac{z}{z-1}\right) = 1 \tag{7.49}$$

The theorems confirm what we know already.

These properties of the z-transform are listed in Table 7.1, while Table 7.2 provides a collection of the z-transforms of common signals.

Example 7.5

Find the z-transform of the pulse function defined by

$$p[n] = \begin{cases} 1, & n = 0, 1, 2, \ldots, m-1 \\ 0, & \text{otherwise} \end{cases}$$

Solution

We can express $p[n]$ in terms of the unit step function.

$$p[n] = u[n] - u[n-m]$$

TABLE 7.1
Properties of the z-Transforms

Property	$x[n]$	$X(z)$
1. Linearity	$ax[n] + by[n]$	$aX(z) + bY(z)$
2. Time-shifting	$x[n-m]$	$z^{-m}X(z) + z^{-m+1}x[-1] + \cdots + z^{-1}x[-m+1] + x[-m]$
3. Frequency scaling	$a^n x[n]$	$X\left(\frac{z}{a}\right)$
4. Time reversal	$x[-n]$	$X\left(\frac{1}{z}\right)$
5. Multiplication by n	$nx[n]$	$-z\frac{d}{dz}X(z)$
6. Multiplication by n^2	$n^n x[n]$	$z\frac{d}{dz}X(z) + z^2\frac{d^2}{dz^2}X(z)$
7. Modulation		
Multiplication by $e^{j\Omega_0 n}$	$e^{j\Omega_0 n}x[n]$	$X(e^{-j\Omega_0}z)$
Multiplication by $\cos\Omega n$	$(\cos\Omega n)x[n]$	$\frac{1}{2}\left[X(e^{j\Omega}z) + X(e^{-j\Omega}z)\right]$
Multiplication by $\sin\Omega n$	$(\sin\Omega n)x[n]$	$\frac{j}{2}\left[X(e^{j\Omega}z) - X(e^{-j\Omega}z)\right]$
8. Accumulation	$\sum_{k=0}^{n} x[k]$	$\frac{z}{z-1}X(z)$
9. Convolution	$x[n] * h[n]$	$X(z)H(z)$
10. Initial value	$x[0] = \lim_{z\to\infty} X(z)$	
11. Final value	$x[\infty] = \lim_{z\to 1}(1 - z^{-1})X(z)$	

TABLE 7.2
Some Common z-Transform Pairs

$x[n]$	$X(z)$	ROC
1. $\delta[n]$	1	All z
2. $\delta[n-\mathrm{m}]$	$\frac{1}{z^m}$	$z \neq 0$
3. $u[n]$	$\frac{z}{z-1}$	$\lvert z\rvert > 1$
4. $a^n u[n]$	$\frac{z}{z-a}$	$\lvert z\rvert > \lvert a\rvert$
5. $nu[n]$	$\frac{z}{(z-1)^2}$	$\lvert z\rvert > 1$
6. $(n+1)u[n]$	$\frac{z^2}{(z-1)^2}$	$\lvert z\rvert > 1$
7. $n^2 u[n]$	$\frac{z(z+1)}{(z-1)^3}$	$\lvert z\rvert > 1$
8. $na^n u[n]$	$\frac{az}{(z-a)^2}$	$\lvert z\rvert > \lvert a\rvert$
9. $(n+1)a^n$	$\frac{z^2}{(z-a)^2}$	$\lvert z\rvert > \lvert a\rvert$
10. $n^2 a^n u[n]$	$\frac{az(z+a)}{(z-a)^3}$	$\lvert z\rvert > \lvert a\rvert$
11. $\exp[-anT]$	$\frac{z}{z-\exp[-aT]}$	$\lvert z\rvert > e^{-aT}$
12. $\cos\Omega n\ u[n]$	$\frac{z^2 - z\cos\Omega}{z^2 - 2z\cos\Omega + 1}$	$\lvert z\rvert > 1$
13. $\sin\Omega n\ u[n]$	$\frac{z\sin\Omega}{z^2 - 2z\cos\Omega + 1}$	$\lvert z\rvert > 1$
14. $a^n(\cos\Omega n)\ u[n]$	$\frac{z^2 - za\cos\Omega}{z^2 - 2za\cos\Omega + a^2}$	$\lvert z\rvert > \lvert a\rvert$
15. $a^n(\sin\Omega n)\ u[n]$	$\frac{za\sin\Omega}{z^2 - 2za\cos\Omega + a^2}$	$\lvert z\rvert > \lvert a\rvert$

Taking the z-transform of each term and applying the time-shifting property, we get

$$P(z) = \frac{z}{z-1} - z^{-m}\frac{z}{z-1} = \frac{z(1-z^{-m})}{z-1}$$

Practice Problem 7.5 Let $p[n]$ be as defined in Example 7.5. Derive the z-transform of

$$q[n] = a^n p[n]$$

Answer: $Q(z) = \dfrac{z(1-a^m z^{-m})}{z-a}$

Example 7.6

Derive the z-transform of $y[n] = na^n u[n]$.

Solution

We let $x[n] = a^n u[n]$. From entry 4 in Table 7.2,

$$X(z) = \frac{z}{z-a}$$

We now apply multiplication-by-n property, which states that

$$Z\{nx[n]\} = -z\frac{d}{dz}X(z)$$

But

$$z\frac{d}{dz}X(z) = \frac{-az}{(z-a)^2}$$

Thus,

$$Y(z) = \mathcal{Z}\{na^n u[n]\} = \frac{az}{(z-a)^2}$$

Practice Problem 7.6 Determine the z-transform of $y[n] = n^2 a^n u[n]$.

Answer: $\dfrac{az(z+a)}{(z-a)^3}$

Example 7.7

Use the z-transform to find the convolution of the following two sequences:

$$x[n] = [1, -2,\ 1,\ 4] \quad \text{and} \quad h[n] = [2,\ 0,\ 1, -3,\ 1]$$

Solution

We use Equation 7.4 to find the z-transform of $x[n]$ and $h[n]$.

$$X(z) = 1 - 2z^{-1} + z^{-2} + 4z^{-3}$$

$$H(z) = 2 + 0z^{-1} + z^{-2} - 3z^{-3} + z^{-4}$$

Applying the convolution property gives

$$\begin{aligned} Y(z) &= X(z)H(z) \\ &= 2 + 0z^{-1} + z^{-2} - 3z^{-3} + z^{-4} \\ &\quad - 4z^{-1} + 0z^{-2} - 2z^{-3} + 6z^{-4} - 2z^{-5} \\ &\quad + 2z^{-2} + 0z^{3} + z^{-4} - 3z^{-5} + z^{-6} \\ &\quad + 8z^{-3} + 0z^{-4} + 4z^{-5} - 12z^{-6} + 4z^{-7} \\ &= 2 - 4z^{-1} + 3z^{-2} + 3z^{-3} + 8z^{-4} - z^{-5} - 11z^{-6} + 4z^{-7} \end{aligned}$$

From this, we obtain $y[n]$ as

$$y[n] = [2, -4, 3, 3, 8, -1, -11, 4]$$

Practice Problem 7.7 Use the z-transform to find the convolution of these sequences:

$$x[n] = [1,\ 1,\ 1,\ 1] \quad \text{and} \quad h[n] = [0,\ 1,\ 0.5,\ 1,\ 1.5]$$

Answer: [0, 1, 1.5, 2.5, 4, 3, 2.5, 1.5]

Example 7.8

Let the z-transform of a signal be

$$X(z) = \frac{z - 0.75}{z(z-1)(z+0.75)}$$

Find the initial and final values.

Solution

Applying the initial and final value theorems, we obtain

$$x[0] = \lim_{z\to\infty} X(z) = \lim_{z\to\infty} \frac{z-0.75}{z(z-1)(z+0.75)}$$

$$= \lim_{z\to\infty} \frac{1/z^2 - 0.75/z^3}{1(1-1/z)(1+0.75/z)} = 0$$

$$x[\infty] = \lim_{z\to1} \frac{z-1}{z} X(z) = \lim_{z\to1}\left(\frac{z-1}{z}\frac{z-0.75}{z(z-1)(z+0.75)}\right)$$

$$= \lim_{z\to1}\left(\frac{z-0.75}{z^2(z+0.75)}\right) = \frac{1-0.75}{1^2(1+0.75)} = \frac{1}{7} = 0.1429$$

Practice Problem 7.8 Obtain the initial and final values of the signal whose z-transform is given by

$$X(z) = \frac{1+2z^{-1}}{1-1.6z^{-1}+0.6z^{-1}}$$

Answer: 1, 7.5

Example 7.9

Find the transfer function $H(z) = Y(z)/X(z)$ for the causal system described by the following difference equation:

$$y[n] - 3y[n-1] + 4y[n-2] = x[n] + x[n-3]$$

Solution

Applying linearity and time-shifting properties, we have

$$Y(z) - 3z^{-1}Y(z) + 4z^{-2}Y(z) = X(z) + z^{-3}X(z)$$

or

$$Y(z)\{1-3z^{-1}+4z^{-2}\} = X(z)\{1+z^{-3}\}$$

Thus,

$$H(z) = \frac{Y(z)}{X(z)} = \frac{1+z^{-3}}{1-3z^{-1}+4z^{-2}} = \frac{z^3+1}{z^3-3z^2+4z}$$

Practice Problem 7.9 Given the difference equation

$$y[n+3] + 2y[n+1] + y[n] = x[n+2] - x[n]$$

Find the transfer function $H(z)$.

Answer: $H(z) = \dfrac{z^2-1}{z^3+2z+1}$

7.5 INVERSE z-TRANSFORM

The problem of finding the inverse z-transform is the problem of finding $x[n]$ for a given $X(z)$. The direct way to do this is to evaluate the contour integral:

$$x[n] = \frac{1}{2\pi j} \oint_{\Gamma} X(z) z^{n-1} dz \tag{7.50}$$

where Γ is a closed contour that encloses the origin and includes all poles of $X(z)$. Γ is usually determined by the region of convergence of the summation in Equation 7.4. It encircles all singularities in the ROC. This direct method of finding the inverse transformation is computationally involved and will not be pursued further. We would rather use two other methods: long division expansion and partial fraction expansion.

7.5.1 Long Division Expansion

To find the inverse z-transform of $X(z)$, we expand $X(z) = N(z)/D(z)$ in a power series in z^{-1} by using long division. The values of $x[n]$ are taken as the coefficients of the series expansion. The long division becomes tedious if more than few values are needed. This major deficiency limits the usefulness of the long division approach.

Example 7.10

Use long division expansion to find $x[n]$ given that

$$X(z) = \frac{z^2 - 2}{z^3 + z + 3}$$

Solution

We carry out the long division as follows:

$$\begin{array}{r l}
 & z^{-1} + 0z^{-2} - 3z^{-3} + 3z^{-4} \\
z^3 + z + 3 \,\big) & z^2 - 2 \\
 & \underline{z^2 + 1 + 3z^{-1}} \\
 & \quad -3 + 3z^{-1} \\
 & \quad \underline{-3 - 3z^{-2} - 9z^{-3}} \\
 & \qquad 3z^{-1} + 3z^{-2} + 9z^{-3} \\
 & \qquad \underline{3z^{-1} + 3z^{-3} + 9z^{4}} \\
 & \qquad\quad 3z^{-2} + 6z^{-3} - 9z^{-4}
\end{array}$$

Hence,

$$X(z) = z^{-1} - 3z^{-3} + 3z^{-4} + \cdots$$

We compare this with the basic definition of the z-transform in Equation 7.4, that is,

$$X(z) = \sum_{n=0}^{\infty} x[n]z^{-n} = x[0] + x[1]z^{-1} + x[2]z^{-2} + x[3]z^{-3} + \cdots$$

Thus,

$$x[0] = 0, \quad x[1] = 1, \quad x[2] = 0, \quad x[3] = -3, \quad x[4] = 3, \ldots$$

Practice Problem 7.10 Using long division expansion, find the inverse transform of

$$X(z) = \frac{z}{z - 0.8}$$

Answer: $(0.8)^n u[n]$.

Example 7.11

Find the inverse z-transform of

$$X(z) = \log\left(\frac{z}{z-a}\right), \quad |z| > |a|$$

Solution

This can be written as

$$X(z) = \log\left(\frac{1}{1 - az^{-1}}\right) = -\log(1 - az^{-1}) \tag{7.11.1}$$

The Maclaurin series expansion for log $(1 - u)$ is given by

$$\log(1 - r) = -\sum_{n=1}^{\infty} \frac{1}{n} r^n \tag{7.11.2}$$

Applying Equation 7.11.2 in Equation 7.11.1 with $r = az^{-1}$ leads to

$$X(z) = \sum_{n=1}^{\infty} \frac{1}{n} a^n z^{-n} \tag{7.11.3}$$

We obtain $x[n]$ as the coefficients of the series in Equation 7.11.3.

$$x[n] = \frac{1}{n} a^n u[n-1]$$

Practice Problem 7.11 By long division, find $x[n]$ corresponding to

$$X(z) = \frac{z+1}{z+2}$$

Answer: $\delta[n] - \delta[n-1] + 2\delta[n-2] - 4\delta[n-3] + \cdots$

7.5.2 Partial Fraction Expansion

We use partial fraction for finding the inverse z-transform in the same way we used it to find the inverse Laplace transform in Section 3.4. In using partial fraction, a function $X(z)$ that does not appear in the z-transform in Table 7.2 is expressed as a sum of functions that are listed in the table. The inverse z-transform is then computed term by term using the appropriate entries in Table 7.2. We may also need to use the z-transform properties in Table 7.1. We will illustrate with examples for the cases of simple poles, repeated poles, and complex poles.

Example 7.12

Find the signal corresponding to

$$X(z) = \frac{4z}{(z+1)(z-0.8)}$$

Solution

This example is on simple poles. If we expand $X(z)$ as

$$X(z) = \frac{4z}{(z+1)(z-0.8)} = \frac{A}{z+1} + \frac{B}{z-0.8}$$

we will find out that the terms $1/(z+1)$ and $1/(z-0.8)$ are not in Table 7.2. Thus, in order to get the inverse of $X(z)$, we need some additional manipulation. If we divide $X(z)$ by z, we get

$$\frac{X(z)}{z} = X_1(z) = \frac{4}{(z+1)(z-0.8)} = \frac{A}{z+1} + \frac{B}{z-0.8}$$

$$A = (z+1)X_1\Big|_{z=-1} = \frac{4}{z-0.8}\Bigg|_{z=-1} = \frac{4}{-1.8} = -2.222$$

$$B = (z-0.8)X_1\Big|_{z=0.8} = \frac{4}{z+1}\Bigg|_{z=0.8} = \frac{4}{1.8} = 2.222$$

Thus,

$$X(z) = \frac{-2.222z}{z+1} + \frac{2.22z}{z-0.8}$$

From Table 7.2, we obtain the inverse as

$$u[n] = -2.222(-1)^n + 2.22(0.8)^n$$

Practice Problem 7.12 Find the inverse of

$$X(z) = \frac{4z-1}{(z-1)(z-2)}$$

Answer: $-0.5\delta[n] - 3u[n] + 3.5(2)^n u[n]$

Example 7.13

Find the inverse z-transform of the function

$$X(z) = \frac{z+1}{(z-0.5)(z-1)^2}$$

Solution

This example is on repeated poles. We let

$$X_1(z) = \frac{X(z)}{z} = \frac{z+1}{z(z-0.5)(z-1)^2} = \frac{A}{z} + \frac{B}{z-0.5} + \frac{C}{z-1} + \frac{D}{(z-1)^2} \quad (7.13.1)$$

We obtain the expansion coefficients as

$$A = zX_1(z)\Big|_{z=0} = \frac{z+1}{(z-0.5)(z-1)^2}\Bigg|_{z=0} = \frac{1}{(-0.5)(-1)^2} = -2$$

$$B = (z-0.5)X_1(z)\Big|_{z=0.5} = \frac{z+1}{z(z-1)^2}\Bigg|_{z=0.5} = \frac{1.5}{0.5(-0.5)^2} = 12$$

$$D = (z-1)^2 X_1(z)\Big|_{z=1} = \frac{z+1}{z(z-0.5)}\Bigg|_{z=1} = \frac{2}{1(0.5)} = 4$$

To find C, we can select any appropriate value of z and substitute in Equation 7.13.1. If we choose $z = 2$, we obtain

$$X_1(z) = \frac{X(z)}{z} = \frac{z+1}{z(z-0.5)(z-1)^2} = \frac{A}{z} + \frac{B}{z-0.5} + \frac{C}{z-1} + \frac{D}{(z-1)^2}$$

$$\frac{2+1}{2(1.5)(1)} = \frac{-2}{2} + \frac{12}{1.5} + \frac{C}{1} + \frac{4}{1}$$

$$\text{or} \quad 1 = -1 + 8 + C + 4 \rightarrow C = -10$$

Thus,

$$X(z) = -2 + \frac{12z}{z-0.5} - \frac{10z}{z-1} + \frac{4z}{(z-1)^2}$$

Taking the inverse z-transform of each term,

$$x[n] = -2\delta[n] + \{12(0.5)^n - 10 + 4n\}u[n]$$

Practice Problem 7.13 Obtain the inverse z-transform for the function

$$X(z) = \frac{z-6}{(z-1)(z-2)^2}$$

Answer: $-1.5\delta[n]-\{5-4.5(2)n + 2n(2)n\}u[n]$

Example 7.14

Find the signal corresponding to the z-transform

$$X(z) = \frac{2z^2 - 3z}{(z-1)(z^2-4z+5)}$$

Solution

This is an example on complex poles. Setting $z^2 - 4z + 5 = (z-2)^2 - 4 + 5 = (z-2)^2 - (j)^2 = (z-2-j)(z-2+j)$ reveals that there are complex poles at $2 \pm j$. We really do not need to know about these poles to find the inverse. We let

$$X_1(z) = \frac{X(z)}{z} = \frac{2z-3}{(z-1)(z^2-4z+5)} = \frac{A}{z-1} + \frac{Bz+C}{z^2-4z+5}$$

or

$$\frac{2z-3}{(z-1)(z^2-4z+5)} = \frac{A(z^2-4z+5)+(z-1)(Bz+C)}{(z-1)(z^2-4z+5)} \quad (7.14.1)$$

Setting the numerators on both sides equal

$$2z-3 = A(z^2-4z+5) + B(z^2-z) + C(z-1)$$

Equating the coefficients

$$z^2 : 0 = A + B \text{ or } B = -A$$

$$z : 2 = -4A - B + C$$

$$\text{Constant:} -3 = 5A - C$$

Solving these gives $A = -1/2$, $B = ½$, and $C = 1/2$.
Thus,

$$X(z) = \frac{-0.5z}{z-1} + 0.5\frac{z^2+z}{z^2-4z+5} \tag{7.14.2}$$

The inverse of the first term is $-0.5u[n]$. We find the inverse of the second term by comparing with entries 14 and 15 in Table 7.2, namely,

$$a^n(\cos\Omega n)u[n] \leftrightarrow \frac{z^2 - za\cos\Omega}{z^2 - 2za\cos\Omega + a^2}$$

$$a^n(\sin\Omega n)u[n] \leftrightarrow \frac{za\sin\Omega}{z^2 - 2za\cos\Omega + a^2}$$

Set $z^2 - 4z + 5 = z^2 - 2az\cos\Omega + a^2$

We notice that $a = \sqrt{5}$ and $2a\cos\Omega = 4 \rightarrow \cos\Omega = \frac{2}{\sqrt{5}}$ and consequently,

$\sin\Omega = \frac{1}{\sqrt{5}}$ and $\Omega = 0.4636$.

Thus,

$$X(z) = \frac{-0.5z}{z-1} + 0.5\frac{z^2-2z}{z^2-4z+5} + 0.5\frac{3z}{z^2-4z+5}$$

The inverse is

$$x[n] = 0.5\left\{-1 + (\sqrt{5})^n\cos(0.4636n) + 3(\sqrt{5})^n\sin(0.4636n)\right\}u[n]$$

Practice Problem 7.14 Obtain the inverse z-transform of

$$X(z) = \frac{2z}{z^2+4z+8}$$

Answer: $\left(\sqrt{8}\right)^n \sin\left(\frac{n\pi}{4}\right)u[n]$

7.6 APPLICATIONS

The z-transform is fundamentally important to digital signal processing, digital communications, and linear control systems. In this section, we consider the use of the z-transform in the analysis of discrete-time linear systems. We will apply the z-transform to two areas: linear difference equation and transfer function.

7.6.1 Linear Difference Equation

Most discrete-time systems of practical interest can be described by linear difference equations. We can use the z-transform to solve the difference equation for $n \geq 0$. To do this we first take the z-transform of the two sides of the difference equation. Next, we solve algebraically the transformed difference equation. At this point, we can evaluate $Y(z)$ and invert it using Table 7.2 to determine $y[n]$.

To obtain the complete solution requires that we know the input signal $x[n]$. For Nth order difference equation, we must also know N initial conditions. Using the initial conditions, we obtain the z-transform of the delayed output signals as follows:

$$\begin{aligned} Z\{y[n-1]\} &= z^{-1}Y(z) + y[-1] \\ Z\{y[n-2]\} &= z^{-2}Y(z) + z^{-1}y[-1] + y[-2] \\ Z\{y[n-3]\} &= z^{-3}Y(z) + z^{-2}y[-1] + z^{-1}y[-2] + y[-3] \end{aligned} \tag{7.51}$$

In general,

$$Z\{y[n-m]\} = z^{-m}Y(z) + z^{-m}\sum_{n=1}^{m} y[-n]z^{n} \tag{7.52}$$

Similarly, the z-transforms of advanced output signals are

$$\begin{aligned} Z\{y[n+1]\} &= zY(z) - zy[0] \\ Z\{y[n+2]\} &= z^{2}Y(z) - z^{2}y[0] - zy[1] \\ Z\{y[n+3]\} &= z^{3}Y(z) - z^{3}y[0] - z^{2}y[1] - zy[2] \end{aligned} \tag{7.53}$$

In general,

$$Z\{y[n+m]\} = z^{m}Y(z) - z^{m}\sum_{n=0}^{m-1} y[n]z^{-n} \tag{7.54}$$

We will illustrate this with an example.

Example 7.15

A second-order discrete-time system is characterized by the difference equation

$$y[n] + 0.2y[n-1] - 0.5y[n-2] = x[n] + x[n-1]$$

Find the complete solution for the initial conditions $y[-1] = 1$, $y[-2] = 2$ and an input $x[n] = u[n]$.

Solution

Taking the z-transform of each term in the difference equation yields

$$Y(z)+0.2\{z^{-1}Y(z)+y[-1]\}-0.5\{z^{-2}Y(z)+z^{-1}y[-1]+y[-2]\}=X(z)+\{z^{-1}X(z)+x[-1]\}$$

We should set $x[-1] = 0$ since $u[n] = 0$ for $n < 0$. Also,

$$X(z)=\frac{1}{1-z^{-1}}$$

Introducing the initial conditions for $y[-1]$ and $y[-2]$ leads to

$$Y(z)\{1+0.2z^{-1}-0.5z^{-2}\}+0.2-0.5z^{-1}-1=X(z)(1+z^{-1})=(1+z^{-1})/(1-z^{-1})$$

$$Y(z)=\frac{0.8+0.5z^{-1}+(1+z^{-1})/(1-z^{-1})}{1+0.2z^{-1}-0.5z^{-2}}=\frac{0.8+0.5z^{-1}-0.8z^{-1}-0.5z^{-2}+1+z^{-1}}{(1-z^{-1})(1+0.2z^{-1}-0.5z^{-2})}$$

$$=\frac{1.8z^{3}+0.7z^{2}-0.5z}{(z-1)(z^{2}+0.2z-0.5)}$$

We use partial fraction to obtain $y[n]$.

$$Y_1(z)=\frac{Y(z)}{z}=\frac{1.8z^{2}+0.7z-0.5}{(z-1)(z+0.8141)(z-0.6141)}=\frac{A}{z-1}+\frac{B}{z+0.8141}+\frac{C}{z-0.6141}$$

$$A=(z-1)Y_1(z)\Big|_{z=1}=\frac{1.8+0.7-0.5}{1.8141\times 0.3859}=2.8569$$

$$B=(z+0.8141)Y_1(z)\Big|_{z=-0.8141}=\frac{1.193-0.5699-0.5}{(-1.8141)(-1.4282)}=0.0475$$

$$C=(z-0.6141)Y_1(z)\Big|_{z=0.6141}=\frac{0.6788+0.4299-0.5}{(-0.3859)(1.4282)}=-1.1044$$

Thus,

$$Y(z)=\frac{2.8569z}{z-1}+\frac{0.0475z}{z+0.8141}-\frac{1.1044z}{z-0.6141}$$

By taking the inverse z-transform of each term, we finally obtain

$$y[n]=\{2.8569+0.0475(-0.8141)^{n}-1.1044(0.6141)^{n}\}u[n]$$

Practice Problem 7.15 Use the z-transform to solve the difference equation

$$y[n]+0.6y[n-1]=x[n]$$

where
$x[n] = 4\delta[n-1]$
$y[-1] = -2.$

Answer: $6.667\delta[n]-5.4667(-0.6)^n u[n]$

7.6.2 Transfer Function

The transfer function plays an important role in the analysis of discrete-time linear systems. In this section, we will define the discrete-time transfer function and how to find the system impulse and step responses. We will also learn how to use the transfer function to determine system performance characteristics such as stability and frequency response.

The relationships between the system input $x[n]$, output $y[n]$, and impulse response and their z-transforms are given in Equation 7.39, namely,

$$y[n]=x[n]*h[n]\leftrightarrow Y(z)=X(z)H(z) \tag{7.55}$$

From this, we obtain the transfer function $H(z)$ as

$$H(z)=\frac{Y(z)}{X(z)} \tag{7.56}$$

This implies that the transfer function is the ratio of the z-transforms of the output and input. We obtain the transfer function by either transforming the difference equation or transforming the system representation.

When the system input is a unit impulse $\delta[n]$ ($\delta[n]\leftrightarrow 1$), the output of the system is equal to the transfer function $H(z)$. Thus, the discrete-time system impulse response is given by

$$h[n]=Z^{-1}\{H(z)\} \tag{7.57}$$

When the input $x[n]$ is a unit step $u[n]$ so that $X(z) = z/(z-1)$, the corresponding output $y[n]$ is called the *step response*.

When systems are interconnected, the rules that apply to continuous-time systems also apply to discrete-time systems. We handle block diagrams such as series, parallel, and feedback interconnection in the same manner as for their corresponding

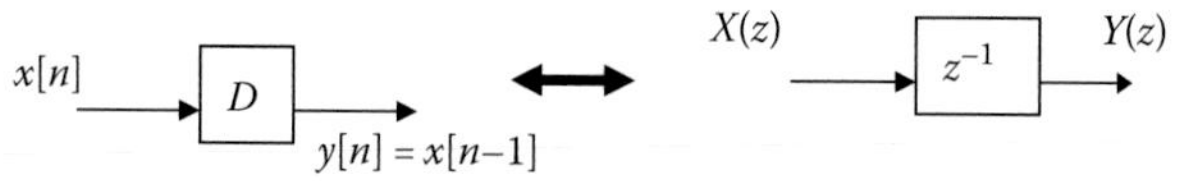

FIGURE 7.3 Unit delay.

continuous-time systems in Section 2.7. For a unit delay, shown in Figure 7.3, the input–output relationship is

$$y[n] = x[n-1] \tag{7.58}$$

or

$$Y(z) = z^{-1}X(z) \tag{7.59}$$

For the series connection, shown in Figure 7.4a, the overall transfer function is

$$H(z) = H_1(z)H_2(z) \tag{7.60}$$

For the parallel connection, shown in Figure 7.4b, the overall transfer function is

$$H(z) = H_1(z) + H_2(z) \tag{7.61}$$

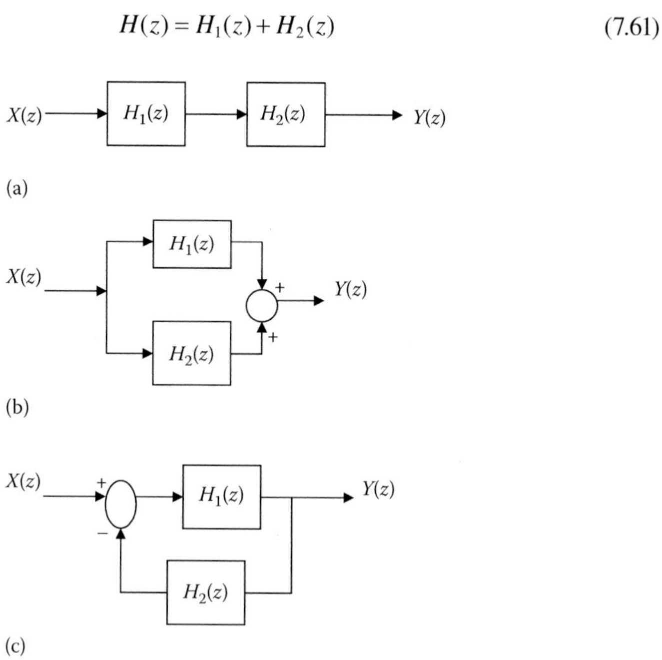

FIGURE 7.4 Transfer functions for (a) series connection, (b) parallel connection, and (c) feedback connection.

For the feedback interconnection, shown in Figure 7.4c, the overall transfer function is

$$H(z) = \frac{Y(z)}{X(z)} = \frac{H_1(z)}{1 + H_1(z)H_2(z)} \tag{7.62}$$

System stability may be addressed in terms of the characteristic roots, the impulse response, and the bounded-input–bounded-output (BIBO) criterion. It may also be deduced from the system transfer function. The system is stable if $h[n]$ is absolutely summable, that is,

$$\sum_{n=-\infty}^{\infty} |h[n]| < \infty \tag{7.63}$$

All the terms in $h[n]$ are decaying exponentials. To achieve this, the magnitudes of all the system poles of $H(z)$ must be less than 1, that is, the poles are inside a unit circle in the z-plane.

Example 7.16

Consider the α-filter, which is described by Figure 7.5. From the figure, it is evident that

$$y[n] + (1 - \alpha)y[n-1] = \alpha x[n]$$

Find $H(z)$ and $h[n]$. You may take $\alpha = 0.2$.

Solution

With $\alpha = 0.2$,

$$y[n] + 0.8y[n-1] = 0.2x[n]$$

We take the z-transform of this to get

$$Y(z) + 0.8z^{-1}Y(z) = 0.2\, X(z)$$

$$H(z) = \frac{Y(z)}{X(z)} = \frac{0.2}{1 + 0.8z^{-1}} = \frac{0.2z}{z + 0.8}$$

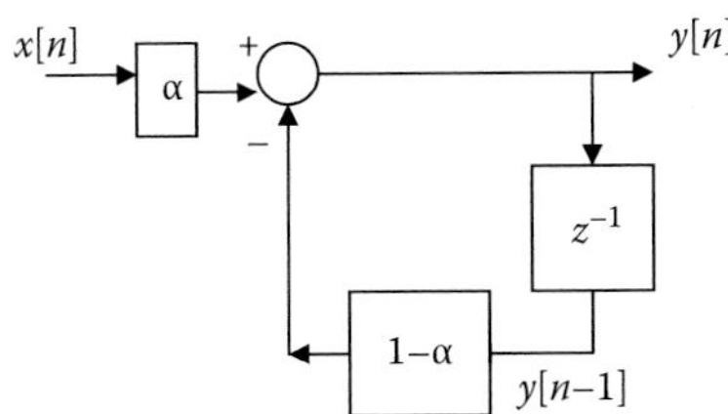

FIGURE 7.5 For Example 7.16.

Taking inverse z-transform gives

$$h[n] = 0.2(-0.8)^n u[n]$$

Practice Problem 7.16 Repeat the problem in Example 7.16 with

$$y[n] - 0.5y[n-1] = x[n]$$

Answers: $H(z) = \dfrac{z}{z-0.5}$, $h[n] = (0.5)^n\, u[n]$

Example 7.17

A discrete-time system is described by the difference equation

$$y[n] - 0.6y[n-1] - 0.05y[n-2] = x[n] - x[n-1]$$

Find the impulse response.

Solution

Since there are not initial conditions, the z-transform of the difference equation is

$$Y(z) - 0.6\{z^{-1}Y(z)\} + 0.05\{\, z^{-2}Y(z)\} = X(z) - \{z^{-1}X(z)\}$$

The transfer function is given by

$$H(z) = \frac{Y(z)}{X(z)} = \frac{1-z^{-1}}{1-0.6z^{-1}+0.05z^{-2}} = \frac{z^2 - z}{z^2 - 0.6z + 0.05}$$

To find the impulse function, we apply partial fraction on $H(z)$

$$H_1(z) = \frac{H(z)}{z} = \frac{z-1}{z^2 - 0.6z + 0.05} = \frac{z-1}{(z-0.5)(z-0.1)} = \frac{A}{z-0.5} + \frac{B}{z-0.1}$$

$$A = (z-0.5)H_1(z)\Big|_{z=0.5} = \frac{0.5-1}{(0.5-0.1)} = -1.25$$

$$B = (z-0.1)H_1(z)\Big|_{z=0.1} = \frac{0.1-1}{(1-0.5)} = -1.8$$

Thus,

$$H(z) = -\frac{1.25z}{z-0.5} - \frac{1.8z}{z-0.1}$$

Taking the inverse z-transform of this yields

$$h[n] = -1.25(0.5)^n u[n] - 1.8(0.1)^n u[n]$$

Practice Problem 7.17 Find the impulse response of the discrete-time system described by

$$y[n] - \frac{1}{12}y[n-1] - \frac{1}{12}y[n-2] = x[n-1] + x[n-2]$$

Answer: $-12\delta[n-1] + \frac{48}{7}\left(\frac{1}{3}\right)^n u[n] - \frac{36}{7}\left(\frac{1}{4}\right)^n u[n]$

7.7 COMPUTING WITH MATLAB®

In this section, we discuss how MATLAB can be used to do the following:

- Find the z-transform of a discrete-time signal using the MATLAB command **ztrans**.
- Determine the inverse z-transform using MATLAB command **iztrans**.
- Determine the poles of a transfer function using MATLAB command **roots** and evaluate the stability of the system.
- Use the MATLAB commands **dstep** and **filter** to find the step response of discrete-time linear system.

Example 7.18

Using MATLAB, find the z-transform of

$$x[n] = n\, u[n]$$

Solution

The MATLAB script for doing this is

```
syms  X x n z
x =  n*heaviside(n);
X = ztrans(x)
```

The program returns

```
X =
z/(z^2 - 2*z + 1)
```

which agrees with what we have in Table 7.2.

Practice Problem 7.18 Use MATLAB to find the z-transform of $y[n] = n^2 u[n]$

Answer: $Y(z) = \dfrac{z^2 + z}{z^3 - 3z^2 + 3z - 1}$

Example 7.19

Given the function

$$X(z) = \frac{6z+4}{z^2 - 3.5z + 3}$$

Find the inverse z-transform $x[n]$.

Solution

We can do this in two ways.

Method 1: Using the **iztrans** command, we have MATLAB script as follows.

```
syms  X x n z
X = (6*z +4)/(z^2 -3.5*z + 3);
x = iztrans(X)
```

MATLAB returns

```
x =
16*2^n - (52*(3/2)^n)/3 + (4*kroneckerDelta(n, 0))/3
```

Method 2: We can also use the **residue** or **residuez** command, which carries out partial-fraction expansion. We illustrate it in the following script.

```
num = [ 6  4]
den = [1  -3.5  3]
[r, p, k] = residuez(num, den)
```

MATLAB responds as follows (r = residues, r = poles, and k = direct terms).

```
r =
    32
   -26
p =
    2.0000
    1.5000
k =
     []
```

This implies that

$$X(z) = \frac{32}{1-2z^{-1}} - \frac{26}{1-1.5z^{-1}} = \frac{32z}{z-2} - \frac{26z}{z-1.5}$$

From this, we obtain the inverse z-transform as:

$$x[n] = 32(2)^n u[n] - 25(1.5)^n u[n]$$

Practice Problem 7.19 Use MATLAB to find the inverse z-transform of

$$X(z) = \frac{z-2}{z^2+3z+2}$$

Answer: $x[n] = 3*(-1)^n - 2*(-2)^n$ –Kronecker Delta (n, 0)

Example 7.20

The transfer function of a system is given by

$$H(z) = \frac{2z-3}{z^3-2z^3-z-0.5}$$

Determine whether the system is stable.

Solution

We only need to find the roots of the denominator to determine the system stability.

```
den = [1  -2 -1  -0.5];
r = roots(den)
```

MATLAB returns

```
r =
   2.4837
  -0.2418 + 0.3779i
  -0.2418 - 0.3779i
```

The poles are located on the complex plane as shown in Figure 7.6. We notice that the magnitude of the first pole (2.4847) is greater than one and therefore lies outside the unit circle. (The other poles lie with the unit circle.) Hence, the system is unstable.

Practice Problem 7.20 Determine the stability of a system whose transfer function is

$$H(z) = \frac{2z-1}{z^3+0.5z^2+0.5z+0.5}$$

Answer: Stable; all poles lie within the unit circle.

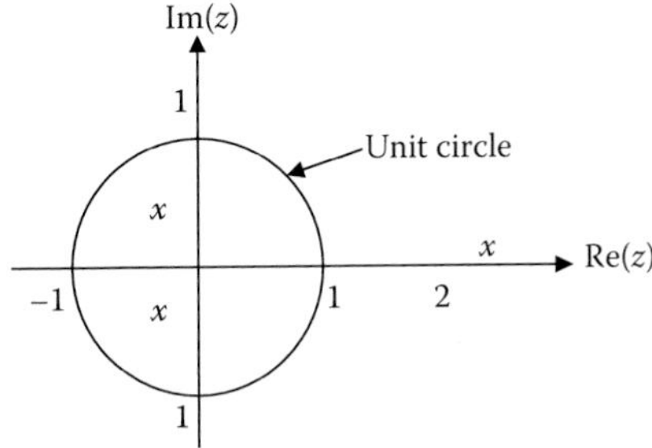

FIGURE 7.6 Pole locations for Example 7.20.

Example 7.21

Obtain the unit step response of a discrete-time system whose transform is

$$H(z) = \frac{3+z}{z^2 - 5z + 2}$$

Solution

The MATLAB script is as follows. We use the MATLAB command **filter** in this example, while the Practice Problem following this example uses **dstep** command. Given the transfer function, the input and initial conditions, filter returns the system output. The step response is shown in Figure 7.7.

```
num = [1  3];
den = [ 1 -5  2];
n = 0:1:30;
x = [1*ones(size(n))]; % unit step input
y = filter(num, den, x);
% d = length(y);
% n = 0:1:d-1
plot(n,y);
xlabel('Sample number n');
ylabel('step response y[n]')
```

Practice Problem 7.21 Change one line of the MATLAB script for Example 7.21 to `y = dstep (num, den)`. You may delete two lines before that statement and add two lines commented after that statement. Plot your result.

Answer: See Figure 7.8.

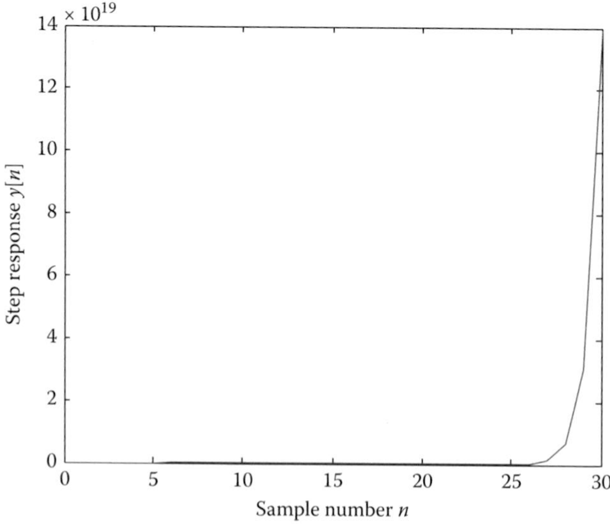

FIGURE 7.7 Unit step response; for Example 7.21.

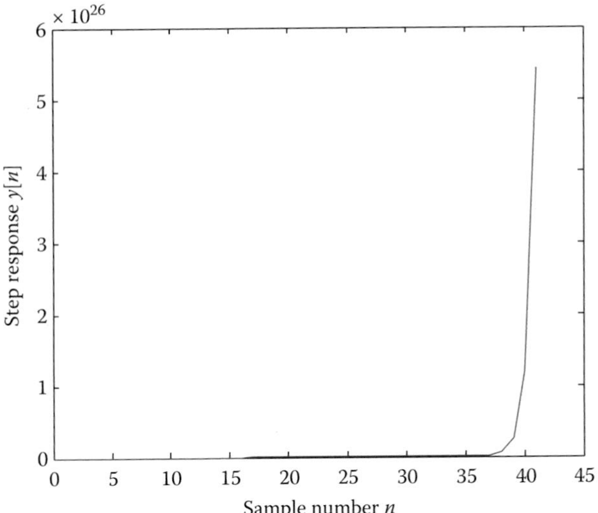

FIGURE 7.8 For Practice Problem 7.21.

7.8 SUMMARY

1. The z-transform is used for analyzing discrete-time signals just as its counterpart Laplace transform is used for analyzing continuous-time signals.
2. The (unilateral) z-transform of a signal $x[n]$ is defined as

$$X(z) = \sum_{n=0}^{\infty} x[n]z^{-n}$$

3. The region of convergence of the z-transform consists of the values of z for $X(z)$ converges.
4. The properties of the z-transform are presented in Table 7.1. Table 7.2 provides the z-transform pairs of common signals.
5. To find the inverse z-transform of $X(z)$, we expand $X(z)$ in partial fractions and identify the inverse of each term using Table 7.2.
6. Among the applications of the z-transform are the solution of difference equation and the system transfer function.
7. The z-transform can be used to solve difference equation of a system with initial conditions.
8. The transfer function $H(z)$ of a discrete-time system is

$$H(z) = Z\{h[n]\} = \frac{Y(z)}{X(z)}$$

where $X(z)$ and $Y(z)$ are, respectively, the z-transform of the input $x[n]$ and output $y[n]$.
The inverse z-transform of $H(z)$ is the impulse response $h[n]$.

REVIEW QUESTIONS

7.1 The z-transform may be regarded as the Laplace transform in disguise.
(a) True, (b) False.

7.2 The z-transform of the sequence shown in Figure 7.9 is
(a) $2z + z^2$, (b) $2z^{-1} + z^{-1}$, (c) $2z^{-1} + z^{-2}$, (d) $z^{-1} + 2z^{-2}$.

7.3 The z-transform of $\delta[n - 1]$ is
(a) 1, (b) z, (c) z^{-1}, (d) e^{-z}, (b) e^z.

7.4 The z-transform for $u[n + 1]$ is

(a) $\dfrac{1}{z(z-1)}$, (b) $\dfrac{1}{(z-1)}$, (c) $\dfrac{z}{(z-1)}$, (d) $\dfrac{z^2}{(z-1)}$.

7.5 Let $X(z) = \dfrac{z+1}{z^2 + 3z + 2}$

The initial value $x[0]$ is
(a) 0, (b) 1/2, (c) 1, (d) ∞.

7.6 Refer to the function $X(z)$ in the previous Review Question. The final value $x[\infty]$ is
(a) 0, (b) 1/2, (c) 1, (d) ∞.

7.7 Given the difference equation

$$y[n] + y[n-1] + y[n-2] = u[n], \quad y[-1] = 2, \quad y[-2] = 1,$$

$y[0]$ is
(a) −2, (b) −1, (c) 0, (d) 1, (e) 2.

7.8 Which MATLAB command is used for finding the inverse z-transform?
(a) ztrans, (b) invztrans, (c) iztrans, (d) filter, (e) dstep.

7.9 Which of these MATLAB command is used for determining the stability of a system?
(a) freqz, (b) residue, (c) residuez, (d) roots.

7.10 The pole plot of a system is shown in Figure 7.10. The system is
(a) stable, (b) unstable, (c) marginally state.

Answers: 7.1a, 7.2c, 7.3c, 7.4d, 7.5a, 7.6a, 7.7a, 7.8c, 7.9d, 7.10a.

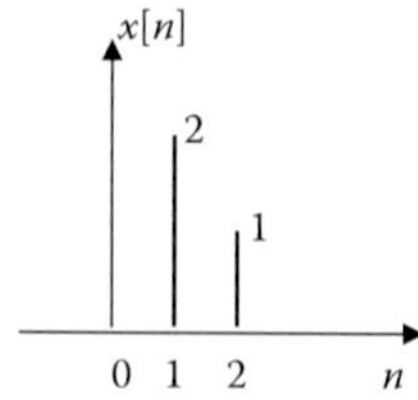

FIGURE 7.9 For Review Question 7.2.

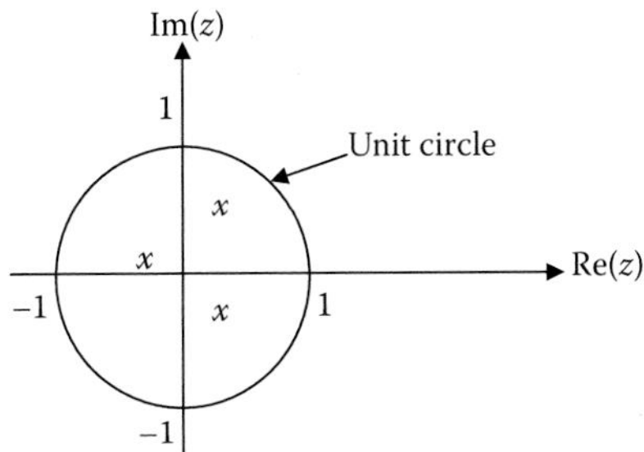

FIGURE 7.10 Pole locations for Review Question 7.10.

PROBLEMS

Sections 7.2 and 7.3—Definition of the z-Transform and Region of Convergence

7.1 Find the z-transform of the sequences shown in Figure 7.11.

7.2 Determine the z-transform of the signal in Figure 7.12.

7.3 Determine the condition for the following summations to converge.

(a) $\sum_{n=1}^{\infty}\left(\frac{2}{3}\right)^{2} z^{n}$

(b) $\sum_{n=1}^{\infty}\left(\frac{1}{5}\right)^{n} z^{-n}$

7.4 Find the z-transform and the ROC for

$$x[n]=2^{n}u[n]+\left(\frac{1}{2}\right)^{n}u[n]$$

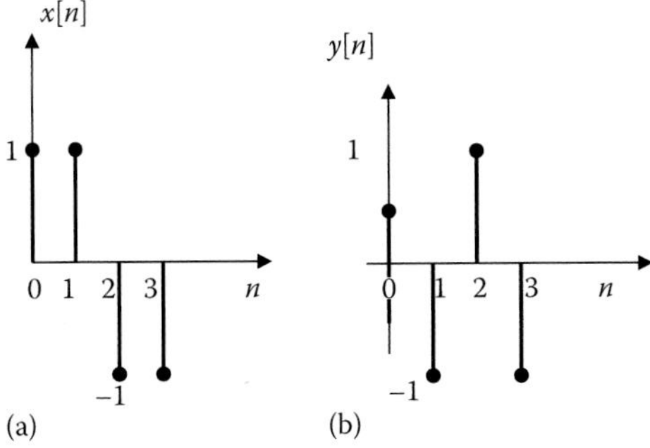

FIGURE 7.11 For Problem 7.1.

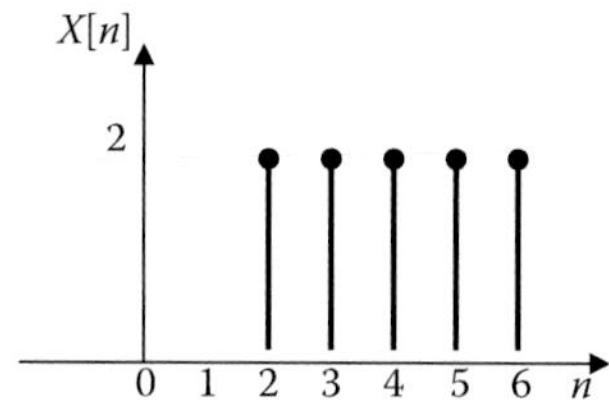

FIGURE 7.12 For Problem 7.2.

7.5 Determine the z-transform and its ROC for

$$x[n]=2\left(\frac{2}{3}\right)^n u[n]-\left(\frac{2}{5}\right)^n u[n]$$

SECTION 7.4—PROPERTIES OF THE Z-TRANSFORM

7.6 Find the z-transform of the following signals:
(a) $u[n-m]$
(b) $na^n u[n]$
(c) $a^n \cos \pi n\, u[n]$

7.7 Determine the z-transform of each of the following signals:
(a) $u[n]-u[n-1]$
(b) $a^{n-1}\, u[n-1]$
(c) $na^n\, u[n-1]$

7.8 Find the z-transform of each of the following sequences:

(a) $x[n]=\left(\frac{1}{2}\right)^2+\left(-\frac{1}{3}\right)^n, \quad n=0,1,2,...$

(b) $x[n]=\begin{cases}\left(\frac{1}{4}\right)^n, & n=0,1,2,\ldots \\ 0, & n=\text{negative}\end{cases}$

(c) $x[n]=\begin{cases}(0.5)^n, & 0\le n\le 5 \\ 0, & \text{otherwise}\end{cases}$

7.9 Find the z-transform of the following sequences:
(a) $x[n]=na^n\cos(\Omega_0 n)u[n]$
(b) $x[n]=n^2u[n]$
(c) $x[n]=e^n\sin(n)u[n]$

7.10 Obtain the z-transform of the following signals:

(a) $x[n]=3\left(\frac{2}{5}\right)^n+4\left(-\frac{1}{3}\right)^n$

(b) $y[n]=1-e^{-an}$

7.11 Prove the following z-transform pairs:

(a) $(n+1)u[n] \Leftrightarrow \dfrac{z^2}{(z-1)^2}$

(b) $n^2 a^n u[n] \Leftrightarrow \dfrac{az(z+a)}{(z-a)^3}$

(c) $a^2(\sin \Omega n)u[n] \Leftrightarrow \dfrac{za \sin \Omega}{z^2 - 2za \cos \Omega + a^2}$

7.12 Show that $\mathcal{Z}\{nx[n]\} = -z\dfrac{dX(z)}{dz}$.

7.13 Assuming that sequence $x[n]$ is real-valued, show that $X(z) = X^*(z^*)$.

7.14 A discrete-time signal $x[n] = A\cos(\Omega n)$ has the z-transform as

$$X(z) = \frac{4z(z-0.8)}{z^2 - 1.6z + 1}$$

Compute the values of A and Ω.

7.15 Given that $x[n] = 3^n u[n]$, find the z-transform of

$$y[n] = nx[n] + x[n-1]u[n-1] + x[n+1]u[n+1]$$

7.16 Determine the initial values of the signals whose z-transforms are given in Problem 7.24.

7.17 Find the initial value in each of the cases in Problem 7.27.

7.18 Find the initial and final values of $x[n]$ for each of the following cases:

(a) $X(z) = \dfrac{2\left(z - \frac{1}{6}\right)}{\left(z - \frac{1}{4}\right)\left(z - \frac{1}{5}\right)}$

(b) $X(z) = \dfrac{z^2 + 1}{z^3 + 2z + 2}$

7.19 The z-transform of a discrete-time signal $x[n]$ is

$$X(z) = \frac{z-2}{z(z-1)}$$

Calculate $x[0]$, $x[1]$, and $x[10^5]$.

7.20 The z-transform of a discrete-time signal is

$$X(z) = \frac{2z}{z^2 + 3z + 1}$$

Find the z-transform of the following signals:

(a) $y[n] = x[n-1]u[n-1]$

(b) $y[n] = \sin(\pi n/4)\, x[n]$

(c) $y[n] = n^n x[n]$

(d) $y[n] = 2x[n]*x[n]$

7.21 Using the z-transform, determine the convolution of these sequences:

$$x[n]=[1,\ -1,\ 3,\ 2],\quad h[n]=[1,\ 0,\ 2,\ 1,\ -3].$$

7.22 Using the convolution property, find the z-transform of $x[n]$, where

$$x[n]=\sin\left(\frac{\pi n}{4}\right)u[n]*u[n]$$

Section 7.5—Inverse z-Transform

7.23 Find the inverse z-transform of

(a) $X_1(z)=\dfrac{1}{z^2+\dfrac{5}{6}z+\dfrac{1}{6}}$

(b) $X_2=\dfrac{z}{(z-1)(z-0.5)}$

(c) $X_3(z)=\dfrac{z^2-8}{(z-1)^2(z-2)}$

7.24 Find the inverse z-transform of the following functions:

(a) $X(z)=\dfrac{2z}{z^2-z+1}$

(b) $Y(z)=\dfrac{z(z+2)}{(z+1)(z-1)}$

(c) $H(z)=\dfrac{4z}{(z^2+z+1)(z+1/2)}$

7.25 Obtain the inverse z-transform of

$$X(z)=\frac{z^2+2z-10}{(z-1)(z+2)(z+3)}$$

7.26 The z-transform of $x[n]$ is given by

$$X(z)=\frac{2z+1}{(z-1)(z+1)}$$

(a) Find the final value $x[\infty]$.

(b) Determine the inverse z-transform of $X(z)$ and check your result in part (a).

7.27 Invert each of the following z-transform:

(a) $X_1(z) = \dfrac{1 - z^{-1}}{1 - z^{-1} - 0.75z^{-2}}$

(b) $X_2(z) = \dfrac{1 + z^{-1}}{1 - 0.8z^{-1} + 0.64z^{-2}}$

7.28 Use long division to find the inverse of

$$X(z) = \frac{z - 2}{1 - z + z^2}$$

Section 7.6—Applications

7.29 A discrete-time system is described by

$$y[n+2] + y[n+1] + 0.36y[n] = 0, \quad y[0] = 2, \quad y[1] = 1.$$

Find the response $y[n]$.

7.30 The difference equation for a system is

$$y[n] + 6y[n-1] + 15y[n-2] = 0, \quad y[-2] = 0, \quad y[-1] = 1.$$

Find $y[n]$.

7.31 Using the z-transform, solve the following difference equation:

$$y[n+1] - 2y[n] = (1.5)^n, \quad y[0] = 1.$$

7.32 Find the transfer function of the discrete-time system represented by

$$y[n+2] + 0.5y[n+1] + y[n] = x[n+1] + 2x[n]$$

7.33 Obtain the complete response of the system represented by

$$y[n] + \frac{1}{3}y[n-1] + \frac{1}{6}y[n-2] = u[n], \quad y[-2] = 0, \quad y[-1] = 1$$

7.34 A digital filter has the impulse response

$$h[n] = \left(\frac{1}{2}\right)^n u[n-1]$$

Find the step response of the filter.

7.34 A discrete-time linear system has impulse response

$$h[n] = 2 + 3(0.6)^n - 5(-0.8)^n$$

Determine the transfer function.

7.36 The transfer function of a discrete-time system is

$$H(z) = \frac{1 + 2z^{-1}}{1 - z^{-1} + z^{-2}}$$

Find the system response $y[n]$ when the input is a unit step function $u[n]$.

7.37 Find the z-transform of the output of the system shown in Figure 7.13.

7.38 Determine the transfer function of the feedback system represented in Figure 7.14.

7.39 Determine the overall transfer function of the system shown in Figure 7.15. Let

$$h_1[n] = \delta[n] + \delta[n-1], \quad h_2[n] = nu[n], \quad h_3[n] = \left(\frac{1}{3}\right)^n u[n]$$

7.40 A discrete-time linear system is shown in Figure 7.16.

(a) Find the transfer function $H(z)$ of the system.

(b) Find the difference equation that describes the system.

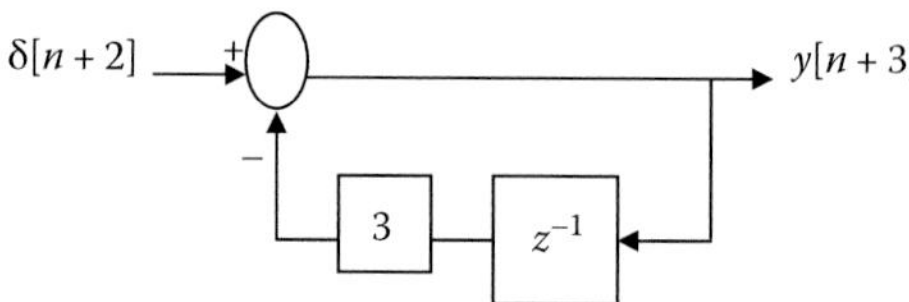

FIGURE 7.13 For Problem 7.37.

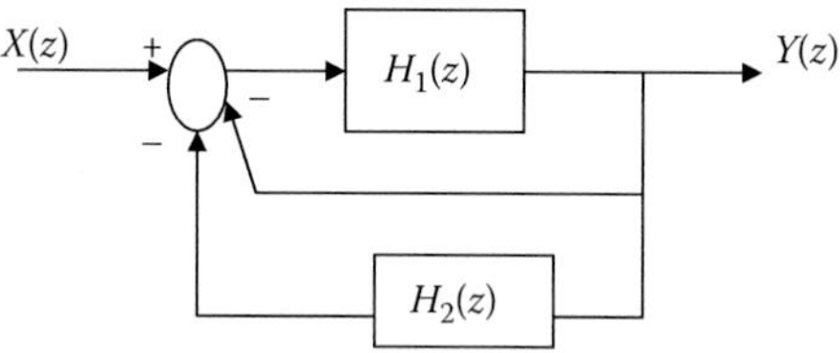

FIGURE 7.14 For Problem 7.38.

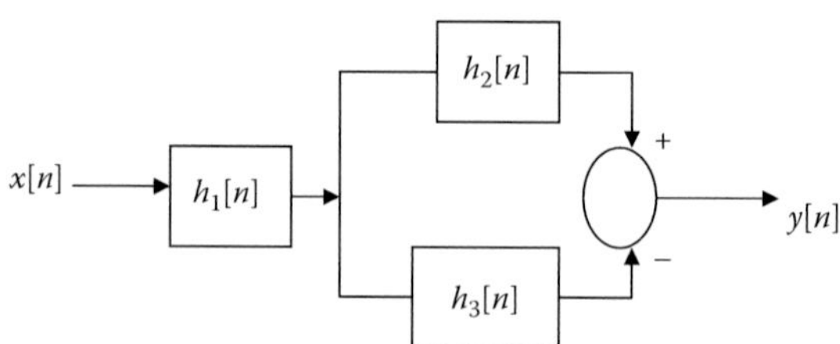

FIGURE 7.15 For Problem 7.39.

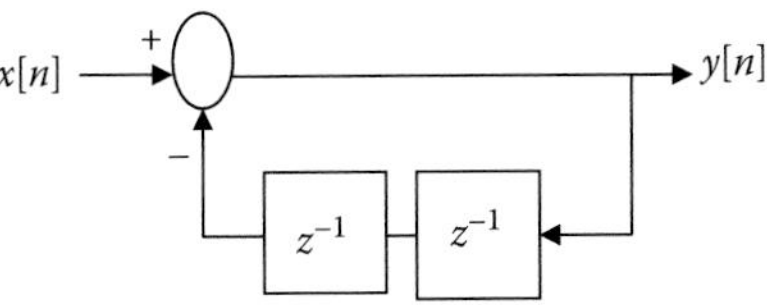

FIGURE 7.16 For Problem 7.40.

7.41 Find the response of a system with a transfer function

$$H(z) = \frac{z - 0.6}{(z + 0.2)(z - 0.8)}$$

and an input $x[n]$ given by
(a) $x[n] = u[n]$
(b) $x[n] = 2^n u[n]$

7.42 A discrete-time system has input $x[n] = \delta[n] + \delta[n-1]$ and output $y[n] = \delta[n] + \delta[n-1] + \delta[n-3]$. Find $H(z)$ and $h[n]$.

7.43 Obtain the impulse and step responses of the discrete-time system with transfer function

$$H(z) = \frac{0.8z}{(z - 0.6)(z - 2)}$$

Section 7.7—Computing with MATLAB®

7.44 Use MATLAB to find the z-transform of the following signals:
(a) $x[n] = (0.8)^n\, u[n]$
(b) $x[n] = n(0.6)^n\, u[n]$

7.45 Use MATLAB to find the inverse z-transform of

$$X(z) = \frac{z}{z - 0.6}$$

7.46 A linear discrete-time system is represented by the transfer function

$$H(z) = \frac{z + 1}{z^3 + 2z^2 + z + 3}$$

Use MATLAB to plot the step response of the system.

7.47 Determine the poles and zeros of the transfer function

$$H(z) = \frac{z^2 + 6z + z}{z^4 + 3z^{-3} + 4z + 10}$$

7.48 The z-transform of a system is

$$X(z) = \frac{z^3 - 0.6z^2 + 0.2z + 1}{z^4 + 0.5z^3 - z^2 + 0.5z + 1}$$

Plot the locations of the poles and zeros of $X(z)$.

7.49 Determine the stability of the systems represented by the following transfer function:

$$H(z) = \frac{z^3 + 3z^2 + z - 1}{z^4 + 1.25z^3 + 0.5z^2 - 0.375z - 0.2}$$

7.50 Check the stability of a system described by the following transfer function:

$$H(z) = \frac{z^{-3} - 2z^2 + 6z + 1}{z^5 - 2z^4 + 5z^2 - z + 4}$$

Selected Bibliography

Allen, D. L. and D. W. Mills, *Signal Analysis*. Hoboken, NJ: John Wiley & Sons, 2004.

Balmer, L., *Signals and Systems: An Introduction*. Upper Saddle River, NJ: Prentice Hall, 1981.

Boulet, B., *Fundamentals of Signals and Systems*. Hingham, MA: Da Vinci Engineering Press, 2006.

Brigham, E. O., *The Fast Fourier Transform and Its Applications*. Englewood Cliffs, NJ: Prentice Hall, 1988.

Buck, J. R., M. M. Daniel, and A. C. Singer, *Computer Explorations in Signals and Systems*, 2nd ed. Upper Saddle River, NJ: Prentice Hall, 2002.

Cadzow, J. A. and H. F. Van Landingham, *Signals, Systems, and Transforms*. Englewood Cliffs, NJ: Prentice Hall, 1985.

Carlson, G. E., *Signal and Linear System Analysis with MATLAB*, 2nd ed. New York: John Wiley & Sons, 1998.

Chapiro, L. F., *Signals and Systems Using MATLAB*. Burlington, MA: Elsevier, 2011.

Chen, C., *Linear System Theory and Design*, 3rd ed. New York: Oxford University Press, 1999.

Chen, C., *Signals and Systems*, 3rd ed. New York: Oxford University Press, 2004.

Denbigh, P., *System Analysis and Signal Processing*. Essex, U.K.: Addison Wesley Longman, 1998.

ElAli, T. S., *Discrete Systems and Digital Signal Processing with MATLAB*. Boca Raton, FL: CRC Press, 2004.

Hsu, H. P., *Signals and System* (*Schaum's Outline*). New York: McGraw-Hill, 1995.

Kailah, T., *Linear Systems*. Englewood Cliffs, NJ: Prentice Hall, 1980.

Kamen, E. W. and B. S. Heck, *Fundamentals of Signals and Systems*, 3rd ed. Upper Saddle River, NJ: Pearson Prentice Hall, 2007.

Karris S. T., *Signals and System: With MATLAB Computing and Simulink Modeling*, 4th ed. Fremont, CA: Orchard Publications, 2008.

Kwakernaak, H. and R. Sivan, *Modern Signals and Systems*. Englewood Cliffs, NJ: Prentice Hall, 1991.

Lathi, B. P., *Linear Systems and Signals*, 3rd ed. New York: Oxford University Press, 2005.

McClellan, J. H., R. W. Schafer, and M. A. Yoder, *Signal Processing First*. Upper Saddle River, NJ: Pearson Education, 2003.

McGillem, C. D. and G. R. Cooper, *Continuous and Discrete Signal and System Analysis*, 2nd ed. New York: Holt, Rinehart, and Winston, 1984.

McMahon, D., *Signals and Systems Demystified*. New York: McGraw-Hill, 2007.

Nemzow, M., *Fast Internet Implementation and Migration Solutions*. New York: McGraw-Hill, 1997.

Oppenheim, A. V. and A. S. Willsky with S. H. Nawab, *Signals and Systems*, 2nd ed. Upper Saddle River, NJ: Prentice Hall, 1997.

Papoulis, A., *Circuits and Systems: A Modern Approach*. New York: Holt, Rinehart and Winston, 1980.

Phillips, C. L. and J. M. Parr, *Signals, Systems, and Transforms*, 2nd ed. Upper Saddle River, NJ: Prentice Hall, 1999.

Poularikas, A. D. and S. Seely, *Signals and Systems*, 2nd ed. Malabar, FL: Krieger Publishing Co., 1994.

Roberts, M. J., *Signals and Systems: Analysis Using Transform Methods and MATLAB*. New York: McGraw-Hill, 2004.

Roberts, M. J., *Fundamentals of Signals and Systems*. New York: McGraw-Hill, 2008.

Semmlow, J. L., *Circuits, Systems, and Signals for Bioengineers: A MATLAB-Based Introduction*. Amsterdam, the Netherlands: Elsevier Academic Press, 2005.

Sherlock, J. *A Guide to Technical Communication*. Boston, MA: Allyn and Bacon, 1985, p. 7.

Sherrick, J. D., *Concepts in Systems and Signals*, 2nd ed. Upper Saddle River, NJ: Prentice Hall, 2005.

Soliman, S. S. and M. D. Srinath, *Continuous and Discrete Signals and Systems*, 2nd ed. Upper Saddle River, NJ: Prentice Hall, 1998.

Strum, R. D., *Contemporary Linear Systems using MATLAB*. Pacific Grove, CA: Brooks/Cole, 2000.

Swisher, G. M., *Introduction to Linear Systems Analysis*. Champaign, IL: Matrix Publishers, 1976.

Ziemer, R. E., W. H. Tranter, and D. R. Fannin, *Signals and Systems: Continuous and Discrete*, 4th ed. Upper Saddle River, NJ: Prentice Hall, 1998.

Appendix A: Mathematical Formulas

This appendix—by no means exhaustive—serves as a handy reference. It does contain all the formulas needed to solve problems in this book.

A.1 QUADRATIC FORMULAS

The roots of the quadratic equation $ax^2 + bx + c = 0$

$$x_1, x_2 = \frac{-b \pm \sqrt{b^2 - 4ac}}{2a}$$

A.2 TRIGONOMETRIC IDENTITIES

$$\sin(-x) = -\sin x$$

$$\cos(-x) = \cos x$$

$$\sec x = \frac{1}{\cos x}, \quad \csc x = \frac{1}{\sin x}$$

$$\tan x = \frac{\sin x}{\cos x}, \quad \cot x = \frac{1}{\tan x}$$

$$\sin(x \pm 90°) = \pm \cos x$$

$$\cos(x \pm 90°) = \mp \sin x$$

$$\sin(x \pm 180°) = -\sin x$$

$$\cos(x \pm 180°) = -\cos x$$

$$\cos^2 x + \sin^2 x = 1$$

$$\frac{a}{\sin A} = \frac{b}{\sin B} = \frac{c}{\sin C} \quad \text{(law of sines)}$$

$$a^2 = b^2 + c^2 - 2bc \cos A \quad \text{(law of cosines)}$$

$$\frac{\tan \frac{1}{2}(A - B)}{\tan \frac{1}{2}(A + B)} = \frac{a - b}{a + b} \quad \text{(law of tangents)}$$

$$\sin(x \pm y) = \sin x \cos y \pm \cos x \sin y$$

$$\cos(x \pm y) = \cos x \cos y \mp \sin x \sin y$$

$$\tan(x \pm y) = \frac{\tan x \pm \tan y}{1 \mp \tan x \tan y}$$

$$2 \sin x \sin y = \cos(x - y) - \cos(x + y)$$

$$2 \sin x \cos y = \sin(x + y) - \sin(x - y)$$

$$2 \cos x \cos y = \cos(x + y) - \cos(x - y)$$

$$\sin 2x = 2 \sin x \cos x$$

$$\cos 2x = \cos^2 x - \sin^2 x = 2\cos^2 x - 1 = 1 - 2\sin^2 x$$

$$\tan 2x = \frac{2 \tan x}{1 - \tan^2 x}$$

$$\sin^2 x = \frac{1}{2}(1 - \cos 2x)$$

$$\cos^2 x = \frac{1}{2}(1 + \cos 2x)$$

$$a\cos x + b\sin x = K\cos(x + \theta), \quad \text{where } K = \sqrt{a^2 + b^2} \quad \text{and} \quad \theta = \tan^{-1}\left(\frac{-b}{a}\right)$$

$$e^{\pm jx} = \cos x \pm j\sin x \quad \text{(Euler's formula)}$$

$$\cos x = \frac{e^{jx} + e^{-jx}}{2}$$

$$\sin x = \frac{e^{jx} - e^{-jx}}{2j}$$

$$1 \text{ rad} = 57.296^\circ$$

A.3 HYPERBOLIC FUNCTIONS

$$\sinh x = \frac{1}{2}(e^x - e^{-x})$$

$$\cosh x = \frac{1}{2}(e^x + e^{-x})$$

$$\tanh x = \frac{\sinh x}{\cosh x}$$

$$\coth x = \frac{1}{\tanh x}$$

$$\operatorname{csc\,h} x = \frac{1}{\sinh x}$$

$$\operatorname{sec\,h} x = \frac{1}{\cosh x}$$

$$\sinh(x \pm y) = \sinh x \cosh y \pm \cosh x \sinh y$$

$$\cosh(x \pm y) = \cosh x \cosh y \pm \sinh x \sinh y$$

$$\tanh(x \pm y) = \frac{\sinh(x \pm y)}{\cosh(x \pm y)}$$

A.4 DERIVATIVES

If $U = U(x)$, $V = V(x)$, and a = constant,

$$\frac{d}{dx}(aU) = a\frac{dU}{dx}$$

$$\frac{d}{dx}(UV) = U\frac{dV}{dx} + V\frac{dU}{dx}$$

$$\frac{d}{dx}\left(\frac{U}{V}\right) = \frac{V\frac{dU}{dx} - U\frac{dV}{dx}}{V^2}$$

$$\frac{d}{dx}(aU^n) = naU^{n-1}$$

$$\frac{d}{dx}(a^U) = a^U \ln a \frac{dU}{dx}$$

$$\frac{d}{dx}(e^U) = e^U \frac{dU}{dx}$$

$$\frac{d}{dx}(\sin U) = \cos U \frac{dU}{dx}$$

$$\frac{d}{dx}(\cos U) = -\sin U \frac{dU}{dx}$$

$$\frac{d}{dx}\tan U = \frac{1}{\cos^2 U}\frac{dU}{dx}$$

A.5 INDEFINITE INTEGRALS

If $U = U(x)$, $V = V(x)$, and a = constant,

$$\int a dx = ax + C$$

$$\int U dV = UV - \int V dU \quad \text{(integration by parts)}$$

$$\int U^n dU = \frac{U^{n+1}}{n+1} + C, \quad n \neq 1$$

$$\int \frac{dU}{U} = \ln U + C$$

$$\int a^U dU = \frac{a^U}{\ln a} + C, \quad a > 0, a \neq 1$$

$$\int e^{ax} dx = \frac{1}{a} e^{ax} + C$$

$$\int x e^{ax} dx = \frac{e^{ax}}{a^2}(ax - 1) + C$$

$$\int x^2 e^{ax} dx = \frac{e^{ax}}{a^3}(a^2x^2 - 2ax + 2) + C$$

$$\int \ln x dx = x \ln x - x + C$$

$$\int \sin ax dx = -\frac{1}{a} \cos ax + C$$

$$\int \cos ax dx = \frac{1}{a} \sin ax + C$$

$$\int \sin^2 ax dx = \frac{x}{2} - \frac{\sin 2ax}{4a} + C$$

$$\int \cos^2 ax dx = \frac{x}{2} + \frac{\sin 2ax}{4a} + C$$

$$\int x \sin ax dx = \frac{1}{a^2}(\sin ax - ax \cos ax) + C$$

$$\int x \cos ax dx = \frac{1}{a^2}(\cos ax + ax \sin ax) + C$$

$$\int x^2 \sin ax dx = \frac{1}{a^3}\left(2ax \sin ax + 2 \cos ax - a^2 x^2 \cos ax\right) + C$$

$$\int x^2 \cos ax dx = \frac{1}{a^3}\left(2ax \cos ax - 2 \sin ax + a^2 x^2 \sin ax\right) + C$$

$$\int e^{ax} \sin bx dx = \frac{e^{ax}}{a^2 + b^2}(a \sin bx - b \cos bx) + C$$

$$\int e^{ax} \cos bx dx = \frac{e^{ax}}{a^2 + b^2}(a \cos bx + b \sin bx) + C$$

$$\int \sin ax \sin bx dx = \frac{\sin(a-b)x}{2(a-b)} - \frac{\sin(a+b)x}{2(a+b)} + C, \quad a^2 \neq b^2$$

$$\int \sin ax \cos bx dx = -\frac{\cos(a-b)x}{2(a-b)} - \frac{\cos(a+b)x}{2(a+b)} + C, \quad a^2 \neq b^2$$

$$\int \cos ax \cos bx dx = \frac{\sin(a-b)x}{2(a-b)} + \frac{\sin(a+b)x}{2(a+b)} + C, \quad a^2 \neq b^2$$

$$\int \frac{dx}{a^2 + x^2} = \frac{1}{a}\tan^{-1}\frac{x}{a} + C$$

$$\int \frac{x^2 dx}{a^2 + x^2} = x - a\tan^{-1}\frac{x}{a} + C$$

$$\int \frac{dx}{\left(a^2 + x^2\right)^2} = \frac{1}{2a^2}\left(\frac{x}{x^2 + a^2} + \frac{1}{a}\tan^{-1}\frac{x}{a}\right) + C$$

A.6 DEFINITE INTEGRALS

If m and n are integers,

$$\int_0^{2\pi} \sin ax \, dx = 0$$

$$\int_0^{2\pi} \cos ax \, dx = 0$$

$$\int_0^{\pi} \sin^2 ax \, dx = \int_0^{\pi} \cos^2 axdx = \frac{\pi}{2}$$

$$\int_0^{\pi} \sin mx \sin nxdx = \int_0^{\pi} \cos mx \cos nxdx = 0, \quad m \neq n$$

$$\int_0^{\pi} \sin mx \cos nxdx = \begin{cases} 0, & m + n = \text{even} \\ \dfrac{2m}{m^2 - n^2}, & m + n = \text{odd} \end{cases}$$

$$\int_0^{2\pi} \sin mx \sin nx dx = \int_{-\pi}^{\pi} \sin mx \sin nx dx = \begin{cases} 0, & m \neq n \\ \pi, & m \neq n \end{cases}$$

$$\int_0^{\infty} \frac{\sin ax}{x} dx = \begin{cases} \frac{\pi}{2}, & a > 0 \\ 0, & a = 0 \\ -\frac{\pi}{2}, & a < 0 \end{cases}$$

$$\int_0^{\infty} \frac{\sin^2 x}{x} dx = \frac{\pi}{2}$$

$$\int_0^{\infty} \frac{\cos bx}{x^2 + a^2} dx = \frac{\pi}{2a} e^{-ab}, \quad a > 0, b > 0$$

$$\int_0^{\infty} \frac{x \sin bx}{x^2 + a^2} dx = \frac{\pi}{2} e^{-ab}, \quad a > 0, b > 0$$

$$\int_0^{\infty} \sin cx dx = \int_0^{\infty} \sin c^2 x dx = \frac{1}{2}$$

$$\int_0^{\pi} \sin^2 nx dx = \int_0^{\pi} \sin^2 x dx = \int_0^{\pi} \cos^2 nx dx = \int_0^{\pi} \cos^2 x dx = \frac{\pi}{2}, \quad n = \text{an integer}$$

$$\int_0^{\pi} \sin mx \sin nx dx = \int_0^{\pi} \cos mx \cos nx dx = 0, \quad m \neq n, m, n \text{ integers}$$

$$\int_0^{\pi} \sin mx \cos nx dx = \begin{cases} \frac{2m}{m^2 - n^2}, & m + n = \text{odd} \\ 0, & m + n = \text{even} \end{cases}$$

$$\int_{-\infty}^{\infty} e^{\pm j2\pi tx} dx = \delta(t)$$

$$\int_{0}^{\infty} x^n e^{-ax} dx = \frac{n!}{a^{n+1}}$$

$$\int_{0}^{\infty} e^{-a^2x^2} dx = \frac{\sqrt{\pi}}{2a}, \qquad a > 0$$

$$\int_{0}^{\infty} x^{2n} e^{-ax^2} dx = \frac{1\cdot 3\cdot 5\cdots(2n-1)}{2^{n+1}a^n}\sqrt{\frac{\pi}{a}}$$

$$\int_{0}^{\infty} x^{2n+1} e^{-ax^2} dx = \frac{n!}{2a^{n+1}}, \qquad a > 0$$

A.7 L'HOPITAL'S RULE

If $f(0) = 0 = h(0)$, then

$$\lim_{x\to 0}\frac{f(x)}{h(x)} = \lim_{x\to 0}\frac{f'(x)}{h'(x)}$$

where the prime indicates differentiation.

A.8 TAYLOR AND MACLAURIN SERIES

$$f(x) = f(a) + \frac{(x-a)}{1!} f'(a) + \frac{(x-a)^2}{2!} f''(a) + \cdots$$

$$f(x) = f(0) + \frac{x}{1!} f'(0) + \frac{x^2}{2!} f''(0) + \cdots$$

where the prime indicates differentiation.

A.9 POWER SERIES

$$e^x = 1 + x + \frac{x^2}{2!} + \frac{x^3}{3!} + \cdots + \frac{x^n}{n!} + \cdots$$

$$\sin x = x - \frac{x^3}{3!} + \frac{x^5}{5!} - \frac{x^7}{7!} + \cdots$$

$$\cos x = 1 - \frac{x^2}{2!} + \frac{x^4}{4!} - \frac{x^6}{6!} + \frac{x^8}{8!} - \cdots$$

$$\tan x = x + \frac{x^3}{3} + \frac{2x^5}{15} + \frac{17x^7}{315} + \cdots$$

$$(1+x)^n = 1 + nx + \frac{n(n+1)}{2!}x^2 + \frac{n(n-1)(n-2)}{3!}x^3 + \cdots + \binom{n}{k}x^k + \cdots + x^n$$

$$\approx 1 + nx, \quad |x| < 1$$

$$\frac{1}{1-x} = 1 + x + x^2 + x^3 + \cdots, \quad |x| < 1$$

$$Q(x) = \frac{e^{-x^2/2}}{x\sqrt{2\pi}}\left(1 - \frac{1}{x^2} + \frac{1\cdot 3}{x^4} - \frac{1\cdot 3\cdot 5}{x^6} + \cdots\right)$$

$$J_n(x) = \frac{1}{n!}\left(\frac{x}{2}\right)^n - \frac{1}{(n+1)!}\left(\frac{x}{2}\right)^{n+2} + \frac{1}{2!(n+2)!}\left(\frac{x}{2}\right)^{n+4} - \cdots$$

$$J_n(x) \approx \sqrt{\frac{2}{\pi x}}\cos\left(x - \frac{\pi}{4} - \frac{n\pi}{2}\right), \quad x \gg 1$$

$$I_0(x) \approx \begin{cases} e^{x^2/4}, & x^2 \ll 1 \\ \dfrac{e}{\sqrt{2\pi x}}, & x \gg 1 \end{cases}$$

A.10 SUMS

$$\sum_{k=1}^{N} k = \frac{1}{2}N(N+1)$$

$$\sum_{k=1}^{N} k^2 = \frac{1}{6}N(N+1)(2N+1)$$

$$\sum_{k=1}^{N} k^3 = \frac{1}{4}N^2(N+1)^2$$

$$\sum_{k=0}^{N} a^k = \frac{a^{N+1}-1}{a-1} \quad a \neq 1$$

$$\sum_{k=M}^{N} a^k = \frac{a^{N+1}-a^M}{a-1} \quad a \neq 1$$

$$\sum_{k=0}^{N} \binom{N}{k} a^{N-k} b^k = (a+b)^N, \quad \text{where} \binom{N}{k} = \frac{N!}{(N-k)!k!}$$

A.11 LOGARITHMIC IDENTITIES

$$\log xy = \log x + \log y$$

$$\log \frac{x}{y} = \log x - \log y$$

$$\log x^n = n \log x$$

$$\log_{10} x = \log x \quad \text{(common logarithm)}$$

$$\log_e x = \ln x \quad \text{(natural logarithm)}$$

A.12 EXPONENTIAL IDENTITIES

$$e^x = 1 + x + \frac{x^2}{2!} + \frac{x^3}{3!} + \frac{x^4}{4!} + \cdots$$

where $e = 2.7182$

$$e^x e^y = e^{x+y}$$

$$(e^x)^n = e^{nx}$$

$$\ln e^x = x$$

A.13 APPROXIMATIONS

If $|x| \ll 1$,

$$(1 \pm x)^n \simeq 1 \pm nx$$

$$e^x \simeq 1 + x$$

$$\ln(1 + x) \simeq x$$

$$\sin x = x \quad \text{or} \quad \lim_{x \to 0} \frac{\sin x}{x} = 1$$

$$\cos x \simeq 1$$

$$\tan x \simeq x$$

Appendix B: Complex Numbers

The ability to handle complex numbers is important in signals and systems. Although calculators and computer software packages, such as MATLAB®, are now available to manipulate complex numbers, it is advisable that students be familiar with how to handle them by hand.

B.1 REPRESENTATION OF COMPLEX NUMBERS

A complex number z may be written in *rectangular form* as

$$z = x + jy \tag{B.1}$$

where $j = \sqrt{-1}$; x is the real part of z while y is the imaginary part, that is,

$$x = \mathrm{Re}(z), \quad y = \mathrm{Im}(z) \tag{B.2}$$

The complex number z is shown plotted in the complex plane in Figure B.1. Since $j = \sqrt{-1}$

$$\begin{aligned}
\frac{1}{j} &= -j \\
j^2 &= -1 \\
j^3 &= j \cdot j^2 = -j \\
j^4 &= j^2 \cdot j^2 = 1 \\
j^5 &= j \cdot j^4 = j \\
&\vdots
\end{aligned} \tag{B.3}$$

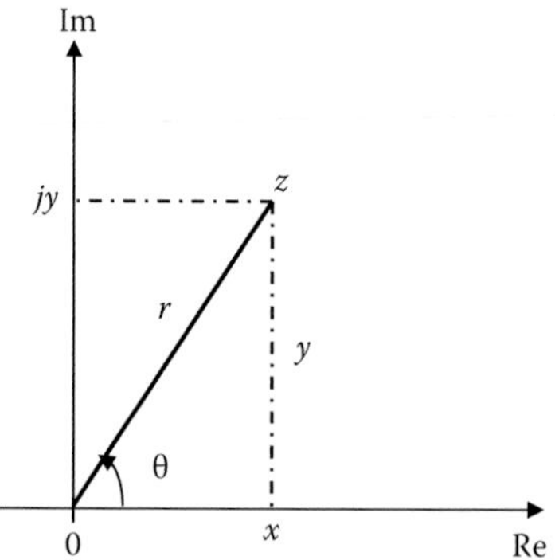

FIGURE B.1 Graphical representation of a complex number.

A second way of representing the complex number z is by specifying its magnitude r and angle θ it makes with the real axis, as shown in Figure B.1. This is known as the **polar form**. It is given by

$$z = |z| \angle\theta = r\angle\theta \tag{B.4}$$

where

$$r = \sqrt{x^2 + y^2}, \quad \theta = \tan^{-1}\frac{y}{x} \tag{B.5}$$

or

$$x = r\cos\theta, \quad y = r\sin\theta \tag{B.6}$$

that is,

$$z = x + jy = r\angle\theta = r\cos\theta + jr\sin\theta \tag{B.7}$$

In converting from rectangular to polar form using Equation B.5, we must exercise care in determining the correct value of θ. These are the four possibilities:

$$\begin{aligned}
z &= x + jy, & \theta &= \tan^{-1}\frac{y}{x} && \text{(First quadrant)} \\
z &= -x + jy, & \theta &= 180° - \tan^{-1}\frac{y}{x} && \text{(Second quadrant)} \\
z &= -x - jy, & \theta &= 180° + \tan^{-1}\frac{y}{x} && \text{(Third quadrant)} \\
z &= x - jy, & \theta &= 360° - \tan^{-1}\frac{y}{x} && \text{(Fourth quadrant)}
\end{aligned} \tag{B.8}$$

assuming that x and y are positive.

The third way of representing the complex number x is the *exponential form*:

$$z = re^{j\theta} \tag{B.9}$$

This is almost the same as the polar form, because we use the same magnitude r and the angle θ.

The three forms of representing a complex number are summarized as follows:

$$\begin{array}{|lll|}\hline z = x + jy, & (x = r\cos\theta, y = r\sin\theta) & \text{Rectangular form} \\ z = r\angle\theta, & \left(r = \sqrt{x^2 + y^2}, \theta = \tan^{-1}\dfrac{y}{x}\right) & \text{Polar form} \\ z = re^{j\theta}, & \left(r = \sqrt{x^2 + y^2}, \theta = \tan^{-1}\dfrac{y}{x}\right) & \text{Exponential form} \\ \hline \end{array} \tag{B.10}$$

B.2 MATHEMATICAL OPERATIONS

Two complex numbers $z_1 = x_1 + jy_1$ and $z_2 = x_2 + jy_2$ are equal if and only, their real and imaginary parts are equal, that is,

$$x_1 = x_2, \quad y_1 = y_2 \tag{B.11}$$

The complex conjugate of the complex number $z = x + jy$ is given as:

$$z^* = x - jy = r\angle -\theta = re^{-j\theta} \tag{B.12}$$

Thus, the complex conjugate of a complex number is found by replacing every j by $-j$.

Given two complex numbers $z_1 = x_1 + jy_1 = r_1\angle\theta_1$ and $z_2 = x_2 + jy_2 = r_2\angle\theta_2$, their sum is

$$z_1 + z_2 = (x_1 + x_2) + j(y_1 + y_2) \tag{B.13}$$

and their difference is

$$z_1 - z_2 = (x_1 - x_2) + j(y_1 - y_2) \tag{B.14}$$

While it is more convenient to perform addition and subtraction of complex numbers in rectangular form, the product and quotient of two complex numbers are best done in polar or exponential form. For their product,

$$z_1 z_2 = r_1 r_2 \angle\theta_1 + \theta_2 \tag{B.15}$$

Alternatively, using the rectangular form

$$z_1 z_2 = (x_1 + jy_1)(x_2 + jy_2)$$
$$= (x_1 x_2 - y_1 y_2) + j(x_1 y_2 + x_2 y_1) \tag{B.16}$$

For their quotient,

$$\frac{z_1}{z_2} = \frac{r_1}{r_2} \angle \theta_1 - \theta_2 \tag{B.17}$$

Alternatively, using the rectangular form as,

$$\frac{z_1}{z_2} = \frac{x_1 + jy_1}{x_2 + jy_2} \tag{B.18}$$

we rationalize the denominator by multiplying both the numerator and denominator by z_2^*.

$$\frac{z_1}{z_2} = \frac{(x_1 + jy_1)(x_2 - jy_2)}{(x_2 + jy_2)(x_2 - jy_2)} = \frac{x_1 x_2 + y_1 y_2}{x_2^2 + y_2^3} + j\frac{x_2 y_1 - x_1 y_2}{x_2^2 + y_2^3} \tag{B.19}$$

B.3 EULER'S FORMULA

Euler's formula is an important result in complex variables. We derive it from the series expansion of e^x, $\cos\theta$, and $\sin\theta$. We know that

$$e^x = 1 + x + \frac{x^2}{2!} + \frac{x^3}{3!} + \frac{x^4}{4!} + \cdots \tag{B.20}$$

Replacing x by $j\theta$ gives

$$e^{j\theta} = 1 + j\theta - \frac{\theta^2}{2!} - j\frac{\theta^3}{3!} + \frac{\theta^4}{4!} + \cdots \tag{B.21}$$

Also,

$$\cos\theta = 1 - \frac{\theta^2}{2!} + \frac{\theta^4}{4!} - \frac{\theta^6}{6!} + \cdots \tag{B.22}$$

$$\sin\theta = \theta - \frac{\theta^3}{3!} + \frac{\theta^5}{5!} - \frac{\theta^7}{7!} + \cdots \tag{B.23}$$

so that

$$\cos\theta + j\sin\theta = 1 + j\theta - \frac{\theta^2}{2!} - j\frac{\theta^3}{3!} + \frac{\theta^4}{4!} + j\frac{\theta^5}{5!} - \cdots \tag{B.24}$$

Comparing Equations B.21 and B.24, we conclude that

$$\boxed{e^{j\theta} = \cos\theta + j\sin\theta} \tag{B.25}$$

This is known as Euler's formula. The exponential form of representing a complex number as in Equation B.9 is based on Euler's formula. From Equation B.25, notice that

$$\boxed{\cos\theta = \text{Re}(e^{j\theta}), \quad \sin\theta = \text{Im}(e^{j\theta})} \tag{B.26}$$

and that

$$\left|e^{j\theta}\right| = \sqrt{\cos^2\theta + \sin^2\theta} = 1 \tag{B.27}$$

Replacing θ by $-\theta$ in Equation B.25 gives

$$e^{-j\theta} = \cos\theta - j\sin\theta \tag{B.28}$$

Adding Equations B.25 and B.28 yields

$$\boxed{\cos\theta = \frac{1}{2}\left(e^{j\theta} + e^{-j\theta}\right)} \tag{B.29}$$

Subtracting Equation B.28 from Equation B.25 yields

$$\boxed{\sin\theta = \frac{1}{2j}\left(e^{j\theta} - e^{-j\theta}\right)} \tag{B.30}$$

The following identities are useful in dealing with complex numbers. If $z = x + jy = r\angle\theta$, then

$$zz^* = |z|^2 = x^2 + y^2 = r^2 \tag{B.31}$$

$$\sqrt{z} = \sqrt{x + jy} = \sqrt{r}e^{j\theta/2} = \sqrt{r}\angle\theta/2 \tag{B.32}$$

$$z^n = (x + jy)^n = r^n\angle n\theta = r^n(\cos n\theta + j\sin n\theta) \tag{B.33}$$

$$z^{1/n} = (x + jy)^{1/n} = r^{1/n}\angle\theta/n + 2\pi k/n, \quad k = 0,1,2,\ldots,n-1 \tag{B.34}$$

$$\ln(re^{j\theta}) = \ln r + \ln e^{j\theta} = \ln r + j\theta + j2\pi k \quad (k = \text{integer}) \tag{B.35}$$

$$\begin{aligned} e^{\pm j\pi} &= -1 \\ e^{\pm j2\pi} &= 1 \\ e^{j\pi/2} &= j \\ e^{-j\pi/2} &= -j \end{aligned} \tag{B.36}$$

$$\begin{aligned} \mathrm{Re}\left(e^{(\alpha + j\omega)t}\right) &= \mathrm{Re}\left(e^{\alpha t}e^{j\omega t}\right) = e^{\alpha t}\cos\omega t \\ \mathrm{Im}\left(e^{(\alpha + j\omega)t}\right) &= \mathrm{Im}\left(e^{\alpha t}e^{j\omega t}\right) = e^{\alpha t}\sin\omega t \end{aligned} \tag{B.37}$$

MATLAB handles complex numbers quite easily as real numbers.

Example B.1

Evaluate the complex numbers:

(a) $z_1 = \dfrac{j(3 - j4)^*}{(-1 + j6)(2 + j)^2}$

(b) $z_2 = \left[4\angle 30° + 2 - j + 6e^{j\pi/4}\right]^{1/2}$

Solution

(a) This can be solved in two ways: working with z in rectangular form or polar form.

Method 1 (working in rectangular form):

Let $z_1 = \dfrac{z_3 z_4}{z_5 z_6}$

where

$$z_3 = j$$

$$z_4 = (3 - j4)^* = \text{the complex conjugate of } (3 - j4) = (3 + j4)$$

$$z_5 = -1 + j6$$

$$z_6 = (2 + j)^2 = 4 + j4 - 1 = 3 + j4$$

Hence,

$$z_1 = \frac{j(3+j4)}{(-1+j6)(3+j4)} = \frac{j3-4}{-3-j4+j18-24}$$

$$= \frac{-4+j3}{-27+j14}$$

Multiplying and dividing by −27 − j14 (rationalization), we have

$$z_1 = \frac{(-4+j3)(-27-j14)}{(-27+j14)(-27-j14)} = \frac{150-j25}{27^2+14^2}$$

$$= 0.1622 - j0.027 = 0.1644\angle -9.46^\circ$$

Method 2 (working in polar form):

$$z_3 = j = 1\angle 90^\circ$$

$$z_4 = (3-j4)^* = (5\angle -53.13^\circ)^* = 5\angle 53.13^\circ$$

$$z_5 = (-1+j6) = \sqrt{37}\angle 99.46^\circ$$

$$z_6 = (2+j)^2 = \left(\sqrt{5}\angle 26.56^\circ\right)^2 = 5\angle 53.13^\circ$$

Hence,

$$z_1 = \frac{z_3 z_4}{z_5 z_6} = \frac{(1\angle 90^\circ)(5\angle 53.13^\circ)}{\left(\sqrt{37}\angle 99.46^\circ\right)(5\angle 53.13^\circ)} = \frac{1}{\sqrt{37}}\angle(90^\circ - 99.46^\circ)$$

$$= 0.1644\angle -9.46^\circ = 0.1622 - j0.027$$

(b) Let

$$z_7 = 4\angle 30^\circ = 4\cos 30^\circ + j4\sin 30^\circ = 3.464 + j2$$

$$z_8 = 2 - j$$

$$z_9 = 6e^{j\pi/4} = 6\cos 45^\circ + j6\sin 45^\circ = 4.243 + j4.243$$

Then,

$$z_2 = \left[z_7 + z_8 + z_9\right]^{1/2}$$

$$= \left[3.464 + j2 + 2 - j + 4.243 + j4.243\right]^{1/2}$$

$$= (9.707 + j5.243)^{1/2} = (11.03\angle 28.374^\circ)^{1/2}$$

$$= 3.32\angle 14.19^\circ$$

Practice Problem B.1 Evaluate the following complex numbers:

(a) $j^3\left[\frac{1+j}{2-j}\right]^2$

(b) $6\angle 30° + j5-3 + ej^{45°}$

Answer: (a) $0.24 + j0.32$, (b) $2.03 + j8.707$

Appendix C: Introduction to MATLAB®

MATLAB® has become a powerful tool of technical professionals worldwide. The term MATLAB is an abbreviation for MATrix LABoratory implying that MATLAB is a computational tool that employs matrices and vectors/arrays to carry out numerical analysis, signal processing, and scientific visualization tasks. Because MATLAB uses matrices as its fundamental building blocks, one can write mathematical expressions involving matrices just as easily as one would on paper. MATLAB is available for Macintosh, Unix, and Windows operating systems. A student version of MATLAB is available for PCs. A copy of MATLAB can be obtained from

The Mathworks, Inc.
3 Apple Hill Drive
Natick, MA 01760-2098
Phone: (508) 647-7000
Website: http://www.mathworks.com

A brief introduction on MATLAB is presented in this appendix. What is presented is sufficient for solving problems in this book. Other information on MATLAB required in this book is provided on a chapter-by-chapter basis as needed. Additional information about MATLAB can be found in MATLAB books and obtained from the online help. The best way to learn MATLAB is to work with it after one has learned the basics.

C.1 MATLAB® FUNDAMENTALS

The Command window is the primary area where you interact with MATLAB. A little later, we learn how to use the text editor to create M-files, which allow executing sequences of commands. For now, we focus on how to work in Command window. We first learn how to use MATLAB as a calculator. We do so by using the algebraic operators in Table C.1.

To begin to use MATLAB, we use these operators. Type commands to MATLAB prompt "≫" in the Command window (correct any mistakes by backspacing) and press the <Enter> key. For example,

```
≫ a = 2; b = 4; c = -6;
≫ dat = b^2 - 4*a*c
dat =
    64
≫ e=sqrt(dat)/10
e =
    0.8000
```

TABLE C.1
Basic Operations

Operation	MATLAB Formula
Addition	a + b
Division (right)	a/b (*means a÷b*)
Division (left)	*a*\b (*means b÷a*)
Multiplication	a*b
Power	a^b
Subtraction	a − b

The first command assigns the values 2, 4, and −6 to the variables *a*, *b*, and *c*, respectively. MATLAB does not respond because this line ends with a colon. The second command sets *dat* to $b^2 - 4ac$ and MATLAB returns the answer as 64. Finally, the third line sets *e* equal to the square root of *dat* and divides by 10. MATLAB prints the answer as 0.8. As function *sqrt* is used here, other mathematical functions listed in Table C.2 can be used. Table C.2 provides just a small sample of MATLAB functions. Others can be obtained from the online help. To get help, type

```
>> help
```

TABLE C.2
Typical Elementary Math Functions

Function	Remark
abs(x)	Absolute value or complex magnitude of x
acos, acosh(x)	Inverse cosine and inverse hyperbolic cosine of x in radians
acot, acoth (x)	Inverse cotangent and inverse hyperbolic cotangent of x in radians
angle(x)	Phase angle (in radian) of a complex number x
asin, asinh(x)	Inverse sine and inverse hyperbolic sine of x in radians
atan, atanh(x)	Inverse tangent and inverse hyperbolic tangent of x in radians
conj(x)	Complex conjugate of x
cos, cosh(x)	Cosine and hyperbolic cosine of x in radian
cot, coth(x)	Cotangent and hyperbolic cotangent of x in radian
exp(x)	Exponential of x
fix	Round toward zero
imag(x)	Imaginary part of a complex number x
log (x)	Natural logarithm of x
log2(x)	Logarithm of x to base 2
log10(x)	Common logarithms (base 10) of x
real (x)	Real part of a complex number x
sin, sinh(x)	Sine and hyperbolic sine of x in radian
sqrt (x)	Square root of x
tan, tanh	Tangent and hyperbolic tangent of x in radian

[a long list of topics come up]

and for a specific topic, type the command name. For example, to get help on *log to base 2*, type

```
>> help log2
```

[a help message on the log function follows]
Note that MATLAB is case sensitive so that sin(*a*) is not the same as sin(*A*).
Try the following examples:

```
>> 3 ^ (log10(25.6))
>> y = 2 * sin(pi/3)
>> exp(y + 4 - 1)
```

In addition to operating on mathematical functions, MATLAB easily allows one to work with vectors and matrices. A vector (or array) is a special matrix with one row or one column. For example,

```
>> a = [1   -3   6   10   -8   11   14];
```

is a row vector. Defining a matrix is similar to defining a vector. For example, a 3 × 3 matrix can be entered as

```
>> A = [1  2  3;  4  5  6;  7  8  9]
```

or as

```
>> A= [1  2  3
       4  5  6
       7  8  9]
```

In addition to the arithmetic operations that can be performed on a matrix, the operations in Table C.3 can be implemented.

TABLE C.3
Matrix Operations

Operation	Remark
A'	Finds the transpose of matrix *A*
det(*A*)	Evaluates the determinant of matrix *A*
inv(*A*)	Calculates the inverse of matrix *A*
eig(*A*)	Determines the eigenvalues of matrix *A*
diag(*A*)	Finds the diagonal elements of matrix *A*
expm(*A*)	Exponential of matrix *A*

Using the operations in Table C.3, we can manipulate matrices as follows:

```
>> B = A'

        1   4   7
    B = 2   5   8
        3   6   9

>> C = A + B

        2    6    10
    C = 6    10   14
        10   14   18

>> D = A ^ 3 - B * C

        372    432    492
    D = 948    1131   1314
        1524   1830   2136

>> e = [1   2;   3   4]

         1   2
    e =
         3   4

>> f = det(e)
    f = - 2

>> g = inv (e)

         -2.0000    1.0000
    g =
          1.5000   -0.5000

>> H = eig (g)

         -2.6861
    H =
          0.1861
```

Note that not all matrices can be inverted. A matrix can be inverted if and only if its determinant is nonzero. Special matrices, variables, and constants are listed in Table C.4. For example, type

```
>>  eye(3)

          1   0   0
    ans = 0   1   0
          0   0   1
```

to get a 3×3 identity matrix.

TABLE C.4
Special Matrices, Variables, and Constants

Matrix/Variable/Constant	Remark
eye	Identity matrix
ones	An array of ones
zeros	An array of zeros
i or *j*	Imaginary unit or sqrt(−1)
pi	3.142
NaN	Not a number
inf	Infinity
eps	A very small number, 2.2e−16
rand	Random element

C.2 USING MATLAB® TO PLOT

To plot using MATLAB is easy. For two-dimensional plot, use the plot command with two arguments as

```
≫plot (xdata, ydata)
```

where *xdata* and *ydata* are vectors of the same length containing the data to be plotted.

For example, suppose we want to plot *y* = 10*sin(2*pi*x) from 0 to 5*pi, we will proceed with the following commands:

```
≫  x = 0:pi/100:5*pi;  % x is a vector, 0 <= x <= 5*pi, increments of
                        pi/100
≫  y = 10*sin(2*pi*x); % create a vector y
≫  plot(x, y);         % create the plot
```

With this, MATLAB responds with the plot in Figure C.1

MATLAB will let you graph multiple plots together and distinguish with different colors.

This is obtained with the command `plot(xdata, ydata, "color")`, where the color is indicated by using a character string from the options listed in Table C.5.

For example,

```
≫ plot(x1, y1, 'r', x2,y2, 'b', x3,y3, '--')
```

will graph data (x1,y1) in red, data (x2,y2) in blue, and data (x3,y3) in dashed line all on the same plot.

MATLAB also allows for logarithm scaling. Rather that the `plot` command, we use

```
loglog      log(y) versus log(x)
semilogx    y versus log(x)
semilogy    log(y) versus x
```

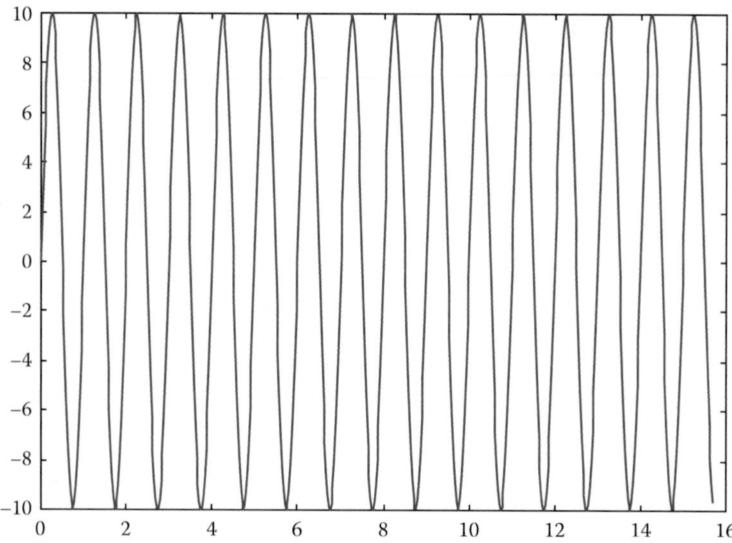

FIGURE C.1 MATLAB plot of y = 10*sin(2*pi*x).

TABLE C.5
Various Color and Line Types

y	Yellow	.	Point
m	Magenta	o	Circle
c	Cyan	x	x-mark
r	Red	+	Plus
g	Green	-	Solid
b	Blue	*	Star
w	White	:	Dotted
k	Black	-.	Dashdot
		--	Dashed

Three-dimensional plots are drawn using the functions *mesh* and *meshdom* (mesh domain). For example, to draw the graph of $z = x^* \, exp(-x^2-y$^2) over the domain $-1 < x, y < 1$, we type the following commands:

```
≫ xx = −1:.1:1;
≫ yy = xx;
≫ [x,y] = meshgrid(xx,yy);
≫ z=x.*exp(-x.^2 -y.^2);
≫ mesh(z);
```

(The dot symbol used in `x.` and `y.` allows element-by-element multiplication.) The result is shown in Figure C.2.

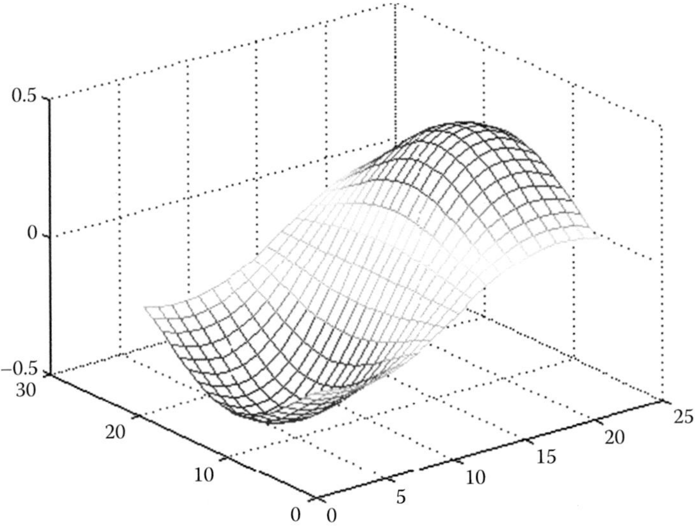

FIGURE C.2 A three-dimensional plot.

TABLE C.6
Other Plotting Commands

Command	Comments
`bar(x,y)`	A bar graph
`contour(z)`	A contour plot
`errorbar (x,y,l,u)`	A plot with error bars
`hist(x)`	A histogram of the data
`plot3(x,y,z)`	A three-dimensional version of plot()
`polar(r, angle)`	A polar coordinate plot
`stairs(x,y)`	A stairstep plot
`stem(n,x)`	Plots the data sequence as stems
`subplot(m,n,p)`	Multiple (*m*-by-*n*) plots per window
`surf(x,y,x,c)`	A plot of 3D colored surface.

Other plotting commands in MATLAB are listed in Table C.6. The `help` command can be used to find out how each of these is used.

C.3 PROGRAMMING WITH MATLAB®

So far MATLAB has been used as a calculator, and you can also use MATLAB to create your own program. The command line editing in MATLAB can be inconvenient if one has several lines to execute. To avoid this problem, one creates a program, which is a sequence of statements to be executed. If you are in Command window, click `File/New/M-files` to open a new file in the MATLAB Editor/Debugger or simple text editor. Type the program and save the program in a file with an extension.m, say filename.m; it is for this reason it is called an M-file. Once the program is saved as an M-file, exit the Debugger window. You are now back in Command window. Type the file without the extension.m to get results. For example, the plot that was made above can be improved by adding title and labels and typed as an M-files called example1.m.

```
x = 0:pi/100:5*pi;              % x is a vector, 0 <= x <= 5*pi,
                                  increments of pi/100
y = 10*sin(2*pi*x);             % create a vector y
plot(x,y);                      % create the plot
xlabel("x (in radians)");       % label the x-axis
ylabel("10*sin(2*pi*x)");       % label the y-axis
title("A sine function");       % title the plot
grid                            % add grid
```

Once it is saved as *example1.m*, we exit text editor, and type

```
≫  example1
```

in the Command window and hit <Enter> to obtain the result shown in Figure C.3.

To allow flow control in a program, certain relational and logical operators are necessary. They are shown in Table C.7. Perhaps, the most commonly used flow

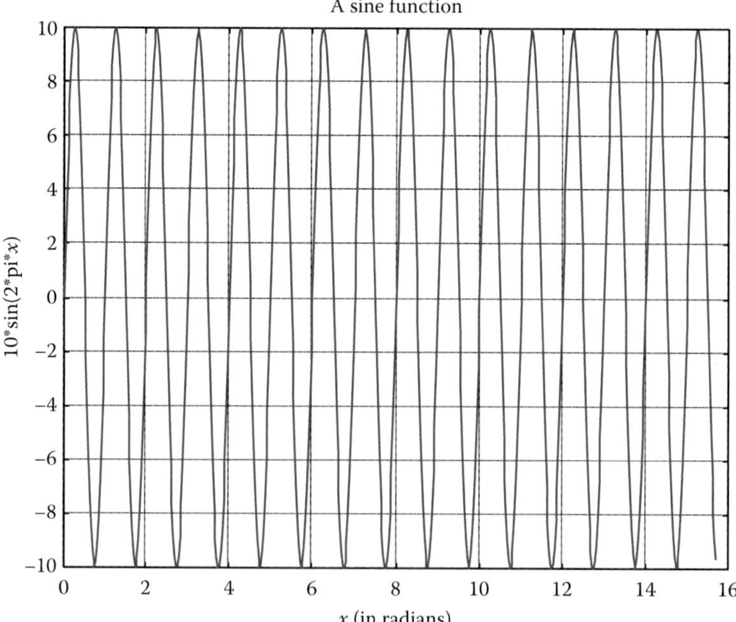

FIGURE C.3 MATLAB plot of $y = 10*\sin(2*\text{pi}*x)$ with title and labels.

TABLE C.7
Relational and Logical Operators

Operator	Remark
<	Less than
≤	Less than or equal
>	Greater than
≥	Greater than or equal
= =	Equal
~ =	Not equal
&	And
|	Or
~	Not

control statements are *for* and *if*. The *for* statement is used to create a loop or a repetitive procedure and has the general form:

```
for x = array
    [commands]
end
```

The *if* statement is used when certain conditions need to be met before an expression is executed. It has the general form:

```
if expression
    [commands if expression is True]
else
    [commands if expression is False]
end
```

For example, suppose we have an array $y(x)$ and we want to determine the minimum value of y and its corresponding index x. This can be done by creating an M-file as shown as follows.

```
% example2.m
% This program finds the minimum y value and its corresponding
x index
x = [1 2 3  4 5 6  7  8 9 10]; %the nth term in y
y = [3 9 15 8 1 0 -2  4 12 5];
min1 = y(1);
for k=1:10
   min2=y(k);
   if(min2 < min1)
      min1 = min2;
      xo = x(k);
   else
      min1 = min1;
   end
```

```
end
diary
min1, xo
diary off
```

Note the use of *for* and *if* statements. When this program is saved as `example2.m`, we execute it in the Command window and obtain the minimum value of y as –2 and the corresponding value of x as 7, as expected.

```
>> example2
min1 =
     -2
xo =
     7
```

If we are not interested in the corresponding index, we could do the same thing using the command

```
>>  min(y)
```

The following tips are helpful in working effectively with MATLAB:

- Comment your M-file by adding lines beginning with a % character.
- To suppress output, end each command with a semicolon (;), you may remove the semicolon when debugging the file.
- Press up and down arrow keys to retrieve previously executed commands.
- If your expression does not fit on one line, use an ellipse (…) at the end of the line and continue on the next line. For example, MATLAB considers

$$y = \sin(x + \log 10(2x+3)) + \cos(x + \ldots$$
$$\log 10(2x+3));$$

  as one line of expression.
- Keep in mind that variable and function names are case sensitive.

C.4 SOLVING EQUATIONS

Consider the general system of n simultaneous equations as

$$\begin{aligned} a_{11}x_1 + a_{12}x_2 + \cdots + a_{1n}x_n &= b_1 \\ a_{21}x_1 + a_{22}x_2 + \cdots + a_{2n}x_n &= b_2 \\ \cdots \quad \cdots \quad \cdots & \\ a_{n1}x_1 + a_{n2}x_2 + \cdots + a_{nn}x_n &= b_n \end{aligned}$$

or in matrix form

$$AX = B$$

where

$$A = \begin{bmatrix} a_{11} & a_{12} & \cdots & a_{1n} \\ a_{21} & a_{22} & \cdots & a_{2n} \\ \cdots & \cdots & \cdots & \cdots \\ a_{n1} & a_{n2} & a_{n3} & a_{nn} \end{bmatrix}, \quad X = \begin{bmatrix} x_1 \\ x_2 \\ \cdots \\ x_n \end{bmatrix}, \quad B = \begin{bmatrix} b_1 \\ b_2 \\ \cdots \\ b_n \end{bmatrix}$$

A is a square matrix and is known as the coefficient matrix, while X and B are vectors. X is the solution vector we are seeking to get. There are two ways to solve for X in MATLAB. First, we can use the backslash operator (\) so that

$$X = A \backslash B$$

Second, we can solve for X as

$$X = A^{-1}B$$

which in MATLAB is the same as

$$X = \text{inv}(A) * B$$

We can also solve equations using the command `solve`. For example, given the quadratic equation $x^2 + 2x-3 = 0$, we obtain the solution using the following MATLAB command:

```
≫  [x]=solve('x^2 + 2*x - 3 =0')
 x =
 [-3]
[1]
```

indicating that the solutions are x=−3 and x=1. Of course, we can use the command `solve` for a case involving two or more variables. We will see that in the following example.

Example C.1

Use MATLAB to solve the following simultaneous equations:

$$\begin{aligned} 25x_1 - 5x_2 - 20x_3 &= 50 \\ -5x_1 + 10x_2 - 4x_3 &= 0 \\ -5x_1 - 4x_2 + 9x_3 &= 0 \end{aligned}$$

Solution

We can use MATLAB to solve this in two ways:

Method 1:
The given set of simultaneous equations could be written as

$$\begin{bmatrix} 25 & -5 & -20 \\ -5 & 10 & -4 \\ -5 & -4 & 9 \end{bmatrix} \begin{bmatrix} x_1 \\ x_2 \\ x_3 \end{bmatrix} = \begin{bmatrix} 50 \\ 0 \\ 0 \end{bmatrix} \quad \text{or} \quad AX = B$$

We obtain matrix A and vector B and enter them in MATLAB as follows:

```
>> A = [25   -5   -20;   -5   10   -4;   -5   -4   9]
        25   -5   -20
   A = -5    10   -4
       -5    -4    9
```

```
>> B = [50   0   0]'
        50
   B = 0
       0
```

```
>> X = inv (A) * B
        29.6000
   X = 26.0000
        28.0000
```

```
>> X = A\B
        29.6000
   X = 26.0000
        28.0000
```

Thus, $x_1 = 29.6$, $x_2 = 26$, and $x_3 = 28$.

Method 2:
Since the equations are not many in this case, we can use the command `solve` to obtain the solution of the simultaneous equations as follows:

```
[x1,x2,x3]=solve('25*x1 - 5*x2 - 20*x3=50', '-5*x1 + 10*x2
- 4*x3 =0', '-5*x1 - 4*x2 + 9*x3=0')
x1 =
148/5
x2 =
26
x3 =
28
```

which is the same as before.

Practice Problem C.1 Solve the following simultaneous equations using MATLAB:

$$
\begin{aligned}
3x_1 - x_2 - 2x_3 &= 1 \\
-x_1 + 6x_2 - 3x_3 &= 0 \\
-2x_1 - 3x_2 + 6x_3 &= 6
\end{aligned}
$$

Answer: $x_1 = 3 = x_3$, $x_2 = 2$.

C.5 PROGRAMMING HINTS

A good program should be well documented, of reasonable size, and capable of performing some computation with reasonable accuracy within a reasonable amount of time. The following are some helpful hints that may make writing and running MATLAB programs easier.

- Use the minimum commands possible and avoid execution of extra commands. This is particularly true of loops.
- Use matrix operations directly as much as possible and avoid *for*, *do*, and/or *while* loops if possible.
- Make effective use of functions for executing a series of commands over several times in a program.
- When unsure about a command, take advantage of the help capabilities of the software.
- It takes much less time running a program using files on the hard disk than on a floppy disk.
- Start each file with comments to help you remember what it is all about later.
- When writing a long program, save frequently. If possible, avoid a long program; break it down into smaller subroutines.

TABLE C.8
Other Useful MATLAB® Commands

Command	Explanation
`angle`	Phase angle
`bode`	Plot bode plot
`conj`	Complex conjugate
`diary`	Save screen display output in text format
`grid`	Provides grid lines for 2D and 3D plots
`mean`	Mean value of a vector
`min(max)`	Minimum (maximum) of a vector
`grid`	Add a grid mark to the graphic window
`poly`	Converts a collection of roots into a polynomial
`polyfit`	Polynomial curve fitting
`roots`	Finds the roots of a polynomial
`sort`	Sort the elements of a vector
`sound`	Play vector as sound
`std`	Standard deviation of a data collection
`step`	Unit step response
`sum`	Sum of elements of a vector
`title`	Plot title

C.6 OTHER USEFUL MATLAB® COMMANDS

Some common useful MATLAB commands, which may be used in this book, are provided in Table C.8.

Appendix D: Answers to Odd-Numbered Problems

Chapter 1

1.1 Proof.

1.3 1/40.

1.5 (a) energy signal, (b) neither power nor energy, (c) neither power nor energy.

1.7 (a) $x(t) = 10u(t) - 15u(t-2) + 5u(t-4)$

(b) $y(t) = 2r(t) - 2r(t-2) - 7u(t-2) + 3u(t-4)$

1.9 (a) $x(t) = 10e^{-t}u(t) + (2.5t - 10 - 10e^{-t})u(t-4) + (5t - 40)u(t-8)$
$-(2.5t - 30)u(t-12)$

(b) See Figure D.1.

1.11 Proof.

1.13 (a) $t - 1$, (b) 4.5, (c) 16, (d) 128.

1.15 See Figure D.2.

1.17 See Figure D.3.

1.19 See Figure D.4.

1.21 (a) 1.353, (b) $27.18e^{3t}$, (c) $27.18e^{-0.75t}$, (d) $545.98\cos^{3t}$.

1.23 See Figure D.5.

1.25 See Figure D.6.

1.27 See Figure D.7.

1.29 See Figure D.8.

1.31 See Figure D.9.

1.33 Proof.

1.35 (a) Linear and time-varying system.

(b) Nonlinear and time-invariant system.

(c) Linear and time-invariant system.

(d) Nonlinear and time-varying system.

1.37 (a) Nonlinear, (b) Nonlinear, (c) Linear.

1.39 (a) Causal and memoryless, (b) Causal and memoryless.

1.41 (a) Not time-invariant, (b) Time-invariant.

1.43 $\dfrac{di(t)}{dt} + \dfrac{i(t)}{RC} = -\dfrac{1}{R}\dfrac{dv_s}{dt}$

1.45 $\dfrac{dy}{dt} + \dfrac{1}{RC}y = \dfrac{1}{RC}x$

1.47 See Figure D.10.

1.49 See Figure D.11.

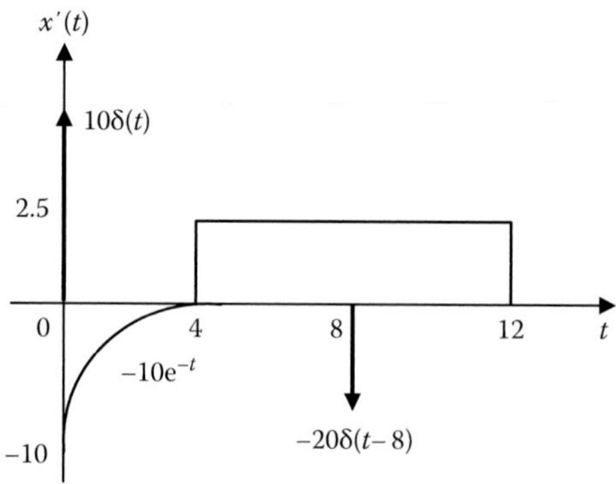

FIGURE D.1 For Problem 1.9(b).

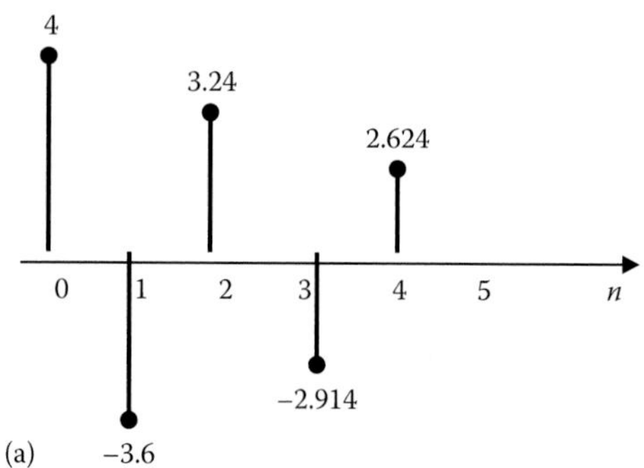

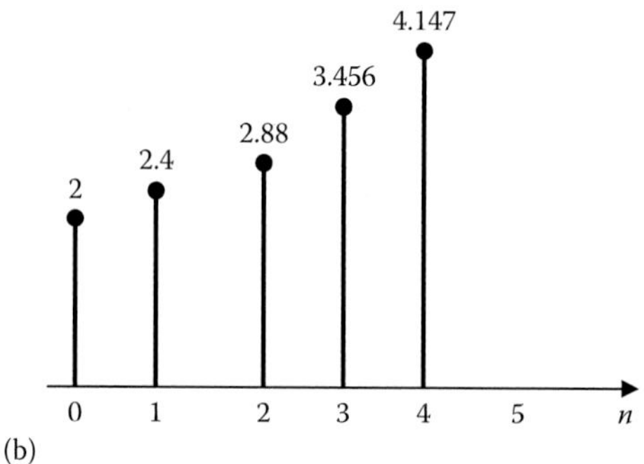

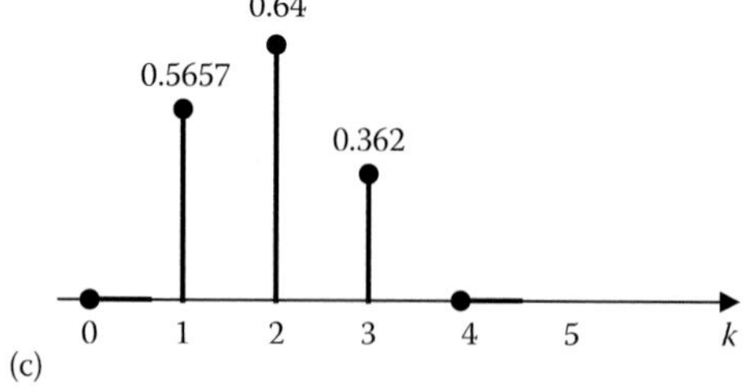

FIGURE D.2 For Problem 1.15.

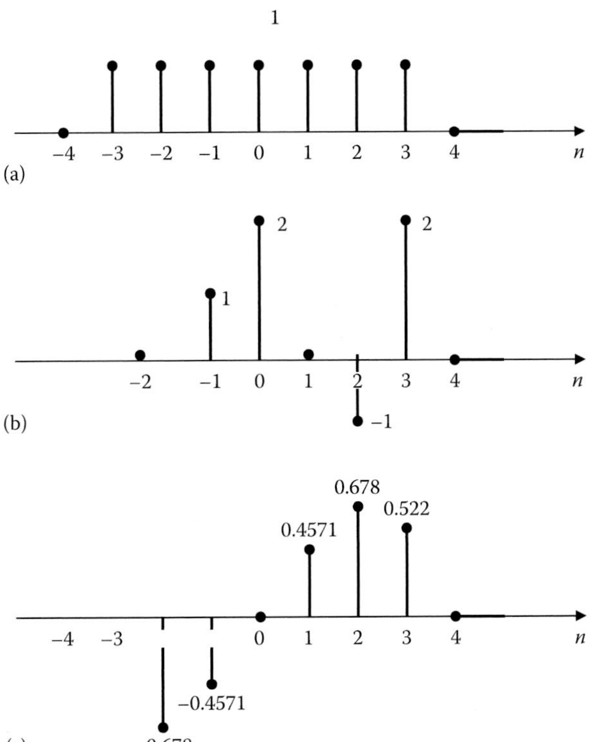

FIGURE D.3 For Problem 1.17.

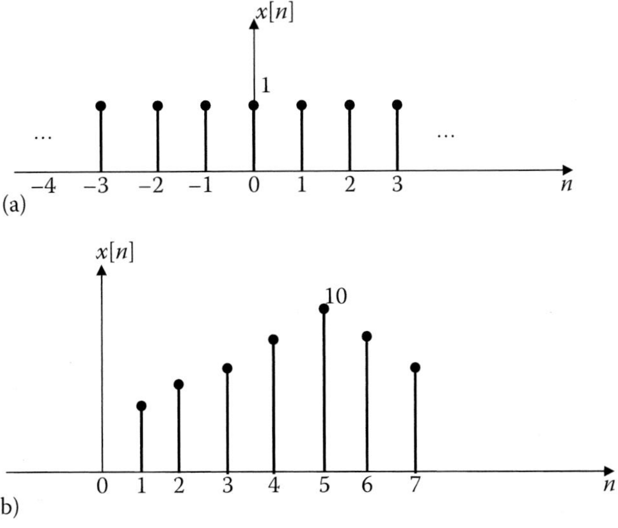

FIGURE D.4 For Problem 1.19.

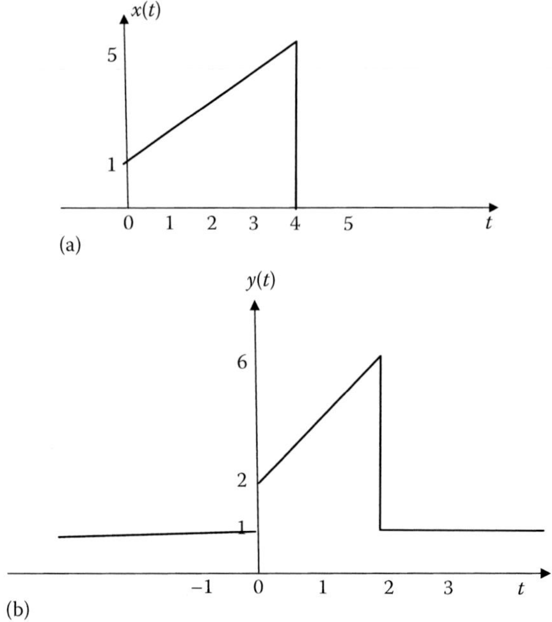

FIGURE D.5 For Problem 1.23.

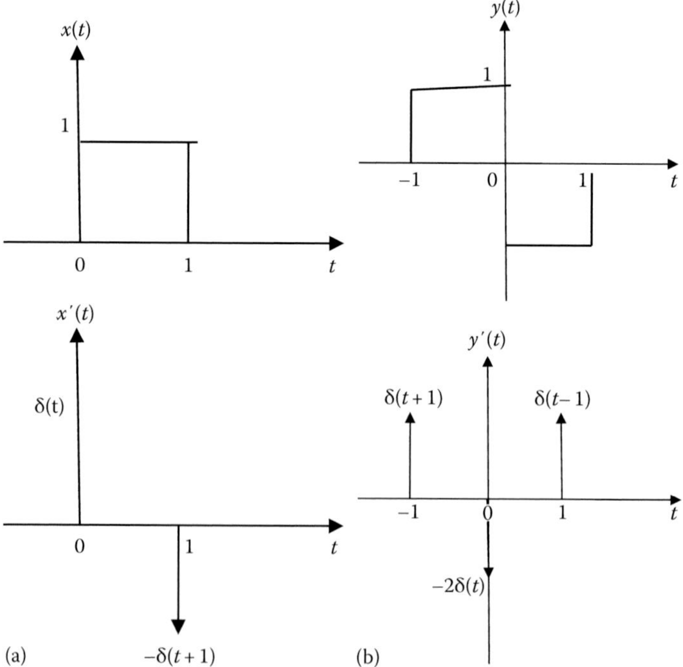

FIGURE D.6 For Problem 1.25.

(Continued)

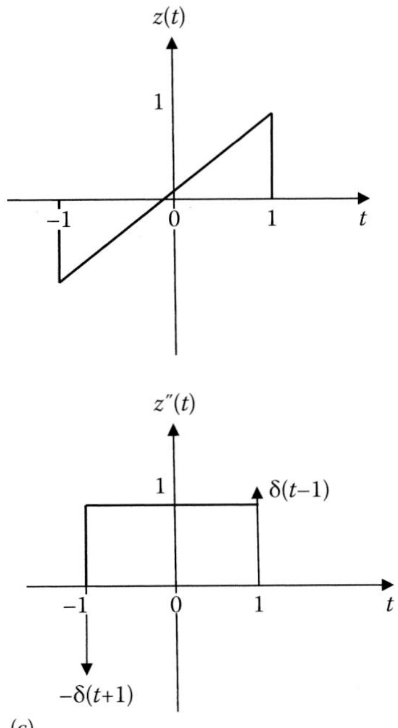

FIGURE D.6 (***Continued***) For Problem 1.25.

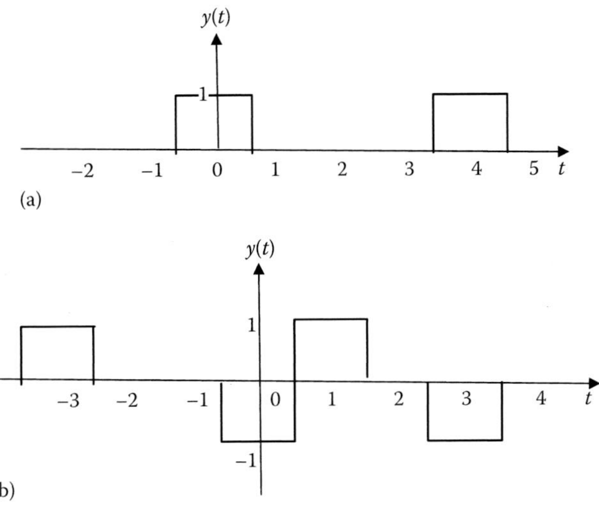

FIGURE D.7 For Problem 1.27.

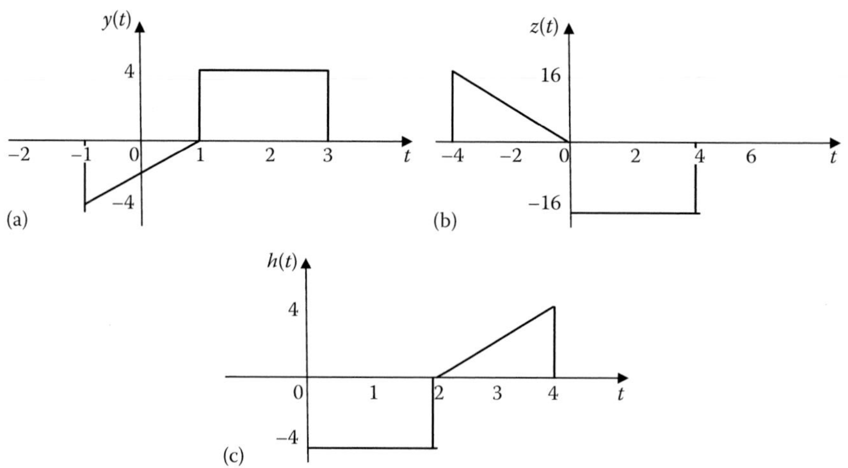

FIGURE D.8 For Problem 1.29.

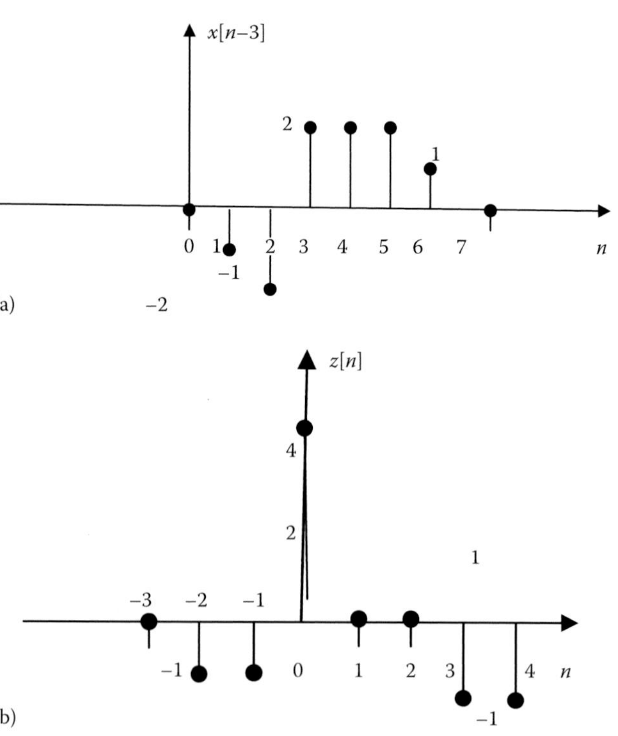

FIGURE D.9 For Problem 1.31

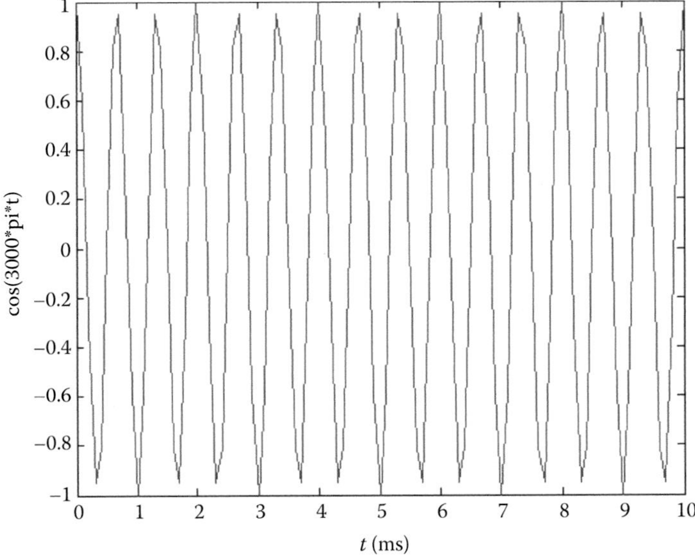

FIGURE D.10 For Problem 1.47.

Chapter 2

2.1 Proof.

2.3 Proof.

2.5 (a) $\frac{t}{a}(e^{at}-1)-\frac{1}{a^2}-\frac{e^{at}}{a^2}(at-1)$

(b) $0.5\cos(t)(t+0.5\sin(2t))-0.5\sin(t)(\cos(t)-1)$.

2.7 See Figure D.12.

2.9 $10(e^{-t}-e^{-2t})u(t)$

2.11 $-e^{-2t}u(t)$

2.13 (a) $\frac{1}{2}\left(1-e^{-2t}\right)u(t)$

(b) $(e^{-2t}-e^{-t})u(t)$

(c) $\frac{1}{4}\left(\cos 2t+\sin 2t-e^{-2t}\right)u(t)$

2.15 (a) unity system, (b) differentiator, (c) integrator.

2.17 Proof.

2.19 $2(1-e^{-t}),\quad t>0$

2.21 $y(t)=\begin{cases} t, & 0<t<2 \\ 4-t, & 2<t<4 \\ 0, & \text{otherwise} \end{cases}$

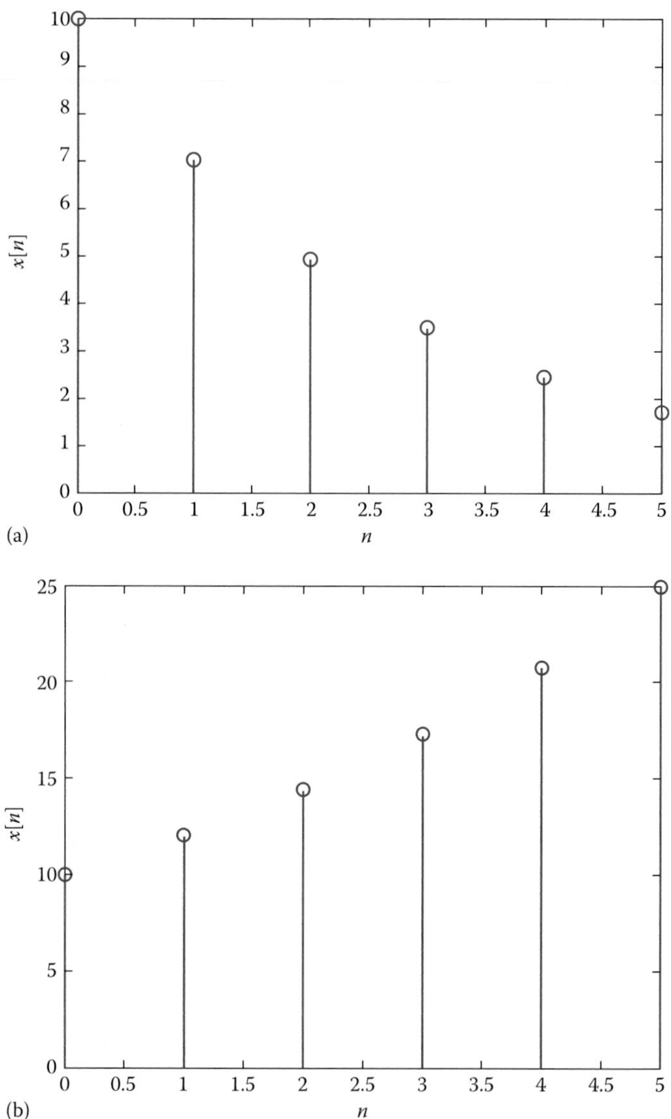

FIGURE D.11 For Problem 1.49.

2.23 $(4e^{-t}-e^{-2t})u(t)$

2.25 Proof.

2.27 (a) $y[n]=\frac{1}{3}\left(4^{n+1}-1\right)u[n]$

(b) $y[n] = 2^n(n + 1)u[n]$

(c) $y[n] = 2^{n+1}(1.5^n-1)u[n]$

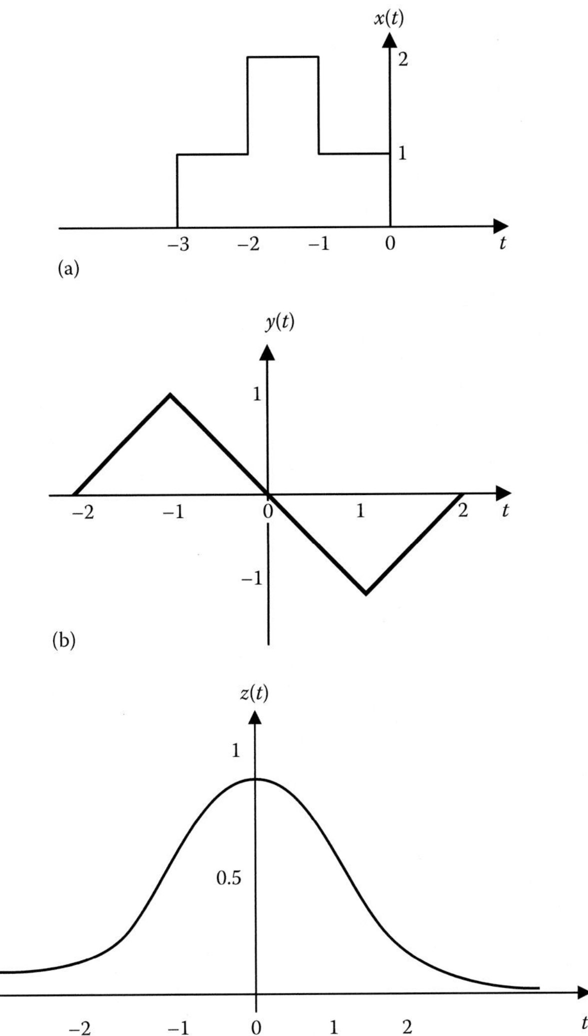

FIGURE D.12 For Problem 2.7.

2.29 $y[n] = \begin{cases} 1, & n = 0 \\ 0, & n = 1 \\ 1, & n = 2 \\ -1, & n = 3 \\ 0, & n = 4 \\ -1, & n = 5 \\ 0, & \text{otherwise} \end{cases}$

2.31 $y[n] = [0 \quad 1 \quad 3 \quad 6 \quad 5 \quad 3]$

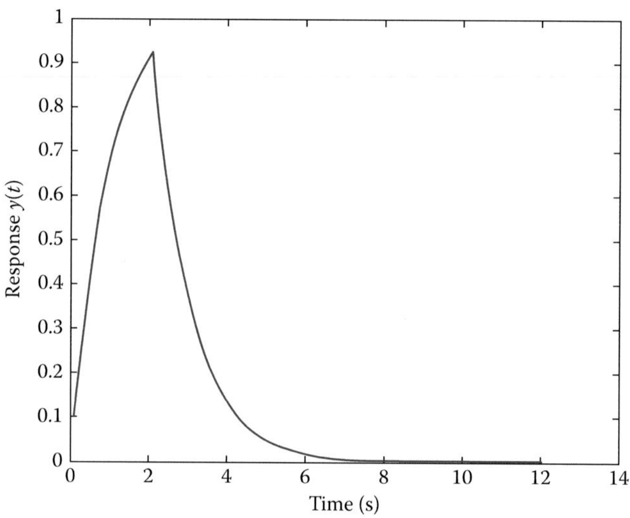

FIGURE D.13 For Problem 2.37(a).

2.33 (a) $y[n] = (n + 2)(0.4)^n u[n] + 1.25(0.4)^n u[n-1]$
(b) $y[n] = (n + 1)(0.4)^n u[n] + n(0.4)^{n-1} u[n-1]$

2.35 [2 3]

2.37 (a) For $y(t)$, see Figure D.13.

(b) $y(t) = \begin{cases} 1 - e^{-t}, & 0 < t < 2 \\ e^{-t}(e^2 - 1), & t > 2 \end{cases}$

2.39 See Figure D.14.

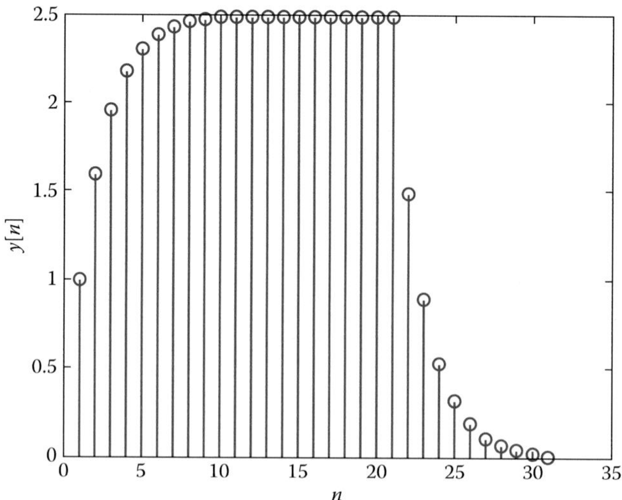

FIGURE D.14 For Problem 2.39.

2.41 $h = [2 \quad 8 \quad 8 \quad -16]$
2.43 Stable.
2.45 Not BIBO stable.
2.47 $v_o(t) = 2(1-e^{-t})u(t)$

Chapter 3

3.1 (a) $\frac{1}{s}\left(5e^{-2s} - 15e^{-s} + 10\right)$

(b) $\frac{1}{s^2}$

3.3 (a) $\left(\frac{2}{s^2} + \frac{8}{s}\right)e^{-4s}$

(b) $5\cos 2e^{-2s}$

(c) $\frac{e^{-\tau(s+1)}}{s+1}$

(d) $\cos 2\tau e^{-\tau s}\frac{2}{s^2+4} + \sin 2\tau e^{-\tau s}\frac{s}{s^2+4}$

3.5 (a) $\frac{1}{s}(1-e^{-2s})$

(b) $\frac{1}{(s+3)^2}$

3.7 $X(s) = \frac{1}{s}\left[2 - e^{-s} - e^{-2s}\right]$

3.9 $f(0) = 0, \qquad f(\infty) = 3$

3.11 (a) $x(t) = u(t) + 2u(t-1)$

(b) $y(t) = \frac{10}{3}(-e^{t} + e^{4t})$

(c) $z(t) = (e^{-t}\cos 3t - e^{-t}\sin 3t)u(t)$

3.13 (a) $x(t) = (3e^{-t}\cos 2t - e^{-t}\sin 2t)u(t)$
(b) $y(t) = (4e^{-t} - e^{-2t})u(t)$
(c) $z(t) = \delta(t) + (e^{-2t} - e^{2t})u(t)$
(d) $h(t) = 2t^3e^{-2t}u(t)$

3.15 $y(t) = 8(t-1)u(t-1) - 4(t-2)y(t-2) - 2u(t-2) - 6u(t-3)$

3.17 (a) $h(t) = (2-e^{-2t})u(t)$
(b) $g(t) = 0.4e^{-3t} + 0.6e^{-t}\cos t + 0.8e^{-t}\sin t$
(c) $f(t) = e^{-2(t-4)}u(t-4)$
(d) $d(t) = \frac{10}{3}\cos t - \frac{10}{3}\cos 2t$

3.19 $h(t) = \delta(t) - 4e^{-5t}u(t)$

3.21 $y(t) = -\frac{1}{4}e^{-2t} + \frac{1}{4}e^{-t}\cos 2t + \frac{3}{4}e^{-t}\sin 2t$

3.23 $H = \dfrac{1}{s^2 + 2as + a^2 + b}$

3.25 $H = \dfrac{G_2G_3(G_1 + F_2)}{1 + G_2F_1}$

3.27 (a) $H = \dfrac{2}{s^2 + s + 0.5}$

(b) $y(t) = 4[1 - e^{-0.5t}\cos 0.5t - e^{-0.5t}\sin 0.5t]u(t)$

3.29 $y(t) = \left(\dfrac{1}{6}e^{-t} - \dfrac{5}{2}e^{-3t} + \dfrac{4}{3}e^{-4t}\right)u(t)$

3.31 $y(t) = (1 + te^{-t})u(t)$

3.33 $i(t) = (2e^{-2t} - 2e^{-2t}\cos t + 2e^{-2t}\sin t)u(t) = 2e^{-2t}(1 - \cos t + \sin t)u(t)$

3.35 $H(s) = \dfrac{s^2 + 2s + 1}{s^2 + 3s + 1}$

3.37 $v_o(t) = (11 - 11e^{-0.5t}\cos 0.866t - 6.351e^{-0.5t}\sin 0.866t)u(t)$

3.39 $I_1 = \dfrac{2(2s+1)}{(s+1)(s^2+3s+1)}, I_2 = \dfrac{2s}{(s+1)(s^2+3s+1)}$

3.41 (a) Unstable, (b) stable, (c) unstable.

3.43 (a) $H(s) = \dfrac{s-3}{s^2 + 2s + 2}$

(b) $h(t) = (e^{-t}\cos t - 4e^{-t}\sin t)u(t)$

3.45 (a) $x =$

heaviside(t − 2) + 1

(b) x =

```
t^2*exp(-t)
```

(c) x =

```
dirac(t) - exp(-2*t)*(cos(2*t) + (5*sin(2*t))/2)
```

3.47 $x =$

heaviside(t − 3)*(exp(6 − 2*t)/2 − exp(3 − t) + 1/2)

− heaviside(t − 2)*(exp(2 − t) − 2* exp(4 − 2*t))

3.49 See Figure D.15.

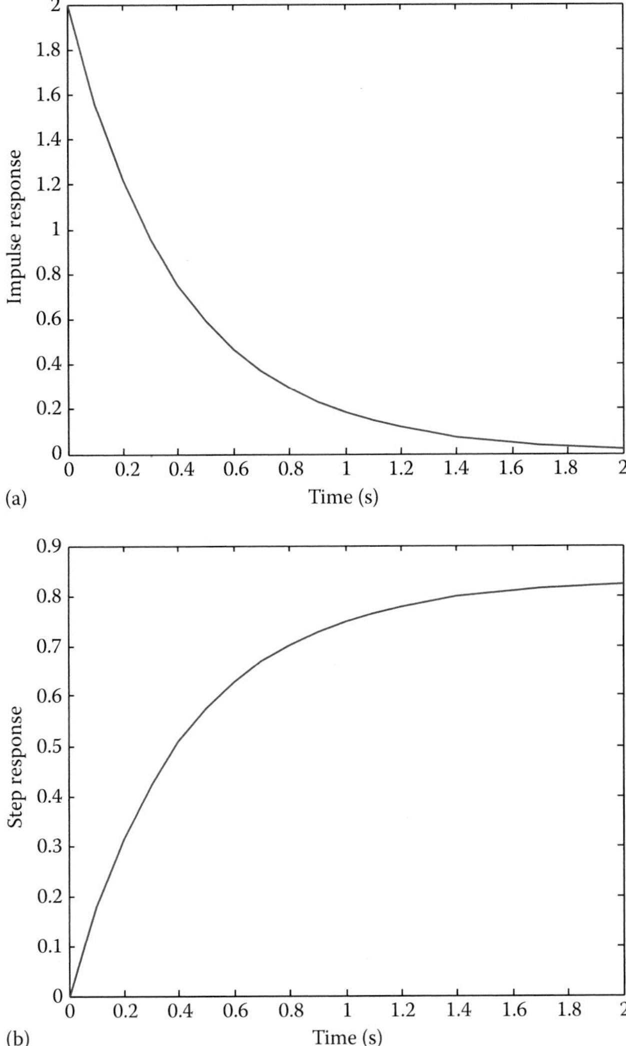

FIGURE D.15 For Problem 3.49 (a) impulse response, (b) step response.

3.51 (a) $z =$

2

$p =$

$$-1.0000 + 3.0000\text{i}$$
$$-1.0000 - 3.0000\text{i}$$

(b) $z =$

$$-1.0000 + 2.0000\text{i}$$
$$-1.0000 - 2.0000\text{i}$$

$p =$

$$\begin{array}{r} 0.0000 + 0.0000i \\ -2.0000 + 3.0000i \\ -2.0000 - 3.0000i \end{array}$$

(c) $z =$

$$\begin{array}{r} -9.4721 \\ -0.5279 \end{array}$$

$p =$

$$\begin{array}{r} -1.5956 + 2.2075i \\ -1.5956 - 2.2075i \\ -0.8087 + 0.0000i \end{array}$$

3.53 See Figure D.16.

Chapter 4

4.1 $x(t) = 7.5 - \sum_{n=\text{odd}}^{\infty} \frac{10}{n\pi} \sin nt$

4.3 (a) $T = 5$ ms, $\omega_o = 1256.64$ rad/s

(b) $a_n = \frac{4}{n} \sin^2\left(\frac{n\pi}{2}\right), \quad b_n = 0$

4.5 See Figure D.17.

4.7 $f(t) = 1 + \sum_{n=1}^{\infty} \left[\frac{3}{n\pi} \sin\frac{4n\pi}{3} \cos\frac{2n\pi t}{3} + \frac{3}{n\pi}\left(1 - \cos\frac{4n\pi}{3}\right) \sin\frac{2n\pi t}{3} \right]$

4.9 $c_2 = \frac{-j}{8}, c_1 = \frac{3j}{8}, c_o = 0, c_{-1} = \frac{-3j}{8}, c_{-2} = \frac{j}{8}$

4.11 $c_n = \frac{\left[e^{(1-jn\pi)} - 1\right]}{2(1 - jn\pi)}$

4.13 (a) 100, (b) 0, (c) 1.

4.15 (a) $T = 6.283s$, $\omega_o = 1$ rad/s

(b) $x(t) = 0.4 + 0.4\cos t + 0.24\cos 2t + 0.16\cos 3t$

4.17 $v(t) = \sum_{n=-\infty}^{\infty} \frac{jV_o}{2n\pi}(\cos n\pi - 1)e^{jnt}$

4.19 (a) $\beta_n = c_{-n}$, (b) $\beta_n = c_n$, (c) $\beta_n = c_n e^{-jn\omega_o 3}$, (d) $\beta_n = jn\omega_o c_n$

4.21 Proof.

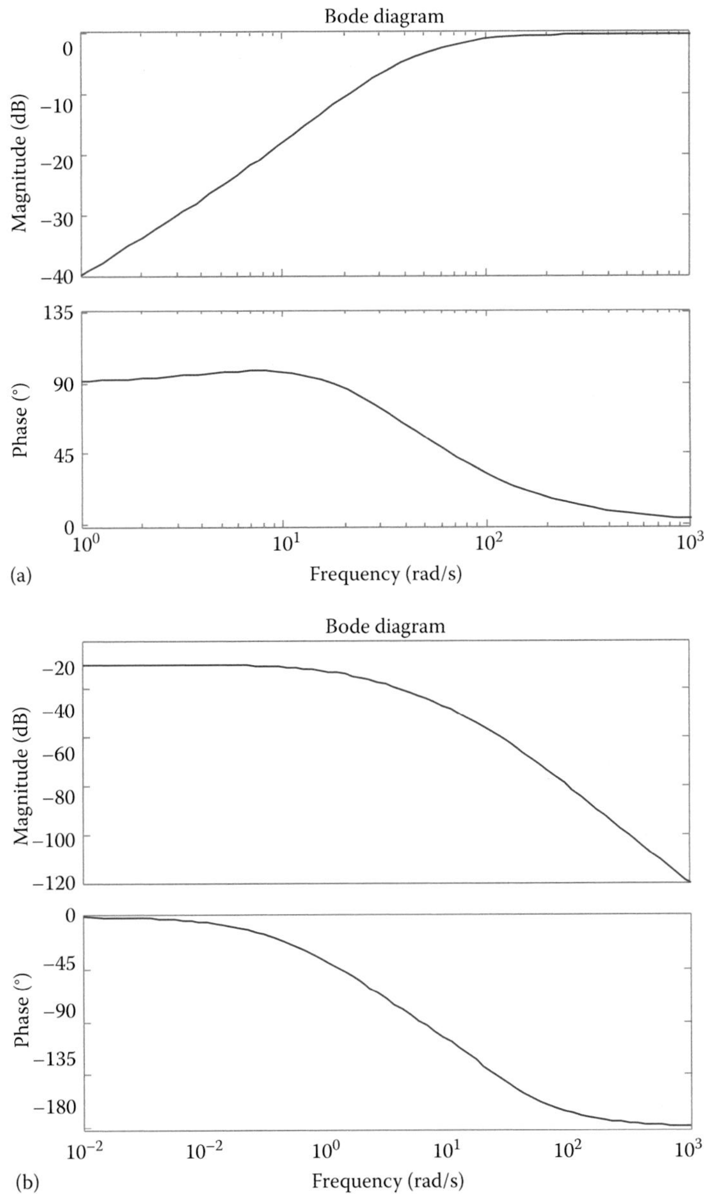

FIGURE D.16 For Problem 3.53.

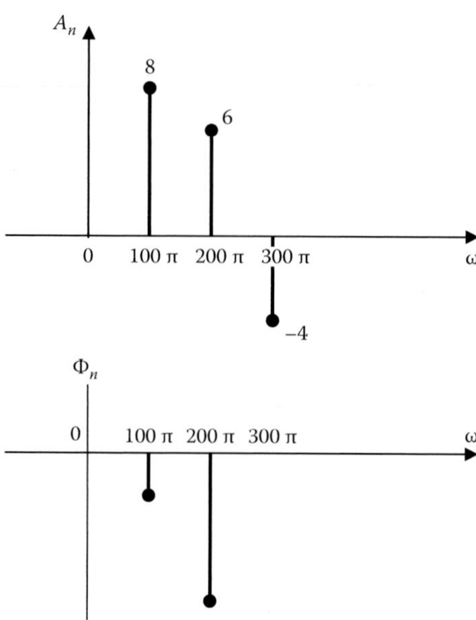

FIGURE D.17 For Problem 4.5.

4.23 (a) Even.
(b) Even.
(c) Even.
(d) Odd.
(e) Odd.
(f) Odd.
(g) Half-wave odd symmetric.
(h) Even.

4.25 Proof.

4.27 109 W.

4.29 $d_n = \dfrac{jV_o \sin^2(n\pi/2)}{n\pi/4}$

4.31 10%.

4.33 $i_o(t) = \sum_{n=1}^{\infty} A_n \sin(2n\pi t + \phi_n)$, where

$$A_n = \frac{10}{\sqrt{25n^2\pi^2 + (2n^2\pi^2 - 50)^2}}, \quad \phi_n = -\tan^{-1}\frac{2n^2\pi^2 - 50}{5n\pi}$$

4.35 See Figure D.18.

4.37 See Figure D.19.

4.39 See Figure D.20.

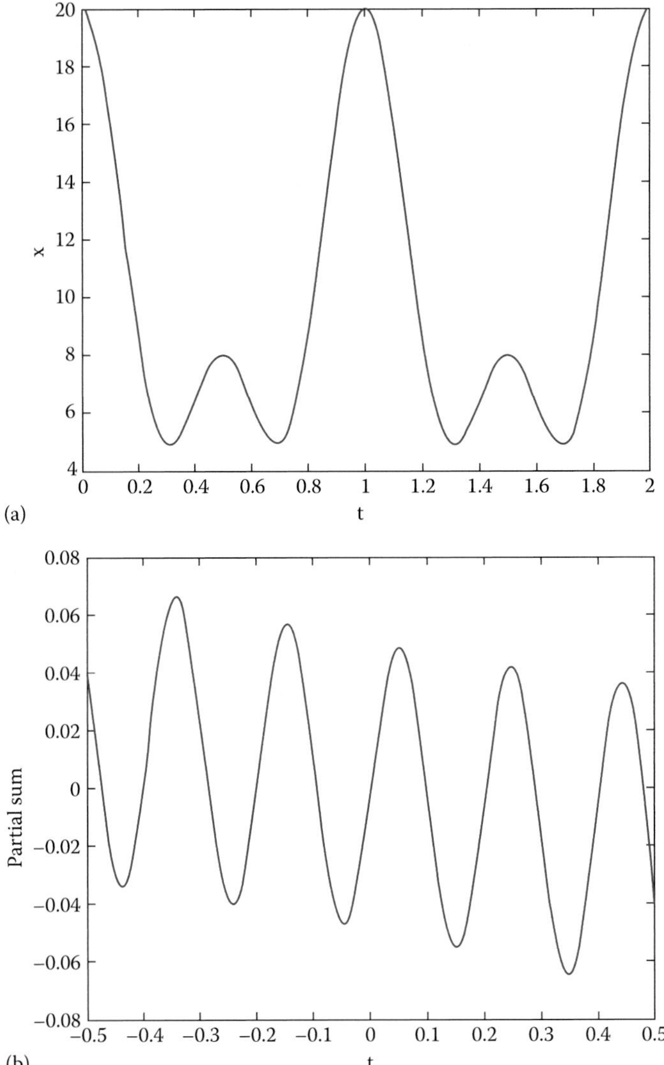

FIGURE D.18 For Problem 4.35.

Chapter 5

5.1 $X(\omega) = \frac{1}{j\omega}\left(1 - e^{-j\omega} - e^{-j2\omega} + e^{-j3\omega}\right)$

5.3 (a) $X(\omega) = \frac{1}{j\omega}\left(1 - 2e^{-j\omega} + e^{-j2\omega}\right)$

(b) $Y(\omega) == \frac{2}{j\omega}\left(\cos\omega - 1\right)$

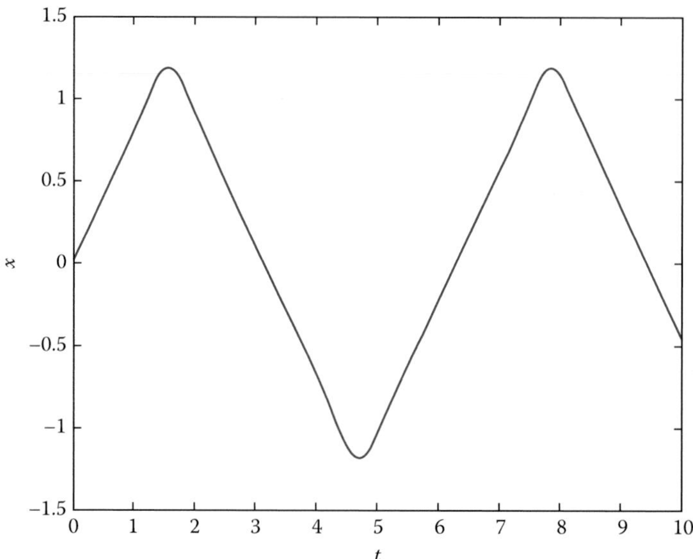

FIGURE D.19 For Problem 4.37.

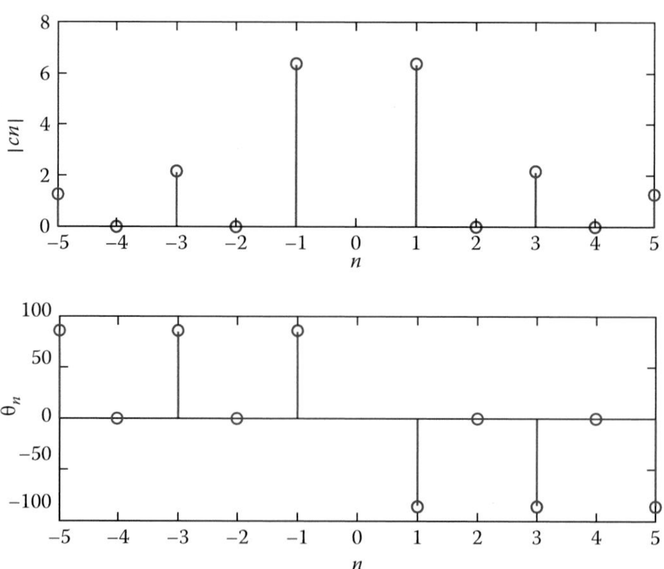

FIGURE D.20 For Problem 4.39.

5.5 $Y(\omega) = \frac{4}{\omega^2}(1 - j\cos\omega)$

5.7 Proof.

5.9 (a) $X(\omega) = 2\pi[\delta(\omega + \omega_o)e^{-j\theta} + \delta(\omega - \omega_o)e^{j\theta}]$

(b) $Y(\omega) = 2\pi[\delta(\omega + 100)e^{j\pi/4} + \delta(\omega - 100\pi)e^{-j\pi/4}]$

5.11 (a) $Y(\omega) = \frac{1}{4}e^{-j2.5\omega}\frac{(10 + j\omega)}{-0.25\omega^2 + j\omega + 1}$

(b) $Y(\omega) = \frac{j(j\omega^2 + 10\omega + j2\omega - j9)}{(-\omega^2 + j2\omega + 1)^2}$

(c) $Y(\omega) = \frac{j}{2}\left[\frac{5 + j(\omega + 2)}{-(\omega + 2)^2 + j2(\omega + 2) + 1} + \frac{5 + j(\omega - 2)}{-(\omega - 2)^2 + j2(\omega - 2) + 1}\right]$

(d) $Y(\omega) = \frac{-\omega^2(5 + j\omega)}{-\omega^2 + j2\omega + 1}$

5.13

(a) $X_{1'}(\omega) = 2\left(\frac{4}{j\omega} + \pi\delta(\omega/2)\right)e^{-j0.5\omega}$

(b) $\mathrm{X}_2(\omega) = \frac{1}{2}\left[\frac{2}{j(\omega + 1)} + \pi\delta(\omega + 1) + \frac{2}{j(\omega - 1)} + \pi\delta(\omega - 1)\right]$

(c) $\mathrm{X}_3(\omega) = 2 + j\omega\pi\delta(\omega)$

5.15 Proof.

5.17 See Figure D.21.

5.19 Proof.

5.21 (a) Proof, (b) (i) $B = \frac{2\pi}{\tau}$, (ii) $B = \frac{2\pi}{\tau}$.

5.23 (a) See Figure D.22.

(b) $X(\omega) = \frac{e^{-j\omega} + j\omega e^{-j\omega} - 1}{\omega^2}$

5.25 Proof.

5.27 $X(\omega) = \frac{e^{-j2\omega} - 1}{\omega^2 - 1}$

5.29 (a) $x(t) = 2\delta(t) - 2e^{-t}u(t)$

(b) $y(t) = \frac{10}{\pi\left(t^2 + 1\right)}$

(c) $z(t) = (-2e^{-2t} + 3e^{-3t})u(t)$

5.31 (a) $y(t) = \frac{j25}{4}\left(\cos 2t - \sin 2t\right)$

(b) $y(t) = 10e^{-2t}u(t)$

(c) $y(t) = (5e^{-2t} - 5e^{-4t})u(t)$

5.33 $y(t) = (e^{-2t} - e^{-4t})u(t)$

5.35 $i(t) = 5(e^{-t} - e^{-2t})u(t)$

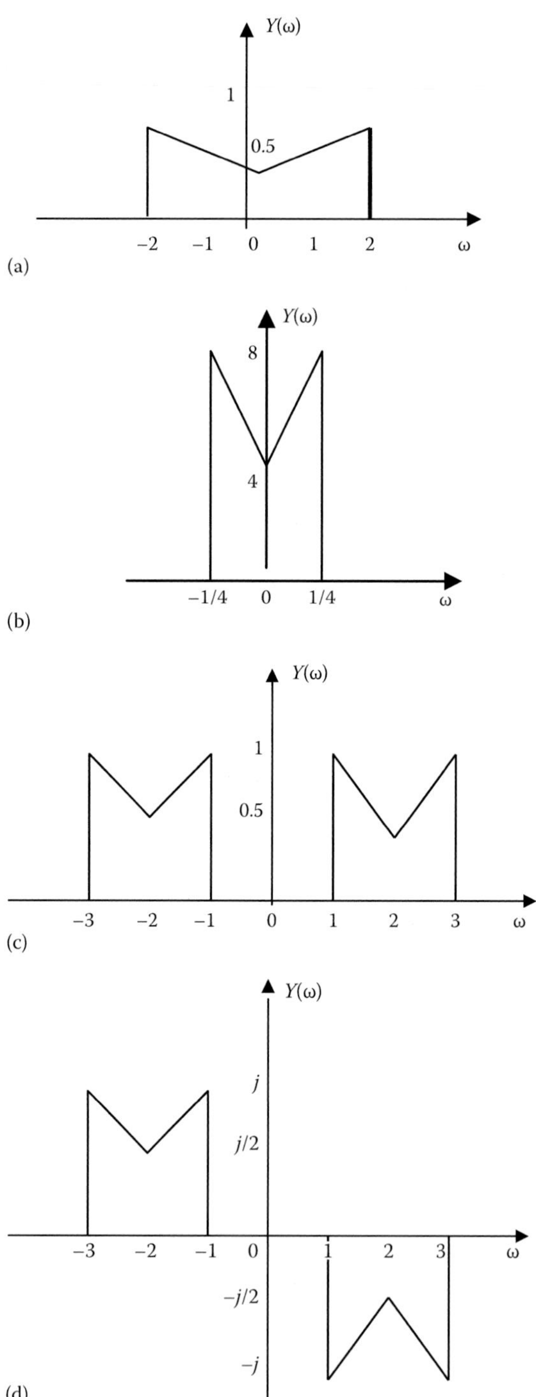

FIGURE D.21 For Problem 5.17.

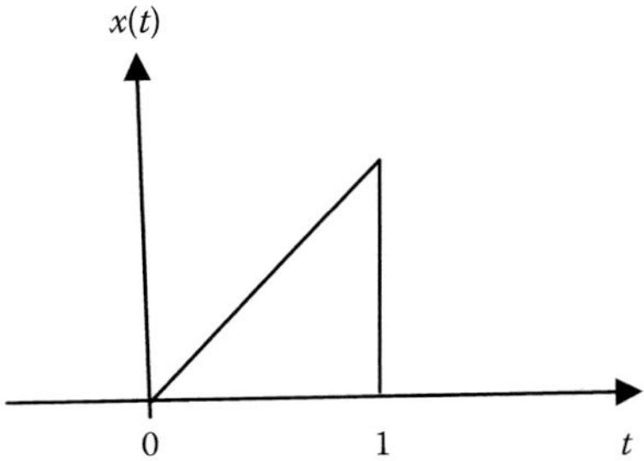

FIGURE D.22 For Problem 5.23(a).

5.37 $H(\omega)=\dfrac{j8\omega}{1+j8\omega}$

5.39 $H(\omega)=\dfrac{j\omega RC}{1+j\omega RC}$, highpass filter.

5.41 $H(\omega)=\dfrac{1+j\omega}{\left[-\omega^2+j5\omega+6\right]}$, $\quad h(t)=(-e^{-2t}+2e^{-3t})u(t)$

5.43 $Y(\omega)=2X(t)+\dfrac{1}{2}X\left(\omega+\omega_o\right)+\dfrac{1}{2}X\left(\omega-\omega_o\right)$

5.45 (a) 60 Hz, (b) 50 Hz
5.47 66.67 J
5.49 2010.62 J
5.51 $X=$

$$\frac{5}{\mathrm{wi}+2-4\mathrm{i}}+\frac{5}{\mathrm{wi}+2+4\mathrm{i}}$$

The plot of $|X(\omega)|$ is in Figure D.23.
5.53 See Figure D.24.
5.55 See Figure D.25.

Chapter 6

6.1 (a) Proof, (b) Proof.
6.3 $j2[\sin\Omega + 2\sin2\Omega + 3\sin3\Omega]$

6.5 (a) $\dfrac{0.36}{1-1.6\cos\Omega}$

(b) $2j\sin\Omega k$

(c) $\dfrac{1}{2}\left[\dfrac{1}{1-0.4e^{-j(\Omega-4)}}+\dfrac{1}{1-0.4e^{-j(\Omega+4)}}\right]$

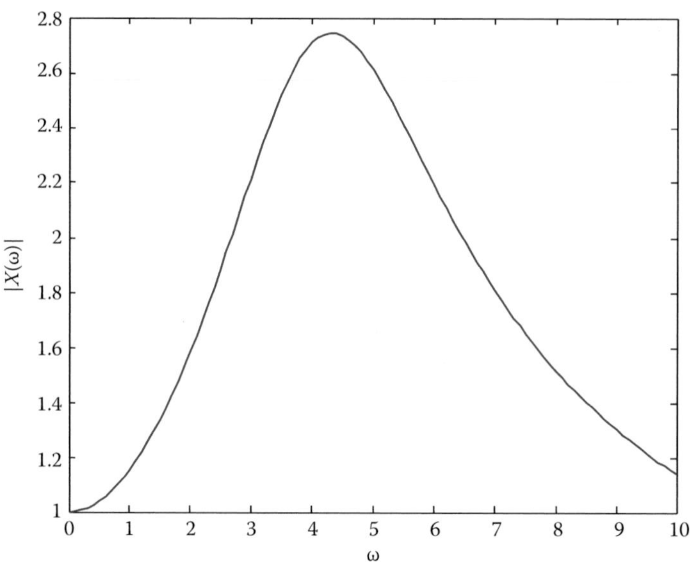

FIGURE D.23 For Problem 5.51.

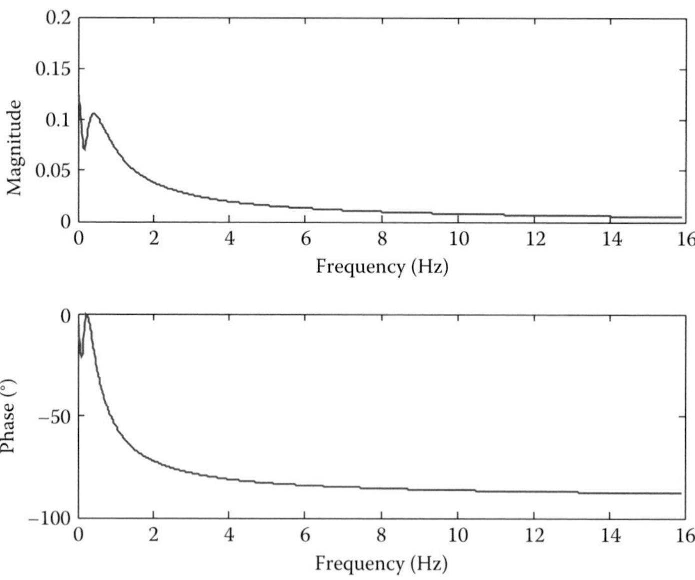

FIGURE D.24 For Problem 5.53.

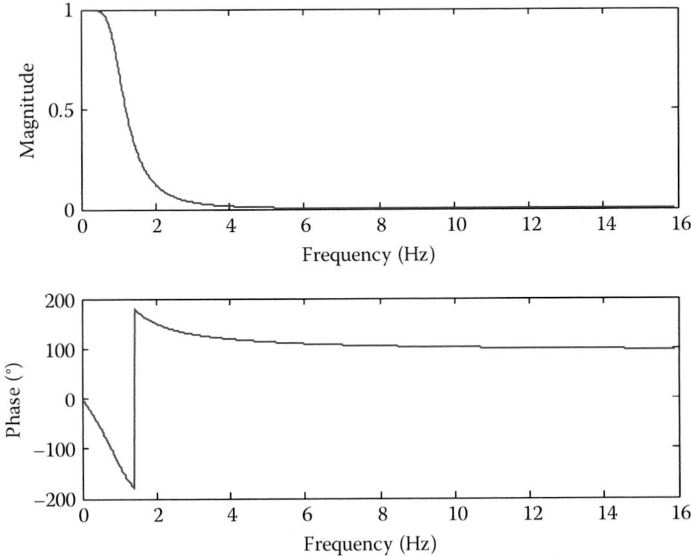

FIGURE D.25 For Problem 5.55.

6.7 (a) $\dfrac{1}{1-0.6e^{-j\Omega}} + j\pi\sum_{k=\infty}^{\infty}\left[\delta(\Omega-2\alpha-2\pi k)-\delta(\Omega+2\alpha-2\pi k)\right]$

(b) $\dfrac{1}{1-0.5e^{j\Omega}} + 2\pi\sum_{k=\infty}^{\infty}\left[\delta(\Omega-2\alpha-2\pi k)+\delta(\Omega+2\alpha-2\pi k)\right]$

6.9 (a) $\dfrac{e^{(1-j3)\Omega}}{e^{j\Omega}+a}$

(b) $\dfrac{e^{j(\Omega-4)}}{e^{j(\Omega-4)}+a}$

(c) $\dfrac{1}{2j}\left[\dfrac{e^{j(\Omega-2)}}{e^{j(\Omega-2)}+a}-\dfrac{e^{j(\Omega+2)}}{e^{j(\Omega+2)}+a}\right]$

(d) $\left(\dfrac{e^{j\Omega}}{e^{j\Omega}+a}\right)^2$

6.11 (a) $y[n] = (1.25)^n u[-n]$
(b) $z[n] = (-1)^n (0.8)^n x[n]$
(c) $w[n] = n(0.8)^n u[n]$

6.13 (a) $x[-2] = 3$, $x[0] = 1$, $x[4] = -2$, $x[5] = 1$

(b) $x[n] = \left(\dfrac{1}{2}\right)^n u[n-1] - \left(\dfrac{1}{2}\right)^{n+1} u[n]$

(c) $x[n] = 3\cos(2n)$

6.15 Proof.

6.17 Proof.

6.19 $\dfrac{1-a^N}{1-ae^{-j2\pi k/N}}$

6.21 (a) $X[k] = [6 \; -2+j2 \; -2 \; -j2]$

(b) $X[k]=[0 \quad 0 \quad 4-j4 \quad 0 \quad 0 \quad 0 \quad 4+j4 \quad 0]$

6.23 Proof.

6.25 $x[0] = 1$, $x[1] = -0.5257$, $x[2] = -0.8507$, $x[3] = 0.8507$, $x[4] = 0.5257$

6.27 For DFT, 4,194,304. For FTT, 11,264.

6.29 The plot is shown in Figure D.26.

6.31 See Figure D.27.

6.33 See Figure D.28.

Chapter 7

7.1 (a) $1 + z^{-1} - z^{-2} - z^{-3}$, (b) $0.5 - z^{-1} + z^{-2} - z^{-3}$

7.3 (a) $|z| < \dfrac{3}{2}$, (b) $|z| > \dfrac{1}{5}$

7.5 $\dfrac{6z}{3z-2} - \dfrac{5z}{5z-2}, \quad |z| > 2/3$

7.7 (a) 1, (b) $\dfrac{1}{z-a}$, (c) $\dfrac{az^2}{z(z-a)^2}$

7.9 (a) $\dfrac{z^2 - za\cos\Omega_o}{z^2 - 2za\cos\Omega_o + a^2}$, (b) $\dfrac{z(z+1)}{(z-1)^2}$, (c) $\dfrac{ze\sin(1)}{z^2 - 2ze\cos(1) + e^2}$

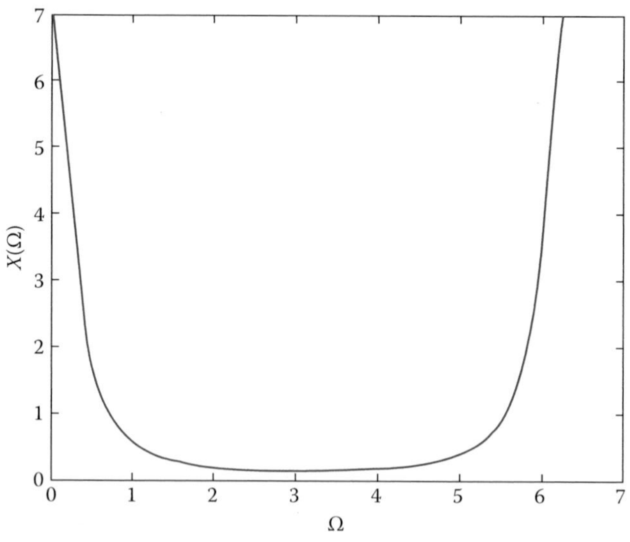

FIGURE D.26 For Problem 6.29.

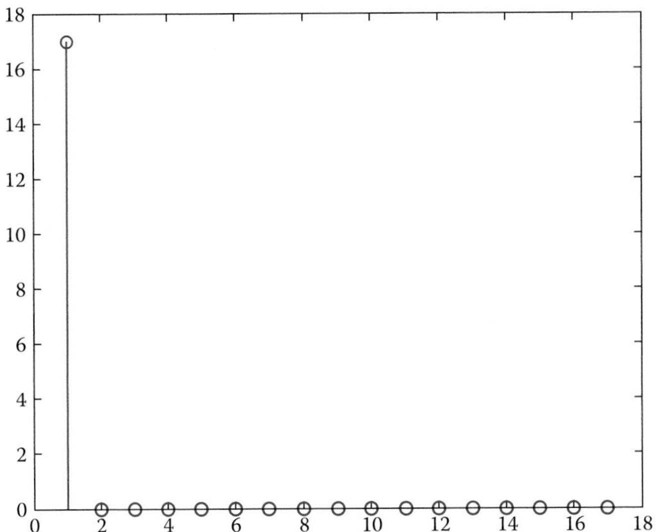

FIGURE D.27 For Problem 6.31.

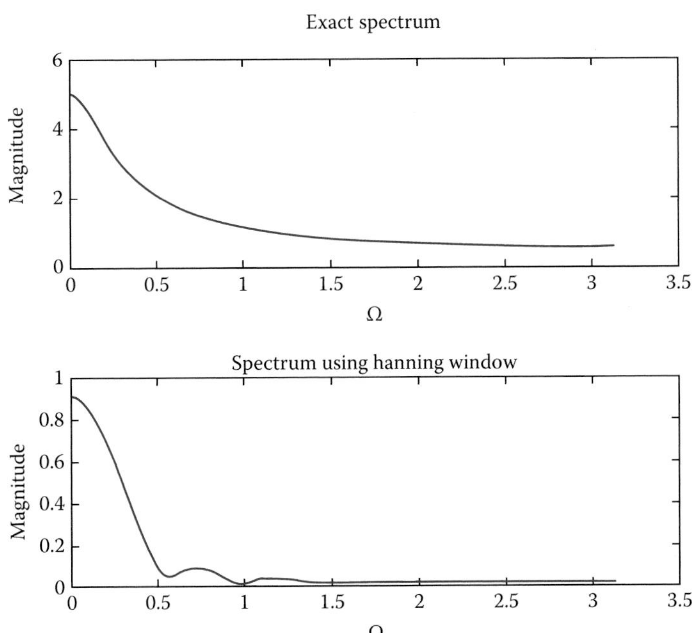

FIGURE D.28 For Problem 6.33.

7.11 Proof.

7.13 Proof.

7.15 $\dfrac{3z}{(z-3)^2}+\dfrac{z+1}{z-3}$

7.17 (a) 1, (b) 1.

7.19 $x[0] = 0$, $x[1] = 1$, $x[10^5] = -1$

7.21 $y[n] = [1, -1, 5, 1, 2, 10, -7, -6]$

7.23 (a) $6\delta[n]+12\left(-\dfrac{1}{2}\right)^n u[n]-18\left(-\dfrac{1}{3}\right)^n u[n]$

(b) $2u[n]-2(0.5)^n u[n]$

(c) $8\delta[n]-10u[n] + 7nu[n]-2(2)^n u[n]$

7.25 $\dfrac{5}{3}\delta[n]-\dfrac{2}{3}u[n]-\dfrac{5}{3}(-2)^n u[n]+\dfrac{7}{12}(-3)^n u[n]$

7.27 (a) $(0.866)^n\cos(54.73^o n)u[n]-0.7072(0.866)^n\sin(58.74^o n)u[n]$

(b) $(0.8)^n\cos(60^o n)u[n] + 2.02(0.8)^n\sin(60^o n)u[n]$

7.29 $y[n] = 2(0.6)^n\cos(146.4n)u[n] + 3.02(0.6)^n\sin(146.4n)u[n]$

7.31 $y[n] = 3(2)^n u[n]-2(1.5)^n u[n]$

7.33 $\text{y[n]}=\dfrac{2}{3}u[n]-0.1491\left(\dfrac{1}{\sqrt{6}}\right)^n \sin(114.1^o n)u[n]$

7.35 $H(z)=2\delta[n]+\dfrac{3z}{z-0.6}-\dfrac{5z}{z+0.8}$

7.37 $Y(z)=\dfrac{z}{z-3}$

7.39 $(z+1)\left[\dfrac{1}{(z-1)^2}-\dfrac{3}{3z-1}\right]$

7.41 (a) $y[n] = 1.667u[n]-0.667(-0.2)^n u[n]-0.8333(0.8)^n u[n]$

(b) $y[n] = 0.5303(2)^n u[n]-0.3636(-0.2)^n u[n]-0.1667(0.8)^n u[n]$

7.43

$$y[n]=0.3076\left[(0.6)^n-(-2)^n\right]u[n],\quad y[n]=\left[0.667+(0.6)^n-0.205(-2)^n\right]u[n]$$

7.45 x =

(3/5)^*n*

7.47 pole =

−5.8284

−0.1716

zero =

$-2.8338+0.0000i$

$-0.0831+1.8767i$

$-0.0831-1.8767i$

7.49 Stable.

Index

G

H

I

N

P

Q

R

S

T

U

Z